ARCHITECTURE IN THE CULTURE
OF EARLY HUMANISM

Architecture in the Culture of Early Humanism

ETHICS, AESTHETICS,
AND ELOQUENCE
1400–1470

Christine Smith

New York   Oxford
OXFORD UNIVERSITY PRESS
1992

Oxford University Press

Oxford   New York   Toronto
Delhi   Bombay   Calcutta   Madras   Karachi
Kuala Lumpur   Singapore   Hong Kong   Tokyo
Nairobi   Dar es Salaam   Cape Town
Melbourne   Auckland

and associated companies in
Berlin   Ibadan

Copyright © 1992 by Oxford University Press, Inc.

Published by Oxford University Press, Inc.,
200 Madison Avenue, New York, New York 10016

Oxford is a registered trademark of Oxford University Press

All rights reserved. No part of this publication may be reproduced,
stored in a retrieval system, or transmitted, in any form or by any means,
electronic, mechanical, photocopying, recording, or otherwise,
without the prior permission of Oxford University Press.

Library of Congress Cataloging-in-Publication Data
Smith, Christine (Christine Hunnikin)
Architecture in the culture of early humanism : ethics,
aesthetics, and eloquence, 1400–1470 / Christine Smith.
p. cm.   Includes bibliographical references and index.
ISBN 0-19-506128-4
1. Architecture, Renaissance.   2. Humanism.   I. Title
NA510.S65   1991
724'.12—dc20   90-20056

1 3 5 7 9 8 6 4 2
Printed in the United States of America
on acid-free paper

ACKNOWLEDGMENTS

The idea of writing this book, and the conviction that such a book was needed, were results of having taught early Renaissance architecture in the buildings themselves, above all in Florence, for ten (now fifteen) years. I had become increasingly uncomfortable with the fact that, with a few important exceptions, little had been done to relate these buildings to their specific intellectual, spiritual, and aesthetic contexts. In particular, the contexts most frequently evoked by art historians did not correspond closely to the way early Humanist culture was understood by other kinds of historians. I came to believe that the first step toward the recuperation of architectural principles in the early Renaissance period required closer examination of some issues within early Humanist culture that had relevance to architecture. I also felt that such a study would be valuable in itself, shedding light on aspects of Humanist thought unlikely to be explored by other historians, and at the same time providing a store of information and concepts from which architectural historians might draw. Thus while the result of my research is not a new interpretation of early Renaissance architecture, but some proposals about the values and concerns of its audience, I believe that the evidence examined in the essays supports the thesis that old assumptions about architecture between 1400 and 1460 need to be revised and that this evidence provides some bases on which a new view may begin to be constructed. This book of essays in intellectual history interprets aspects of both Western and Byzantine culture; many of the arguments rest on analysis of literary sources. Since I am an architectural historian who specializes in the Western medieval and Renaissance periods, I am grateful to those specialists in Byzantine and Renaissance intellectual, cultural, and literary history who have shared their knowledge with me and who helped equip me to write the kind of book I believed was needed.

I am indebted to many for the intellectual and methodological position expressed in this work, but the decisive influences are those of Eugenio Garin and Alexander Kazhdan. Eugenio Garin's approach to the early Renaissance has been fundamental. His interpretation of what were the main themes of concern to the early Humanists (a point on which there is some disagreement among scholars) underlies many of the questions I raise in regard to their attitudes toward architecture. His insights persuaded me that early Humanist culture was characterized by dynamic tensions and contradictions, and that no true picture could be sketched which did not include opposing views; my essays delineate some of these antitheses in regard to questions affecting attitudes toward architecture.

Alexander Kazhdan's influence is most evident in my method. Working closely with him, I saw how the kind of intellectual history that he practiced could be applied to the material which interested me. I learned from his interpretation of Byzantine texts a method of approaching literary sources that could elucidate problems in Renaissance architecture. I am grateful to this extraordinary scholar for teaching me, and to the National Endowment for the Humanities, Dumbarton Oaks, and the Center for Advanced Study in the Visual Arts for grants in 1984 and 1985 that enabled me to study with him for the better part of a year. Henry Millon, dean of the Center for Advanced Study, was one of the first to encourage this exchange between a student of Renaissance architecture and a specialist in Byzantine literature. I have benefited from his breadth of vision and imagination.

Another important influence has been the Villa I Tatti, the Harvard University Center for Renaissance Studies in Florence, which has been my intellectual home since 1971. I am grateful to Craig Smyth, former director of I Tatti, for his encouragement, advice, and support during the past two decades. My understanding of the Italian Renaissance was shaped by the lectures I have heard at I Tatti and by the scholars I have met there, no less than by the Berenson Library itself. Above all, the insight and immense learning of Salvatore Camporeale have illuminated my understanding of the early Humanists. I am happy to thank Father Camporeale and the present director, Walter Kaiser, for reading the completed manuscript and for offering welcome criticism. It is a great satisfaction to me that a grant from the John Simon Guggenheim Memorial Foundation for 1989/1990 enabled me to complete my book as visiting scholar at I Tatti.

Research was facilitated, in addition to the support already mentioned, by an appointment at the Institute for Advanced Study in Princeton in 1988/1989. The opportunity to study without interruption for a year and to converse with scholars from a wide spectrum of specializations enabled to me to broaden and deepen my cultural preparation and contributed to making this a better book. I thank the permanent members of the School of Historical Studies at the institute for including me, and especially Irving Lavin. Both Irving and Marilyn Lavin have been supportive of my work and I am grateful to them.

A final campaign of research in the fall of 1989 was facilitated by an appointment as Visiting Scholar at the American Academy in Rome; I thank the director, Joseph Connors, and the staff of the Academy library.

I would also like to thank Charles Trinkaus and William Loerke, whose bold thought and subtle interpretations set my mind on fire, and Stella Rudolph, whose brilliant imagination is rivaled only by her clarity of thought. It has been my privilege to be influenced by these scholars and to call them my friends. I thank those who read portions of the manuscript at various stages of completion, especially Debra Pincus, whose fine logic caught lapses on my part and whose enthusiasm persuaded me that I could overcome them. Charles Trinkaus, Pauline Watts Trinkaus, and Richard Krautheimer read chapters 1 and 2, and John Barker read chapters 7 and 8. Their criticism has been of great value to me. Earlier versions of chapters 1 and 2 were published in the *Journal of Medieval and Renaissance Studies* (1989) and in *Rinascimento* (1988).

As this book was nearing completion, I began to collaborate on a different, but related, project with the Classicist Joseph O'Connor. Our early discussions touched on some of the problems I was exploring for this book. Although my essays were too near completion to benefit fully from his learning, imagination, and intelligence, they gained polish and elegance from his suggestions. Dr. O'Connor was generous enough to review my translations from Greek and Latin as well as to offer new translations for some exceptionally difficult passages. I should say that all of the translations published here are the product of our collaboration.

Florence
February 1991

C. S.

CONTENTS

Introduction, xi

I ETHICS

1. *Profugiorum ab aerumna libri III*: Architectural Allegories of Virtue in a Dialogue by Leon Battista Alberti, 3
2. Originality and Cultural Progress: Brunelleschi's Dome and a Letter by Alberti, 19
3. Foul Enormity or Grandiose Achievement? The Moral Problem of Size, 40

II AESTHETICS

4. Norm Versus Innovation: The Problem of the Middle Ages, 57
5. Alberti's Description of Florence Cathedral as Architectural Criticism, 80
6. *Varietas* and the Design of Pienza, 98

III ELOQUENCE

7. Byzantine Learning and Renaissance Eloquence, 133
8. Manuel Chrysoloras's *Comparison of Old and New Rome*, 150
9. The Recovery of Eloquence and a Problem in Historiography, 171

Appendix: Manuel Chrysoloras's *Comparison of Old and New Rome*, 199

Notes, 217

Bibliography, 269

Index, 291

INTRODUCTION

These essays on architectural descriptions written in Italy between 1400 and about 1460 are studies of literary analysis and interpretation; they do not constitute a history of early Renaissance architecture, nor are they directly concerned with the principles or practice of early Renaissance architects. My primary interest is the place of architecture in the culture of early Humanism, rather than the influence of that culture on architectural theory and practice. As an architectural historian, however, I am also alert to the implications of my findings for an understanding of early Renaissance architecture, some of which I will suggest in this book. Since I am writing about what the early Humanists thought, and not about what actually happened in the Quattrocento, I consider only written evidence.[1] My approach to this body of evidence is fundamentally different from that usually adopted by historians and architectural historians, who have valued these texts chiefly for the factual information they provide.[2] From their point of view, the rhetorical character of architectural ekphrases has seemed regrettable. Goldbrunner, for example, warned against trying to draw conclusions about specific cities from such texts on account of their "klischeehafter, von panegyrischen und toposhaften Elementen durchsetzter Charakter."[3] Typical of this approach is Storman's assessment of Bessarion's "Encomion to the City of Trebizond":

> It is still too dominated in part by rhetorical conventions to commend itself entirely as good literature, good history, or good description, all of which it sets out to be. However, in spite of the occasional toying with figures of speech or other displays of verbal virtuosity, and in spite, too, of the obvious idealization of Trebizond, past and present, a good deal of factual information is conveyed.[4]

In contrast to these attitudes, I am not concerned with the amount or accuracy of factual information in these texts. For my purposes, the use of topoi and imitation of rhetorical models is meaningful and informative, since they identify the literary tradition within which the work may be interpreted. As I will show in many different ways in this book, not only individual evaluations of phenomena, but perceptions themselves are limited and shaped by available categories of historical periodization (chapter 4), of stylistic excellence (chapter 5), of moral worth (chapter 1), and of specialized vocabulary (chapters 4 and 7). In evaluating the texts, I have assumed that the authors meant what they said (although not, of course, that what they said was true).

The nine essays are concerned with what contemporaries saw or, more precisely, what they thought was important enough in the built environment to deserve comment. Because the vast majority of observers never wrote about architecture at all, the evidence I present does not serve to reconstruct the attitudes of a whole society; this is not a work of social or contextual history. Those who did write, a small elite, were not primarily interested in recording for posterity their impressions of the development of architectural style, and in fact they did not do so. That Brunelleschi reinstated the Orders as the basic vocabulary of design remains true, even though not one of his contemporaries seems to have noticed it. A major focus of my studies is the revival of the primacy of sensory experience in epistemology. The relevance of this theme for Renaissance painting from Assisi on is well known. But its implications in regard to architecture, which is not a mimetic art, are quite different. Since very few of the authors considered here were especially interested in architecture for itself, it seemed useful to ask why they wrote about it. How did description of the built environment further their own real interests? If the built environment was evidence of something, what was it evidence of? How did the Humanists use what they saw around them to understand what they could not see, like the nature of the soul, their position in history, and their relation to the Classical past? I found that architectural imagery could be used to make moral discussion more vivid (chapters 1, 3, 7, and 9). Buildings could also serve as evidence for cultural status (chapters 2, 7, and 9), for the dignity of man (chapters 2, 5, and 7), or as historical documents useful for evaluating the past or transmitting an image of the present to the future (chapters 7 and 9).

More than twenty years ago, Creighton Gilbert called for the study of the "history of aesthetic psychology" in regard to early Renaissance art.[5] His comment occurred in the context of a discussion of Goro Dati's description of Florence (1423), which, although not useful as a source of factual information, could serve to answer certain questions. How did the Florentines respond to their art? Did they understand its exceptional character? How did they link it to their ordinary lives and interests? Did they see it only as functional, or was there interest also in its aesthetic quality? His questions impressed me, since at the time that I read his article, I had already begun to study descriptions of architecture from Classical Antiquity to the Renaissance, using them as instruments for determining the values of those who wrote. Exploring the early Humanists' relation to the built environment in order to learn about their values meant experimenting in a kind of hands-on intellectual history of the kind advocated by Huizinga: "The specific forms of the thought of an epoch should not only be studied as they reveal themselves in theological and philosophic speculations, or in the conceptions of creeds, but also as they appear in practical wisdom and everyday life."[6] While anyone can avoid looking at pictures and statues, it really was not possible for any Florentine to ignore Brunelleschi's dome: it was an important element of daily life to which meaning might or might not be attached. I am interested in those contexts in which architecture was understood as having *connotative* as well as *denotative* qualities. The processes of association and interpretation occur through the interaction of memory, imagination, and reason with the sensory and factual data stored in the mind, and it is these processes that confer cultural significance on architectural monuments. Not what is seen, but the understanding and interpretation of perception constitute the history of ideas. In a recent book, John Onians expresses a negative attitude toward architectural theory in particular and writing about architecture in general.[7] He argues that architectural theory sets out to persuade the spectator that his reaction to what he sees, or its meaning, is different from what he naturally thought and that "each writer [of theory] relates his formulations not to what happened in real buildings but to notions current in other forms of literature." What Onians complains of—the transformation of sensation and of the judgment of sense by associations drawn from other, usually literary, contexts—is the subject of this book, since it is by such associations that architecture acquires its expressive charge.

In order to interpret the descriptions—that is, to draw from them the values of their authors—they first have to be treated as literary works within a well-defined tradition (or traditions). Rarely, perhaps never, did an author of a full-blown description simply record in his own words what he saw. The texts are always in a dynamic relation with other literary texts, especially with other descriptions of architecture, not only in regard to their aesthetic categories and vocabulary, but also in the larger moral or cultural issues that they address. This is especially true of the early Humanist and late Byzantine texts examined in this book. The literary production of the early Humanists, in particular, is a kind of dialogue with works by Classical authors and by other Humanists—an ongoing debate about the significance of their own culture and about right action in the present. Since the texts are not self-enclosed works but replies within a dialogue, one has to reconstruct both sides of the argument in order to recapture their full meaning. Thus since the authors rarely identify the literary works to which they are responding, my first task was to identify their sources. To do this, I collected texts in which the early Humanists described architecture or used architectural imagery, and at the same time I assembled a similar body of evidence from ancient and medieval authors. Then, having examined the availability of these earlier works in the early Renaissance through inventories and other sources, I eliminated works that were not known to the authors under consideration. I also excluded reports about early Renaissance attitudes from later phases of the Renaissance. These controls served to guarantee that the ideas and values I attributed to early Renaissance culture were demonstrably held by people living at the time. Once the boundaries were established within which discourse could take place, themes and problems became apparent. These then determined the topics for the essays, since I wanted to permit the early Humanists to speak in their own voices about the issues that most concerned them. It is also true, however, that my perception of which problems were important was informed by my experience as an architectural historian and that I was more interested in writing about what was less well known than in repeating what was common knowledge.

The essays are organized in relation to three themes: (1) the use of architecture in ethical discourse; (2) the critical criteria with which the early Humanists did, and did not, approach architectural experience; and (3) the development of architectural description as it relates to the recovery of eloquence. The essays in parts I and II focus on Alberti and his circle and on the architectural descriptions of Florence Cathedral in his *Profugiorum ab aerumna, libri III* and his letter to Brunelleschi, which prefaces *Della Pittura*; the essays in part I take up the place of architecture in ethical argument, the context within which it was most frequently discussed in early Humanist culture. Chapter 1 examines Alberti's use of architectural imagery as allegory for the ideal state of the soul in *Profugiorum ab aerumna*. In this case, a work of architecture served to focus Alberti's exploration of the problem of how the demands of personal salvation can be reconciled with those of active participation in society. Chapter 2 examines what Alberti and his contemporaries believed to be important about Brunelleschi's dome. They showed no interest in its shape, style, or relation to Classical models. Instead, what struck them as important was its originality and difficulty, especially as demonstrated in its engineering. For Alberti, in his letter to Brunelleschi, these qualities suggested that human civilization could be understood as progressing rather than declining and that the achievements of his own times might surpass those of the Ancients. In chapter 3, I investigate Alberti's contradictory attitude toward great size. Given the strong Classical and medieval tradition of condemnation of enormous structures, and Alberti's own condemnation of them in his treatise on architecture, how did he justify his praise of Brunelleschi's dome? In fact, his attitude and that of others in the Renaissance was deeply ambivalent and remained unresolved.

The three essays in part II explore the categories, criteria, and vocabulary with which architecture was described and evaluated in Alberti's circle. Chapter 4 returns to Alberti's description of Florence Cathedral, asking how it is that

Alberti, who certainly admired Classical architecture, chose the most fully Gothic structure in Florence as his image for the ideal state of the soul. It emerges that his categories of architectural criticism were style-neutral, and that principles of excellence such as beauty and utility could be embodied in Gothic as well as Classical works. A closer look at the vocabulary and literary structure of the same description occupies chapter 5. For Alberti, the principles of good architecture are not different from those that govern good speech. The equilibrium of opposites that he claims to see embodied in the cathedral reveals an ideal relation whose analogues are in oratory, music, and the nature of God. The last chapter in this section offers a rereading of Pienza, one of the most important projects of early Renaissance urban planning. Its example serves to challenge assumptions about early Renaissance preference for Classical style and about the currency of Platonic ideals. The complex, I suggest, is more fruitfully understood as inspired by the quite different values of change, variety, and movement.

The essays in part III center on a problem that has long troubled me, the contribution of Byzantium to the Italian Renaissance. While art historians acknowledge the importance of frequent contact with, and influence from, Byzantine art in Italian medieval art and architecture, in Italian Renaissance studies "Byzantine" is most frequently identified with all that is *retardataire*. Yet historians of the Renaissance, unlike art historians, attribute an important and positive role to Byzantine culture in this same moment. It seemed odd to me that the very century that saw the Ferrara/Florence Council, the fall of Constantinople, and the presence of more Byzantines in Italy than at any time in the preceding thousand years should have ignored the Byzantine artistic tradition. This problem can be approached in many ways, only one of which is represented in this book. The last three chapters assess the contribution of Manuel Chrysoloras to the early Humanists' ability to write about the built environment. As in the first two parts, although I concentrate on that side of the argument that most requires development, I do not mean to lessen the validity of that about which we already know. Without question, Latin, not Greek, authors were the main inspiration for Renaissance authors. Chapter 7 examines Chrysoloras's education, the Byzantine tradition of architectural description, and the Byzantine scholar's interaction with his students in Florence in the last years of the Trecento. Then chapter 8 offers an analysis of his *Comparison of Old and New Rome*, a work that I believe is far more innovative than has been recognized and that demonstrates the criteria, language, and sources that influenced early Humanist descriptions. The analysis highlights those skills Chrysoloras possessed in a more highly developed form than did contemporary Italians, especially the ability to perceive connections between various disciplines and activities. This permitted the Byzantine scholar to implement a descriptive language that made evident the wider cultural associations of the built environment. My translation of the work, making it available in English for the first time, forms an appendix. Chapter 8 parallels my discussions of Alberti's descriptions in earlier chapters, using the texts as vehicles for an investigation of the authors' literary culture and their moral and practical concerns. The last chapter begins with literary analyses of architectural descriptions in the circle of Chrysoloras's friends and students, above all Leonardo Bruni's *Laudatio florentinae urbis*. It then examines the negative and positive responses to the new eloquence of these rhetorical works and ends with a suggestion about their contribution to the new historiography that developed just after midcentury.

Searching through volumes of letters and histories for chance references to architecture has been a time-consuming task—one that architectural historians, for whom this book is primarily intended, may not wish to repeat. Consequently I have included as much of the literary evidence as possible. I also felt that since my work was interpretive the reader should have the evidence at hand with which to agree or disagree with my reading. For historians of the Renaissance, for whom this book is also intended, I hope the connections I have drawn between literary and historical matters and the art of architecture will form a useful bridge between diverse cultural manifestations. I do not claim to have read all that was written in the Quattrocento; in particular, I have not used any unpublished sources. A select bibliography of primary sources is provided for those interested in the genre. Nor have I read the entire literary production of Antiquity and the Middle Ages; therefore, my suggestions about sources can be only partial. References to this literature are in the notes but not in the bibliography; wherever possible, I have used those editions most easily found in libraries. The core of my work consists in establishing relations among original texts; therefore, the secondary literature was useful either in establishing specific facts about the works or in suggesting ideas that I could apply to my material. Because the subjects of my essays have received little attention, and because they are not only interdisciplinary but also cross-cultural, a large number of secondary works had to be consulted, each of which contributed some fragment to my understanding. The bibliography of these does not include all works cited in the notes, but those that helped form my ideas, whether they were then cited or not. The complexity of the issues, and the need to document both the Humanists' opinions and their Classical sources in order to make this book as useful as possible to the reader, suggested that it was better to treat selected problems in some depth rather than provide a general view of the early Humanists' writings on architecture. Some questions had to be put aside entirely: my study of writings about the Roman ruins was simply too lengthy to be included; my investigation of the role of fantasy in description is only touched on in the last chapter. Nevertheless, I hope to have put together the written evidence on some of the central problems in early Renaissance architecture and early Humanist culture; obviously, I have not treated problems in which these two phenomena do not overlap. For each problem, I developed the less well-known side of the issue more than that for which the arguments are already known. There was no need, for example, to lay out in detail the relation between the Ciceronian position on language and the imitation of authoritative Classical models by architects in chapter 4. By contrast, the relevance of the Eclectic position on language to architectural theory and practice had never been argued and therefore required extended demonstration. This method was not meant to replace the Ciceronian position with its opposite, but to confer an equal status on each as alternatives between which early Humanists chose.

Whereas historians have long recognized that there were at least two well-developed, and opposing, points of view on practically every question of importance to the early Humanists, architectural historians (and historians when they have written about architecture) have explicated the new architecture in terms of a unified cultural attitude. Thus Holmes, who was one of the first to show how early Humanist culture differed from that of later phases of the Renaissance, and how one of its salient characteristics was diversity, when he came to describing the new architecture wrote:

> The humanists and their disciples were acutely conscious of the difference between the gothic architecture which had become normal in fourteenth-century Italy and the style of classical architecture which they could see in the remains of the Roman buildings. The main impulse to the creation of a new style was given by the desire to imitate and recreate the glories of the classical past.[8]

It seems to me that this has come about, in part, because the architecture of early Humanism has been seen as an early phase of a development that culminated in the sixteenth century. Generalizations about Renaissance architecture, which includes both the fifteenth and sixteenth centuries, have tended to privilege the theoretical positions of the more developed phase. This position was favored partially because almost all contemporary writing on the subject dates from after 1450. It was natural to assume that since Brunelleschi wrote almost nothing, Manetti's biography could be used as evidence for his attitudes, especially since Manetti knew the architect and wrote a few decades after his death. Yet I will show that it was precisely in those intervening decades that the whole system of early Humanist values, within which Brunelleschi operated, gave way to a new culture, within which Manetti wrote. The use of later values to explain earlier events has been widespread precisely because it was unconscious. When Wittkower set out, in his revolutionary *Architectural Principles in the Age of Humanism*, "to dispose, once and for all, of the hedonist, or purely aesthetic, theory of Renaissance architecture," he did not differentiate, in essential ways, between early and later Renaissance architecture.[9] Zurko attacked Wittkower's treatment of Alberti's thought, claiming (rightly, I believe) that spatial mathematics "is not the distinguishing feature or integrating factor of humanist Renaissance architecture," nor does it display a hierarchy of values "because hierarchy implies fixed principles and authoritative organization," whereas what distinguished fifteenth-century developments were its freedom of approach and an open mind.[10] The open-mindedness of the early Humanists did not mean that they had no values. Rather, their culture was characterized by a vivid confrontation between different beliefs, polemic, and shifts in position. The early Humanists were not serene in their belief that good style entailed the imitation of Classical models, but embattled over the question. As for spatial mathematics, no matter how important this may have been in architectural practice, there is little evidence of interest in the matter on the part of the Humanists up to about 1460. This aspect of architectural history, therefore, cannot be interpreted within the culture of *early* Humanism, as opposed to later Renaissance thought or currents in the early Renaissance apart from Humanism.

Generalizations about Renaissance architecture as a whole encouraged the study of some aspects of early Renaissance architecture over others. Skillfully delineated, especially in an elegant article by Gombrich, a certain set of ideas has come to dominate our understanding of the intellectual bases of early Renaissance architecture.[11] Thus Peter Murray, whose survey of Renaissance architecture is one of the most frequently used for teaching, tells us about the Florentine dome that "Brunelleschi, like every other classically-minded architect, would have wished to build a hemispherical dome because of its perfection of shape and because the great Roman domes, above all the Pantheon, are hemispherical."[12] Murray's text disseminated assumptions that have not been challenged in any systematic way by architectural historians: (1) that early Renaissance architecture was inspired by a Pythagorean–Platonic aesthetic; (2) that early Renaissance architects strove to imitate Classical models; (3) that they rejected Gothic architecture; and (4) that the Pantheon was considered to be the paradigm of Classical architecture. Some of these attitudes do indeed correspond to one strand of early Humanist culture, although the emphasis on Plato and appreciation of the Pantheon are typical of that strand at a later moment in its development, in the late Quattrocento and not in 1420. But Murray's formulation, and others like it, is ultimately misleading because reductive. It gives no hint that, as is clear from the following essays, the early

Humanists read a good deal more Cicero, Aristotle, and Augustine than Plato, whose works were only beginning to be known when Brunelleschi designed the dome. It ignores the vivid controversy among the Humanists about the imitation of Classical models, and it dismisses the admiration some Humanists felt for the products of "modern" culture, including the Gothic (discussed in chapters 2 and 4). And it applies to a work of around 1420 values that came to the fore half a century later (chapter 6).

Of course, a more nuanced picture has been presented of individual problems in the scholarly literature, and no specialist would totally agree with the necessarily simplified view of a general survey. Howard Burns warned, for example, that "it is necessary, before one can understand the architectural aesthetic of the Quattrocento, to examine the whole intellectual climate within which architects sought to understand and imitate the antique."[3] Yet most work on architectural principles in the age of Humanism—and I quote Wittkower's title advisedly—has served to flesh out the concepts Murray selected, rather than consider the relevance of other, opposed concepts equally important in the culture. Thus it has not been possible to connect many of the new observations of recent research with the general attitudes and values of early Humanism. For example, we know that Brunelleschi, Michelozzo, and Alberti all included Gothic motifs in their works, but not what theoretical justifications there could have been for doing so. Specialists realize that Pienza is not an ideal city plan, but have not been able to propose alternative aesthetic values. We know a good deal about how Brunelleschi's architecture relates to one strand of Humanism, but not to other strands within the same culture. In regard to Michelozzo's work, in many ways antithetical to Brunelleschi's, we have no positive critical apparatus with which to assess it. My essays are a first step in exploring the early Humanists' ideas about architecture in the context of their thought as a whole. It will be seen that some of the early Humanists, including Alberti (at least on some issues), held views that not only were different from those of the later Renaissance, but were opposed to them.

PART I

Ethics

ONE

Profugiorum ab aerumna, libri III: Architectural Allegories of Virtue in a Dialogue by Leon Battista Alberti

Profugiorum ab aerumna, libri III, also known as *Della Tranquillità dell'animo*, was written in the vernacular by Leon Battista Alberti in 1441 or 1442 at the end of a lengthy stay in Florence.[1] As is suggested by its Latin title, the moral essay is somber, even pessimistic, in tone: the desired state of spiritual tranquillity is but briefly glimpsed, primarily through architectural allegories at the beginning and end of the work; most of the dialogue concerns "perturbazioni"—disturbances of the soul caused by adverse fortune, man's injustice to man, or the variability of fortune. Each book of the treatise examines a different spiritual state: the first considers what tranquillity is and its potential acquisition; the second, how equilibrium can be maintained under adverse circumstances; the last by what efforts man can combat despair and painful inner turmoil.

All the participants in the dialogue are historical figures. The Florentine statesman Agnolo Pandolfini is the main speaker.[2] In real life, his disillusionment with the successive exiles of the Medici and Strozzi families had prompted him to withdraw from public affairs (in 1434) and retire to his villa. This personal experience explains why Alberti cast him as spokesman for the perils of the active life and the ways in which they may be remedied. Agnolo's religious piety, attested to by his contemporaries, turns Alberti's dialogue toward a specifically Christian, rather than generally ethical, discussion. Agnolo's career and the theme of the essay invite comparison with Classical works known to Alberti: with the *Tusculanarum disputationum*, written when Cicero had retired from public life; and with Seneca's *De Tranquillitate animi*, in which the Stoic advises withdrawal from political activity when such activity becomes morally untenable. But although some of Alberti's arguments are drawn from these Classical texts, his ideal is quite different: spiritual tranguillity is sought not as a good in itself, but within the context of Christian salvation.

Agnolo's interlocutor, Nicola di Vieri de'Medici, was a student of Leonardi Bruni, a friend of Poggio Bracciolini, and one of the executors of Niccolò Niccoli's will.[3] He was, perhaps, more successful as a student of the humanities than as a businessman: by 1433 his bank had incurred such debt, due to his inexperience in practical affairs, that he lost his inheritance and was reduced to poverty. Thus the real Nicola had good reason to seek guidance from Agnolo in Alberti's dialogue. He proposes a wide variety of philosophical doctrines that Agnolo, drawing on his practical experience and religious faith, evaluates.

The third figure, and silent partner in the dialogue, is Alberti himself. He does not reveal what circumstances in his life led him to treat this subject. We know, however, from the prefatory letter to *Theogenius*, of about 1439, that Alberti was depressed by being subject to what he perceived as maltreatment from his fellow man and sought consolation from his bad fortune by writing.[4] Mancini suggested that Alberti might have been distressed by the exile of the Strozzi, Peruzzi, and Albizzi families in 1434;[5] if so, then his choice of Agnolo for the main speaker in *Profugiorum* acquires additional significance. The polemic over the *certame coronario* (contest for vernacular poetry) of 1440–1441 may also have played a role.[6] The contest, which took place in the cathedral, the setting also of *Profugiorum*, ignited a polemic concerning the possibility of creating great poetry in the vernacular. Alberti vehemently defends the expressive possibilities of modern language, a position he underscores by opening his treatise with the praise of Florence's Gothic cathedral, as we shall see in chapter 4.

Unique to Alberti's treatise, within its literary genre, is its vernacular language and use of architectural imagery. His enumeration of human ills shows his absorption of not only the works by Classical authors already mentioned, but also those paradigms of the literature of pessimism: the books of Ecclesiastes and Job, and Innocent III's *De Miseria humanae vitae*. Yet his is no treatise on human suffering and the vanity of human effort; Alberti believes that beatitude is also within human experience and that man's works deserve praise. Thus his essay is an important forerunner of Renaissance treatises on the dignity of man, which, beginning with Bartolomeo Facio's *Dialogus de vitae felicitate*

(1445–1446), and developed in Antonio da Barga's *Libellus de dignitate et excellentia humanae vitae* (1447) and Giannozzo Manetti's *De Dignitate et excellentia hominis libri quattuor* (1452 or 1453), culminated in Pico della Mirandola's *De Hominis dignitate* (1486).[7]

Profugiorum is a precious source for the architectural historian, showing how the experience of architecture could be used for moral argument in the culture of early Humanism and revealing Alberti's own aesthetic response to architectural style. That the structure singled out for this unique status in his work was the Gothic Cathedral of Florence is most unexpected, given Alberti's admiration for Classical architecture. (Reflections on this choice are the subject of chapter 4.) Alberti's description of Florence Cathedral in *Profugiorum* elucidates the principles of good architectural design as they are felt by an individual in relation to a specific building. (These principles, and Alberti's critical vocabulary, will be examined in Chapter 5.) Here, my task is to investigate the moral and literary implications of the description. In this, I follow Alberti's own priorities: for had he not valued the cathedral primarily as a rhetorical vehicle for a spiritual message, he would not have bothered to describe it at all. It was the necessity of stimulating the imagination of the reader, and thus enabling him to think about spiritual tranquillity, that prompted Alberti to bring the cathedral before his eyes.

Architectural imagery serves a variety of purposes in *Profugiorum*: in Book I, it illustrates the ideal harmony of the soul and its ability to withstand misfortune; in Book III, architectural design is the means by which the troubled soul regains peace. But it also provides a physical setting for each of the three books, which, by symbolizing the various conditions of the soul, strengthens the conceptual structure of the literary work. Book I takes place inside Florence Cathedral, the allegory for spiritual well-being; in the second book, where the soul successfully battles adverse fortune, the speakers meet in the cathedral, walk through the city, and return to the security of the cathedral at the end; in Book III, a discussion of the means by which a soul in profound melancholy may raise itself up takes place a few steps outside the cathedral.

The image of Florence Cathedral as allegory for the tranquillity of the soul in Book I is drawn in the opening paragraphs of the treatise. Alberti and Nicola de'Medici, strolling in the cathedral, are joined by Agnolo Pandolfini, who, commending them for being there, explains his approval by describing the building:

> And certainly this temple has in itself grace and majesty; and, as I have often thought, I delight to see joined together here a charming slenderness with a robust and full solidity so that, on the one hand, each of its parts seems designed for pleasure, while on the other, one understands that it has all been built for perpetuity. I would add that here is the constant home of temper-

ateness, as of springtime: outside, wind, ice and frost; here inside one is protected from the wind, here mild air and quiet. Outside, the heat of summer and autumn; inside, coolness. And if, as they say, delight is felt when our senses perceive what, and how much, they require by nature, who could hesitate to call this temple the nest of delights? Here, wherever you look, you see the expression of happiness and gaiety; here it is always fragrant; and, that which I prize above all, here you listen to the voices during mass, during that which the ancients called the mysteries, with their marvelous beauty. (p. 107)

(E certo questo tempio ha in sè grazia e maiestà: e quello ch'io spesso considerai, mi diletta ch'io veggo in questo tempio iunto insieme una gracilità vezzosa con una sodezza robusta e piena, tale che da una parte ogni suo membro pare posto ad amenità, e dall'altra parte comprendo che ogni cosa qui è fatta e offirmata a perpetuità. Aggiugni che qui abita continuo la temperie, si può dire, della primavera: fuori vento, gelo, brina; qui entro socchiuso da'venti, qui tiepido aere e quiéto: fuori vampe estive e autunnali; qui entro temperatissimo refrigerio. E s'egl'è, come è dicono che le delizie sono quando a' nostri sensi s'aggiungono le cose quanto e quali le richiede la natura, chi dubiterà appellare questo tempio nido delle delizie? Qui dovunque tu miri, vedi ogni parte esposte a giocondità e letizia; qui sempre odoratissimo; e, quel ch'io sopra tutto stimo, qui senti in queste voci al sacrificio, e in questi quali gli antichi chiamano misteri, una soavità maravigliosa.)

Agnolo's words express his direct visual observations and communicate his personal feelings. But, at the same time, they describe the soul's equilibrium through the allegory of the architectural style of the church and the interior ambient it encloses. Its style is presented as pairs of opposite qualities in equilibrium: grace and majesty; fragility and strength; beauty and utility. In a second series of images, Agnolo depicts an antithesis between the constancy of the cathedral interior and the extremes of nature outside its walls—that is, the opposition between spiritual tranquillity and the variability of fortune.

These images are no mere literary digression or mise-en-scène: Nicola immediately recognizes that Agnolo is addressing a moral question that concerns him. Let us leave aside the praise of the building's beauty and analysis of its engineering, he suggests, in order to focus on its spiritual significance (p. 109). Agnolo is quite ready to do this. Drawing on Plato's *Phaedrus*, he explains that the soul is composed of two parts: one is tranquil, constant, and industrious; the other is lazy, pleasure-loving, and subject to passions.[8] Agnolo represents as moral qualities the stylistic pairs used to describe the cathedral: that which gives pleasure and that which is useful. It would be better, continues Agnolo, if we could entirely dominate that part of the soul subject to disturbance, but this is not possible to us in our frailty (p. 111). Nonetheless, we can strive to keep the aspects in equilibrium, fortifying ourselves against the blows of fortune.[9] Here he explicates his antithesis between the temperate interior of the cathedral and the changes in nature that buffet its walls. To make his point more clearly, Agnolo adds another architectural example: just as a column can sustain any amount of weight placed on it as long as it is upright, so the soul opposes adverse fortune with the rectitude of virtue (p. 113). This ethical interpretation of a column as a symbol of balance, purity, and incorruptibility may derive from Proclus's commentary on Euclid's *Elements*.[10]

Nicola doubts that it is within human power to repel the "fluctuations and tempests of the soul," to which we are subject by our mortal nature and from which only death can free us (p. 115). To this, Agnolo makes a remarkable reply: just as we protect ourselves from the cold of winter with warm clothing, walls, and places of refuge, so the soul protects us from the tempests of fortune (p. 121). Indeed, he goes on, although the soul itself is subject to change and variety, we must construct ("edificare") another nature within ourselves. Agnolo's philosophical bases require some comment. In part, what he says reflects Aristotle's view that man possesses neither intellectual nor moral virtue by nature, but only the capacity for them.[11] Spiritual tranquillity, then, is a human creation, built up as a moral edifice through human intention, knowledge, and skill. This constructive approach to virtue begins to explain Alberti's choice of an architectural monument for his allegory. Also important is Augustine's discussion of the tranquillity of the soul in the *City of God* (IX.1–7), where, after reviewing Platonic, Stoic, and Peripatetic views, he emphasizes the importance of reason in controlling the passions.

But if human virtue opposes nature, why did Agnolo seem to accept nature as good, claiming that the cathedral interior offered his senses "what, and how much, they require by nature"? Are we to live in harmony with nature, as the Stoics thought, or opposed to it, as Aristotle's explanation of virtue implies? The question must be considered within the context of ethics—that is, human life after the Fall. For although a harmonious relation between man and nature might have prevailed in the Garden of Eden, or during the golden age, when man did no work and nature provided for all his needs, that era is not only past, but beyond recovery.[12] Even so, nature as we experience it can be thought of as perfect, since it is God's handiwork, in which case it is virtuous to accept it. An example of this approach is Horace's tenth epistle, uniting the Stoic precept "to live agreeably to nature" with the topos of the *locus amoenus*:

> If "to live agreeably to nature" is our duty, and first we must choose a site for building a house, do you know any place to be preferred to the blissful country? Is there any where winters are milder, where a more grateful breeze tempers the Dog-star's fury and the Lion's onset, when once in frenzy he has caught the sun's piercing shafts?[13]

Horace's house is but the accessory means permitting him to remain in the midst of a temperate nature. For Alberti, instead, nature is fierce and inimicable: in *De Re aedificatoria*, he explained the origins of architecture as man's attempt to defend himself against nature, and Agnolo, too, said that we fortify ourselves against nature by building houses.[14] To investigate the moral implications of this position, we need to go beyond the comparable Vitruvian passage to Cicero.[15] For the latter, the re-formation of nature is an achievement of human cooperation, and thus of civilization: "How could houses ever have been provided in the first place for the human race, to keep out the rigours of the cold and alleviate the discomforts of the heat... had not the bonds of social life taught men in such events to look to their fellow-men for help?"[16] Man's great task in the world is "the fashioning of another world, as it were, within the bounds and precincts of the one we have."[17] Cicero believed that history can be understood as human progress from a state of misery and deprivation toward civilization through the mastery of nature, a view shared by Lucretius (*De Rerum natura* II.16-19, V.195-234, V.925-1457). Although such Classical discussions of man as second creator focused on manual skills and physical advantage, Agnolo's injunction to build another nature within ourselves extends this concept to morality, where, of course, it finds support in Aristotle. But, more than that, Agnolo's approach must be understood within his own Christian concerns for personal salvation and participation in the redemptive process of the created world.

It is not Alberti who first expanded the notion of second creation in a spiritual direction, thus questioning whether God's creation was sufficient and good, and allocating to man the task of bringing it to perfection.[18] Boccaccio laid special emphasis on the role of intellect, enabling man to invent not only customs, arts, and sciences, but also virtues:

> Indeed, those who are produced by nature are rude and ignorant, unless they are taught—dirty, rustic and beasts—among whom a second Prometheus arises, that is, the educated man, and taking them up like stones, creates them almost as if new. He teaches and instructs them, and through his demonstrations from natural men makes them civil and noteworthy in their customs, sciences and virtues; so much so that it is crystal clear that nature had produced one set of people, and learning re-produced the others.[19]

The moral and theological implications of this view were explored by Alberti's contemporaries. Lorenzo Valla extolled the value of work as the means by which man, transforming nature, fulfills himself;[20] Nicolas of Cusa likened human invention to God's creative activity.[21]

In choosing Florence Cathedral, a work of human ingenuity, as his image of spiritual equilibrium, Alberti asserted that we should live in accord with nature, not as we find it, but as we re-create it. Thus while flatly contradicting Seneca's advice to "Follow nature and you will not need the skills of a craftsman," he accepted the Stoic dictum to live agreeably to nature—meaning re-created nature.[22] Alberti's choice opposes Innocent III's pessimistic assessment of architectural activity as an affliction of the soul, having as its aim riches, honors, offices, and power; and it utterly rejects that disparagement of human works (evoked as evidence by Innocent) in Ecclesiastes (2:4–9): "Then I considered all that my hands had done and the toil I spent in doing it, and behold, all was vanity and a striving after wind, and there was nothing to be gained under the sun."[23] Much that has been defined as new in Alberti's architectural thought by Westfall and Bialostocki finds its explanation in the relationship between man and nature examined here.[24]

Man "finds pleasure in what is his own, such as his works and words," said Aristotle; in *Profugiorum*, what gives pleasure is man-made.[25] The idea would be further developed in Manetti's treatise on the dignity of man:

> After that first, new and rude creation of the world, everything seems to have been discovered, constructed and completed by us out of some singular and outstanding acuteness of the human mind. For those things are ours, that is, they are human, which are seen to be produced by men: all homes, all towns, all cities, finally all buildings in the world which certainly are so many and of such a nature that they ought rather to be regarded as the works of angels than of men.[26]

Agnolo calls the cathedral a "nest of delights," describing his happiness at human capacity to bring opposed qualities into harmonious relation. In *De Re aedificatoria*, begun the year after *Profugiorum*, Alberti tells us that good architecture elicits pleasurable emotion in the spectator who, enchanted by wonder at the excellence of the work, is stimulated to express an intellectual judgment about its qualities (VII.iii, pp. 543–45). But examples of such responses are absent from *De Re aedificatoria*: the treatise, concerned with truth and precept rather than opinion, abstains from subjective expression.[27] Alberti's exposition of the principles of good architecture does, to be sure, include abundant examples, but they are qualified by, at most, one or two adjectives signifying approval. His recommendations regarding good design are either practical, and limited to single aspects of a building, or abstract and theoretical. Since there is no extended description of any building in *De Re aedificatoria*, it is not clear—as it is, instead, in *Profugiorum*—how Alberti would have applied his precepts to the architecture familiar to him. Yet Alberti tells us in *De Re aedificatoria* that the spectator who has been moved by the beauty of a building will seek to put his feelings into words. In *Profugiorum*, recognition urges communication. That the perception of beauty should be shared is an unusual idea found, as far as I know, only in Lucian. Whereas vulgar people look at beauty in silence and point, he says, "When a man of culture beholds beautiful things, he will not be content, I am sure, to harvest their charm with his eyes alone, and will do all he can to linger there and make some return for the spectacle in speech."[28] Agnolo speaks as a cultured man, but it is his perception of virtue rather than his appreciation of beauty that causes him joy and moves him to speech. In describing the Cathedral of Florence, he shares his moral insight with Nicola and Alberti. Book I of *Profugiorum* explores the philosophical implications of Agnolo's initial reaction to an architectural experience.

What has been said about the conceptual and literary structure of Book I raises important questions about Alberti's theory of knowledge, in which the initial impulse for thought is sensory experience. Here again, his approach is rooted in Aristotle. "The soul never thinks without a mental picture," he says in *On the Soul*, and "no one could ever learn or understand anything, if he had not the faculty of perception; even when he thinks speculatively, he must have some mental picture with which to think."[29] And, in *On Memory and Recollection*, Aristotle notes that "memory and the knowledge of things thought cannot exist without a mental picture. So that they would seem to belong incidentally to the thinking faculty, but in themselves to the first sense perception."[30] Alberti's description of the cathedral furnishes such an image, but not in the way intended by the philosopher. For the architectural example in *Profugiorum* is not only an image, but an allegory in which is condensed the entire logical structure and emotional resonance of Alberti's meaning. Yet his allegory is of a fundamentally different kind from the extended architectural antithesis of Augustine's *City of God*. Augustine's cities have no material presence as urban structures, no buildings, and therefore no stylistic qualities. Indeed, there is not a single architectural description in the entire work. Contrary to Aristotle's belief, imaginative mental and spiritual activity can take place without the stimulus of sensory perception, or so we must suppose from claims, like that of Cicero, that "the thought of the wise man scarcely ever calls in the support of the eyes to aid his researches."[31] St. Bernard, who having spent a year secluded in his cell could not remember whether it was vaulted, was evidently not dependent on visual stimuli for his thoughts.[32]

Alberti's decision to use an image to convey his meaning may suggest his acceptance of Aristotelian epistemology over other Classical and medieval approaches to learning. In any case, his approach needs to be seen in the context of a growing tendency to rely on such imagery in early Humanist literature. Salutati, for example, considered precept alone to be insufficient for argument. Frontinus, he says, "was not content to impart his science by means of precepts and rules but placed it before the eye, as it were, by means of infinite examples."³³ Examples, if not descriptions, abound in Alberti's *De Re aedificatoria*; and the example of Florence Cathedral's dome is a metaphor for the revival of the arts in his preface to *Della Pittura* (chapter 2). But if Alberti's imagery is, perhaps, the most eloquent and telling, he was not the first Humanist to employ architectural metaphor. I think particularly of Leonardo Bruni's *Ad Petrum Paulum Histrum*, which debates the value and achievement of recent Florentine culture.³⁴ Each of the two dialogues is preceded by a description of the city. The first, in Bruni's *proemio*, serves to introduce the subject, suggesting that the high quality of the city's architecture illustrates the intellectual level of its citizens. Salutati's observations about the magnificence of Florence's architecture that preface the second dialogue lead into his panegyric of the cultural attainment of the city's inhabitants.

The new desire to understand through sight is one of the most important concerns of the early Humanists, and a central theme of this book. (The use of the built environment as evidence for the nature and status of the city is explored in chapters 2, 8, and 9; its importance for understanding the past is explored in chapter 8.) The early Humanist tendency to supplement words with images—whether mental, as in *Profugiorum*, or physical—is reflected in an interesting addition made to manuscripts of the *City of God* in the fourteenth and fifteenth centuries: Augustine's omission of a description of the physical appearance of the two cities was amended by illustrations of the saint's vision.³⁵ The assumption was, of course, that Augustine could not have thought without the stimulus of sight. This tendency to translate conceptual matter into imagery may explain an illumination in a late-fifteenth-century manuscript of the *Nicomachean Ethics*: a city with owls on its walls and towers has, in its center, a domed building topped by a figure of Athena.³⁶ Such visualization of virtue in an urban setting is radically different from Aristotle's own use of architectural imagery, but it is analogous to, if simpler and more naive than, Alberti's description of Florence Cathedral.

In Bruni's dialogue, and in the illustrated manuscripts, architectural imagery is a subsidiary aid to understanding the conceptual matter of the work. This is not the case in Alberti's moral treatise. Since Alberti nowhere defines the tranquillity of the soul in words, the architectural allegory is its only vehicle of expression. All that we know about the ideal state of the soul is what Agnolo tells us about the cathedral, verbalizing his emotional response. If, as Aristotle thought and Alberti seems to agree, states of the soul are emotional conditions, they may elude objective definition and even be difficult to verbalize (*Nicomachean Ethics* II.v.i). Thus Agnolo's description helps the reader to understand spiritual tranquillity by eliciting his imaginative participation in a visual experience. Florence Cathedral, the largest and most ambitious project of the time, was a structure known to most, or even all, of the dialogue's readers. The meaning of Book I depends on the reader's ability to perceive spiritual equilibrium and fortitude through visual memory, guided by Agnolo's verbal comments and his explication of the cathedral's moral significance. That the cathedral does indeed provide an experience of spiritual tranquillity is assumed by the speakers; the evidence of an individual's feeling is not questioned. Alberti's is not a philosophical discussion of the nature of virtue, for which logical forms of expression are necessary. His dialogue is evidence for the Humanists' preference for conversation, based on rhetoric and aiming at persuasion, over disputation as practiced by the Scholastics (chapter 7). Virtue is examined in its moral dimension as a disposition of the soul, and therefore as an emotion, perceived in a particular time and place by an individual. Direct knowledge of virtue is circumstantial and subjective, but its structural configuration can be abstracted, objectified, and embodied in human works. Architecture presents, as in perpetual equilibrium, those opposed forces that tranquillity reconciles.

Antithetically to Book I, Book III shows human suffering resulting from such unhappiness that the goal of spiritual equilibrium seems entirely beyond reach. Nicola says to Agnolo:

> Tell me, what virtue will relieve us from our oppression of bad luck and these ruinous times? Those wise men will say, pay no attention to your pain. Easily said, easily said! But that man who has lost his servants, family and friends; lost all that makes his life convenient and satisfactory; lost his family inheritance, his grandeur, his public authority and place of honor; and who now finds himself alone, in need, abject, destitute, and perhaps infirm in mind and body—how can that man come to his own aid and strengthen himself? (p. 163)

> (Dite, qual virtù sara qualla che noi sollievi oppressi da e'nostri casi avversi e dalle ruine de'nostri tempi? Diranno que'savi: non curare e' tuoi dolori. Facile precetto a dirlo, facile a dirlo. Ma colui el quale perdette e' noti a sè, domestici, coniunti, amici, e perdette l'altre sue commodità e onestamenti, e perdette sue fortune domestiche, amplitudine, autorità publica e luogo di dignità, e ora si truova in solitudine, assediato da ogni necessità, abietto, destituto, e forse malfermo e poco intero in suoi nervi e membra, come aiterà e soverrà a sè stessi?)

Even though these may be fragile and transient goods, says Nicola, I desire them intensely, and it gives me great pain to be deprived of them. Agnolo agrees that, being human, we should not deny the truth of our feelings. But it is no use crying like spoiled children over what we cannot have. Which seems worse, he asks: to lose what can never be regained, or to lose what is born to be lost and can be found again (pp. 168–70)? After all, we can only hope for good fortune, but we can require virtue from ourselves, acquire wisdom from our studies, and earn praise for a well-balanced mind (p. 173). Agnolo passes in review various remedies for unhappiness—wine and song, women, conversation with friends, sports—before coming to his own solution.

When all else fails to raise his spirits, he seeks to regain tranquillity of the soul through intellectual activity—that is, through study (p. 181). He memorizes poetry or prose and writes literary criticism. But more effective than these pursuits is the design of mechanical devices and the exploration of architectural design:

> I am accustomed, most of all at night, when the agitation of my soul fills me with cares and I seek relief from these bitter worries and sad thoughts, to think about and construct in my mind some unheard-of machine to move and carry weights, making it possible to create great and wonderful things. And sometimes it happens that I not only calm the agitation of my soul, but invent something excellent and worthy of being remembered. And at other times, instead of pursuing these kinds of thoughts, I compose in my mind and construct some well-designed building, arranging various orders and numbers of columns with diverse capitals and unusual bases, and linking these with cornices and marble plaques which give the whole convenience, and a new grace. (pp. 181–82)

> (Soglio, massime la notte, quando e' miei stimoli d'animo mi tengono sollecito e desto, per distormi da mie acerbe cure e triste sollicitudini, soglio fra me investigare e construere in mente qualche inaudita macchina da muovere e portare, da fermare e statuire cose grandissime e inestimabili. E qualche volta m'avenne che non solo me acquetai in mie agitazioni d'animo ma e ancora giunsi cose rare e degnissime di memoria. E talora, mancandomi simili investigazioni, composi a mente e coedificai qualche compositissimo edficio, e desposivi più ordini e numeri di colonne con vari capitelli e base inusitate, e collega'vi conveniente e nuova grazia di cornici e tavolati.)

When even this does not succeed in raising his spirits, he turns to mathematics, particularly those theoretical problems that might have some practical application. In this, says Agnolo, he is like Alberti, who not only explained the principles of painting on the basis of mathematics, but set down incredible propositions in his treatise *De Motibus ponderis*.[37] Indeed, continues Agnolo, I become so deeply involved in mathematical questions that I no longer see or hear anything that goes on around me. Imagine, then, what tranquillity those with greater mathematical ability must possess. Agnolo closes his long discourse (and Alberti, his treatise) with a final example.

During the seige of Syracuse, Archimedes had defended his city with never-before-seen machines and instruments of war that postponed, but could not prevent, its capture. He was so completely absorbed in studying geometrical problems, and drawing them on the floor of his home, that he did not hear the noise of weapons or the cries of the dying citizens; nor was he aware that the city was in flames and its buildings collapsing. Nothing at all could intrude on his mathematical investigation. Therefore, concludes Agnolo, have no doubt that if we pursue virtue, act rightly, and study worthy subjects we will be perfectly tranquil in the face of all pain, all adversity of the times or of fortune, and fortified against the perfidy and iniquity of our fellow men.

The treatise closes, as it opened, with an image of pleasurable absorption and tranquillity of the soul. In Book I, this joy was a passive response to the stylistic qualities and moral significance of a great edifice; by the end of Book III, man is actively seeking to master the principles on which such an edifice is built. Although both are images of the contemplative life, Alberti's assumption throughout the essay is that human ambitions and desires are realized only in human interaction. As Agnolo says, God gave us two things—reason and society (p. 124). But if Alberti's moral ideal is the application of reason to problems of social utility, this solution encounters all manner of difficulties in its application. Indeed, his essay is not a debate about the relative merits of the active and contemplative lives, but about the moral problems that arise in public life. As Agnolo puts it, "in a crowd you can neither stand still nor walk about without someone knocking into you" (p. 124).

At first glance, the solutions presented in Book III seem to follow Seneca in *De Tranquillitate animi*: when public life is untenable, it is wise to withdraw into solitude (III.1–3). Seneca suggests this as only a temporary strategy: "the two things must be combined and resorted to alternately—solitude and the crowd" (XVII.3). He furnishes many examples of important public figures and philosophers who, like Archimedes, displayed perfect tranquillity when they were threatened with death (XIV).

Alberti's substitution of the geometer for the statesman or philosopher is extremely significant, and will be investigated shortly. I want first to focus on a more general and also more fundamental difference between these similarly titled works. Alberti cannot share Seneca's conclusion that the joys and sorrows of human life are not to be taken seriously, since we are born from and return to nothingness (XV.4). As we saw, Alberti's discussion takes place within a Christian dimension: man, made in God's image, must fulfill his task as second creator. Since he defines his being through his actions, he is morally obliged to pursue the active life.[38] But he is also under an obligation to seek spiritual tranquillity for the health of his soul, which will come to divine judgment. Thus the first concern of a Christian, salvation, may be compromised in the exercise of an equally Christian moral obligation toward his fellow man and toward the realization of his own nature.

This complex moral and philosophical problem was of the greatest concern to many of Alberti's contemporaries. Bruni, Salutati, and Valla are among those who debated the respective values of the active and contemplative lives, usually concluding in favor of the former.[39] Their attitude can be, and has been, seen as a Renaissance rejection of medieval withdrawal from life.[40] But what has not been noted, although it is very striking in *Profugiorum*, is the persistence of medieval goals within the Quattrocento's new framework of action. All the reevaluation of man's nature, position, and role in the world did not (to the early Humanists) suggest the renunciation of traditional spiritual aims, such as the tranquillity of the soul. Alberti's essay centers on the dynamic conflict between two kinds of virtue that seem so thoroughly incompatible as to be mutually exclusive. Spiritual disturbance appears to be an inevitable consequence of engagement in the active life.

Alberti is not able to resolve this clash between conflicting virtues; nonetheless, his final examples merit close consideration as an ingenious approach to the problem. Agnolo's studies have a hierarchical sequence, beginning with letters and ending with mathematics. The more profound the spiritual trouble, the more abstract the subject of study must be. This sequence, culminating in the mathematical sciences, suggests Aristotelian sources, such as *Metaphysics* I.i.14–16 and *Nicomachean Ethics* X.viii.7, where the highest form of learning, mathematics, is said to relate to neither the pleasures nor the necessities of life, and intellectual activity is recommended as most closely emulating God's contemplation. As Aristotle himself admitted, if man fulfills himself in intellectual endeavor of an abstract nature, then the life of moral virtue can be happy only in a secondary degree (*Nicomachean Ethics* X.viii.1). The course of action most desirable from an ontological point of view is opposed to that which is good according to ethical criteria. But this is not an acceptable solution for Alberti, and he does not adopt it.

Cicero shows the way out in Book V of the *Tusculanarum disputationum*: if intellectual investigation has potential utility and application, then contemplation is not a withdrawal from social responsibility. His argument is developed as follows. Philosophy, which is the contemplation and study of nature, is the best of human activities: it cures disturbances of the soul and calms the tempest of life. The wise man, then, spends his days and nights trying to understand the movements and forces of the universe. His contemplation leads to self-knowledge and to a perception of union with the divine mind. From this foundation begins the comprehension of moral goodness, which in turn leads to happiness, or spiritual tranquillity. Therefore, mental excellence is identical with moral excellence. And if the disposition of a man's soul is laudable, the same must be true of his life, for his intelligence and his integrity will serve his fellow citizens.

What Cicero does, it seems to me, is to break down fundamental distinctions between private and public, active and contemplative, by applying a single criterion of moral virtue to both. Everything that is morally good, fine, and distinguished, he says, is infinitely productive of joy, and this is what is meant by happiness. Instead of projecting public life and contemplation as mutually exclusive opposites, he sees them as continuous: right behavior is but the actualization of contemplation. This formulation was accepted by Augustine, who praised both the active and the contemplative life in the *City of God* (XIX.19). The life of leisure, providing opportunity for the investigation and discovery of truth, is good as long as we share our discoveries with others; the active life offers the occasion for promoting the well-being of the common people. The linkage between virtue and action also applies in the case of knowledge and application; indeed, Cicero adduces as his example Archimedes.

The choice of Archimedes as exemplar of moral virtue provides a solution to the problematic relation between the active and contemplative lives. Since his subject of study was mathematics, he meets Aristotle's requirements regarding the levels of intellectual pursuit; his invention of war machines for his city satisfies all possible demands of service to one's fellow man. Not only have few historical figures so admirably exemplified both the active and the contemplative life, but the story of his death shows that he possessed, to an extraordinary degree, tranquillity of the soul. Thus in the *Tusculanarum disputationum*, Cicero proposed a synthesis between the Aristotelian ideal of abstract speculation and Roman *utilitas*. Clagett has gathered most of the accounts of Archimedes from antiquity to the Renaissance.[41] It is interesting that, for most medieval biographers, Archimedes was a philosopher or an astronomer, not a mathematician.[42] But only as the last could he serve to synthesize the active and contemplative lives, and only if, within the field of mathematics, he was a student of mechanics, which clearly unites theoretical speculation with practical application.

At this point we may return to the last pages of *Profugiorum*. Agnolo's activity, alone at night, is the study of mechanics. He investigates mathematical problems that might have some utility, such as, in his own example, the invention of unheard-of machines to move and carry weights, making it possible to create great and wonderful things. Thus his contemplation has potential utility, even though his aim is to restore his own spiritual tranquillity. The conception of the contemplative life as potentially active is not new with Alberti. As Kristeller points out, Plato had presented theory and practice as complementary activities in the *Republic*, remarking that knowledge is the prerequisite for action and that action has no merit unless it is guided by knowledge.[43] Of special interest is Valla's observation that the original word behind the Latin Vulgate *virtus* is *dunamis*, implying that virtue carries a potential for action.

And Valla specifically connected the progress of the arts with the activity of contemplation in *De Vero bono*:

> Contemplation is nothing except the progress of learning, that is, what we call reflection as well as invention, and is the concern of men, not of the gods. Whence it has happened that also the most important arts have been carried to the height of perfection by the more profound contemplation of men.[44]

The example of Archimedes shows that the importance of creative intellectual activity (or pursuit of virtue) is not diminished, even though an inventor may live in a world where his works are in vain or where they are destroyed, or he may die before carrying them out. Thus Alberti refuses to judge the individual on the basis of his external accomplishments, the results of virtue rather than virtue itself.[45] Virtue alone can oppose fortune; only virtue is entirely within our control.[46]

The image of virtue in Book I, the Cathedral of Florence, was described as maintaining opposed qualities in equilibrium: that which gives pleasure and that which is useful were equally present; one did not exclude the other. At the end of the treatise, Alberti gives an example of how this can be actuated in human life: the joy of contemplation produces, quite naturally and without distortion, works of utility. The same attitude underlies a passage in *De Re aedificatoria*:

> How much the activity of construction is pleasant to us and deeply rooted in our soul is shown by, among other things, the fact that whoever can allow it always feels the strongest desire to build something, and if he makes some discovery he tells others about it and, as if urged by a natural need, he spreads word of it so that it may serve men. (Prologue, p. 10)

It seems likely that Alberti was influenced, both in *De Re aedificatoria* and in *Profugiorum*, by Augustine's discussion in XIX.19 of the *City of God*, where, having said that it makes no difference whether one follows the active or contemplative life, as long as one meets one's obligations of Christian love, he adds: "The attraction of a life of leisure ought not to be the prospect of lazy inactivity, but the chance for the investigation and discovery of truth, on the understanding that each person makes some progress in this, and does not grudgingly withhold his discoveries from another."[47]

Alberti's argument in *Profugiorum* is a narrow one, for it applies to few disciplines except mechanics, where, undoubtedly, the study of nature enables man to master and transform the created world. But if his solution is flawed because it cannot be applied generally enough, it is of special interest to the architectural historian. For the profession in which Agnolo dabbles at night, and for which Archimedes died, is architecture. Mastery of the science of mechanics is the first element by which Alberti defines the architect in *De Re aedificatoria*: "I will call an architect he who, with a secure and perfect method rationally plans and executes by means of the moving of weights and the joining of bodies, works which are best suited for the most important needs of man" (Prologue, p. 7).

In this definition, the architect is no longer a manual laborer and craftsman. Rather, as we will see in chapter 2, he is a man whose knowledge of a mathematical science—mechanics—is used to serve society. As the examples given in *Profugiorum* argue, the activity of architecture bridges the gap between the active and contemplative lives, and the architect himself may be raised to the status of moral exemplar. Shortly after these reflections on the moral significance of architecture and the architect, Alberti began writing the first treatise on the art since Classical Antiquity.

TWO

Originality and Cultural Progress: Brunelleschi's Dome and a Letter by Alberti

ALBERTI'S LETTER to Brunelleschi that prefaces the 1436 vernacular version of his treatise on painting, *Della Pittura*, has long puzzled scholars.[1] Why should a treatise on painting be dedicated to an architect? Why is Alberti's only specific example of the revival of culture in Quattrocento Florence Brunelleschi's dome for the cathedral (Figure 1)?

Krautheimer's suggestion that the theoretician wished thus to acknowledge Brunelleschi's contribution to the working-out of a correct linear-perspective construction is an attractive explanation, particularly considering that in Remigio dei Girolami's division of the sciences, both perspective and the invention of machines are subaltern aspects of the science of geometry.[2] Yet the letter itself affords no direct support for such an interpretation. Salmi's observation that Alberti must have been strongly impressed by the dome—completed during the translation of the treatise in 1435—is surely accurate, yet fails to explain the purpose of the letter.[3] Argan, seeking a more theoretical explanation, suggested that the dome, spatial center of Florence, served Alberti as the image of what he considered to be an ideal political and social system.[4] But again, the letter provides little evidence in support of the thesis. Benevolo's intuition that the dome was "the visible result of the new interest in engineering which Alberti found when he returned from exile" and that it stood for the status of the city as a whole is nearer the mark.[5]

Figure 1. Dome, Florence Cathedral.

I believe that a closer reading of the letter reveals its content to be quite different from what has been thought until now. Its aim, I will show, is to argue for an understanding of historical process as cultural progress. The dome is the specific proof with which Alberti rejects both the organic model for historical progress, in which birth and maturation are followed by decline and death, and the cyclical, in which periods of greatness and decadence alternate in a recurring pattern.[6] Yet Alberti's view is only partly historical: he is concerned with defining the relationship of the present to the past, not with historical analysis of the past itself. His approach to the past is ethical, endeavoring to understand man's place in the world—the contemporary world of early Quattrocento Florence—and recommending a certain mode of action in this present. The letter's argument is consonant with a Christian understanding of history as the progressive unfolding of revelation, and as a linear process in regard to salvation. In my interpretation, Alberti's letter is an important text for the understanding of early Humanist thought. A highly contrived literary piece, its meaning becomes clear only when read together with other contemporary discussions of the same questions. Thus although Alberti's evidence for man's ability to progress is drawn from architecture, the true subject of the letter is human creativity, the capacity to invent. Cultural progress is the fruit of the interaction of individual *ingegno* with scientific knowledge. The paradigm for this notion is mechanical engineering, but its validity extends to painting, which, says Alberti, is based on mathematics.

Gombrich has shown that the letter is not the direct expression of Alberti's personal feelings and that it utilizes at least one literary topos—the exhaustion of nature.[7] Yet much more can and should be observed about its literary composition. Above all, it needs to be recognized that the letter is not a piece of art criticism, much less art history; nor does it open with a lament on the state of the arts, as is so often claimed. Alberti says that he used to mourn the lack of musicians, geometers, rhetoricians, and augurs—representative of the liberal arts—as well as of painters, sculptors, and architects—practitioners of the mechanical arts.[8] He does not find any dearth of grammarians, arithmeticians, and philosophers (in the liberal arts) or of weavers, farmers, merchants, hunters, and physicians (in the mechanical arts). Not all the arts, then, but some cultural manifestations only, seemed to be in a state of neglect; not all, but a certain kind of culture was lacking.

Overlooked by scholars is Alberti's concern with individuals, as opposed to abstractions and institutions: he speaks of artists and scientists, rather than of arts and sciences. To be precise, his letter is not about the state of the arts and sciences at all, but about the quality and intentions of individuals. Although it might be argued that the achievements of individuals, taken collectively, do in fact constitute history (at least on the Classical model of historiography), I believe that Alberti's insistence on discussion from the individual is not only a quality that he shares with other Humanist writers, but also evidence of his ethical approach. For it is individual action, not the overall condition of society, that is relevant to a discussion of moral virtue.

Before coming to Florence, says Alberti, he believed that Nature no longer produced great intellects ("ingegni"), just as it no longer produced giants. Now he sees, and most clearly in Brunelleschi, that there are individuals possessing the same talent and ambition as in Classical times. Gombrich suggested that Alberti's image of the exhaustion of nature was a topos derived from Pliny's letter to Caninius Rufus (*Letters* VI.xxi), which had its source, in turn, in a Greek epigram about Homer.[9]

> It is not true that the world is too tired and exhausted to be able to produce anything worth praising: on the contrary, I have just heard Vergilius Romanus reading to a small audience a comedy which was so skilfully modelled on the lines of the Old Comedy that one day it might serve as a model itself.[10]

I do not doubt that Alberti had in mind this text, only recently discovered, in 1419, by Guarino da Verona. Before Guarino's discovery, Pliny's letters had been known in a recension that ended with Book V. Thus this letter, in Book VI, was new to the Humanists.[11] But I believe its influence on his letter to Brunelleschi to be of a different, and more profound, nature than Gombrich claimed.

There is a crucial difference, for instance, in Pliny's praise of Vergilius Romanus and Alberti's of Brunelleschi. Pliny awarded approval insofar as contemporary achievement approached the stature of authoritative models, complimenting Vergilius for having "also written comedies in imitation of Menander and his contemporaries, which can be classed with those of Plautus and Terence."[12] Alberti, instead, claims that the merit of the dome lies in its absolute originality, describing it as a feat of engineering "that people did not believe possible these days and was probably equally unknown and unimaginable among the Ancients." As I will show, this radical difference is essential to Alberti's message.

Very important for Alberti was the first sentence of Pliny's letter, of which the topos of the exhaustion of nature is an extension and clarification. This first sentence defines the theme of Pliny's argument: "I am an admirer of the Ancients, but, not like some people, so as to despise the talent of our own times."[13] Pliny's concern, then, as also Alberti's, is to define cultural status. Yet the tone of Pliny's panegyric is slightly playful, unlike Alberti's very serious discussion. Both of these letters would seem to be (rather different) responses to the same source, Lucretius's *De Rerum natura*. Lucretius asserted that everything grows old and dies: "Even now indeed the power of life is broken, and the earth exhausted scarce produces tiny creatures, she who once produced all kinds and gave birth to huge bodies of wild beasts."[14] That Alberti used this specific text, discovered by Poggio Bracciolini in 1417, for his image of the exhaustion of nature is suggested by the explicit mention of enormous creatures in both, a detail absent in Pliny.[15] I suggest that it was Alberti's meditation on this important new source that lent significance to his reading of Pliny's letter, and that the latter work showed the Humanist how Lucretius's belief in progressive degeneration might be refuted through the evidence of one's own experience.[16]

Lucretius's discussion has serious ethical and cultural implications: those born late in time can achieve nothing great. So, he says, the plowman sighs that the labor of his hand comes to nothing and, comparing the present and the past, envies the fortune of his ancestors.[17] Here again, Alberti follows the literary structure of the original source rather than Pliny, comparing the abundance of practitioners of the arts and sciences in antiquity with their present scarcity. And again, Pliny's letter illuminated his reading of Lucretius, demonstrating the direct relevance of this view of historical process to the contemporary problem of the authority of the Ancients. For Pliny, this may have been a cultural problem only, devoid of moral implications, as the tone of his letter suggests. But for the Christian Alberti, it had profound ethical significance. Thus the Humanist sets out not only to find something praiseworthy in his own times, but to propose an exemplar of man's continuing capacity for achievement. Brunelleschi's dome, evidence of the human power to invent, serves to discredit Lucretius's degenerative concept of history, affirming instead the idea of history as progress. At stake for Alberti, but not for Lucretius or Pliny, was the definition of human participation in the process of salvation. It was intolerable for the Humanist to concede that the divinely ordained gifts of memory, intellect, and will could decline in the natural course of things, or that man's ability to perfect himself, and the world given into his care, should be diminished.[18]

That this is the kernel of Alberti's intention is supported by the dialectical relation between the letter to Brunelleschi and other Humanist discussions about the possibility of cultural progress. It may be read as a continuation of, and response to, Leonardo Bruni's 1401 dialogue, *Ad Petrum Paulum Histrum*.[19] Bruni foretold, there, the coming revival of the arts, although he was unable to claim that it had yet occurred: "In it [Florence] the seeds of noble arts and all human qualities remained; and these seeds, which seemed to have died out entirely, grow stronger every day, and very soon, I believe, will give out a great light."[20] Like Alberti, Bruni chose the fecundity of nature for his image, although he speaks of potential rather than actual fertility. And, again like Alberti, Bruni proposed the example of Florence's architecture as proof of the excellence of its citizens.[21] But he did not, of course, mention Brunelleschi's dome, since it had not yet been begun.

In Bruni's dialogue, Niccolò Niccoli utterly rejects an optimistic assessment of the times. Florentine culture can never achieve anything truly great, he says:

> Even if there is someone of great intellect and very anxious to learn, he is prevented from arriving where he would like by the difficulty of the situation. In fact, without culture, without teachers, without books, no one can show his excellent capacities for study. Who therefore will be surprised if, deprived as we are of every chance, no one of us for some time, and not even on rare occasions, has approached the divinity of the Ancients?[22]

Alberti was caught between these conflicting assessments. In *Profugiorum ab aerumna*, his speakers touch on the same topic.[23] One of them, Nicola de'Medici, follows Niccoli's view. The Asians and the Greeks, he says, invented all the arts and disciplines: their discoveries erected a temple whose walls were the investigation of truth; whose columns, the observation of nature; and whose roof, protecting the whole, was ethical theory and precept. From this noble and public edifice, modern writers remove such small fragments as serve their needs. It is true, he concludes, that "nihil dictum quin prius dictum."[24] Here is the Humanists' dilemma. On the one hand, a sense that the cultural foundations that permit great achievement had been undermined by the loss of Classical knowledge: texts were either missing or corrupt, and the transmission of their contents through teaching had been interrupted. On the other, there was an equally depressing perception that even the fragmentary heritage of Antiquity that they possessed was of such ample scope, such profundity and completeness of thought and expression, as to leave no room for original contribution in the present.[25] Powerful arguments, these, which might well foster an attitude of subservience to the past, suggesting that recuperation of what had been lost, and emulation of what had once been, were the highest ambitions to which Quattrocento man might aspire.

But if this was, as we saw, the conclusion of Niccolò Niccoli, it is not the solution Alberti accepts either in his letter to Brunelleschi or in *Profugiorum*.[26] The ability to achieve great things, he says in the letter, lies in our own industry and diligence rather than in the times, and "our fame should be all the greater if without preceptors and without any model to imitate we discover arts and sciences hitherto unheard of and unseen."[27] In Bruni's dialogue, *Ad Petrum Paulum Histrum*, even Niccolò Niccoli was made to concede that an exceptional individual (Coluccio Salutati) was somehow able to overcome the deficiencies of modern times by virtue of his extraordinary, and almost divine, "ingenio".[28] Quint has shown that this has as its literary source Cicero's *De Oratore*, in which Antonio is praised for his almost unique and divine power of *ingenio*, which enables him to win court cases without deep study of the law.[29] The capacity for achievement, then, lies within the creative intelligence of the individual. The implication of this, important for the Humanists' self-definition, is that cultural status need not always be defined in terms of the mastery of Classical knowledge and precedent. Not the return to authoritative models but the capacity to make entirely new things would serve as proof that the process of history is one of progress. Although in Bruni's dialogue this possibility is not so positively stated as it is in Alberti's letter, in both, the assessment of cultural achievement is firmly linked to fertility, the power to create.

But was this power the possession of only a few exceptional individuals or within potential reach of many? Alberti, in his letter, inclines toward the latter view: "I recognized in many, but above all in you, Filippo, and in our great friend the sculptor Donatello and in the others, Nencio, Luca and Masaccio, a genius for every laudable enterprise in no way inferior to any of the Ancients who gained fame in these arts." Alberti's contemporary Lorenzo Valla agreed with him. Having in mind, I believe, that worrisome passage in Lucretius, he asserted that "the natural capacities of man are not, certainly, exhausted; it can scarcely be doubted that our merits will soon approach the skill of the Ancients in many, if not all, honorable and useful matters."[30] And in the opening pages of his *Elegantiae*, Valla described this renewal as dependent on only human will to achieve it:

> Certainly, the more wretched those earlier ages, with not one learned man to be found in them, the more we should rejoice in our own age in which, if we strive a little more, I am sure that the Roman tongue will soon flourish more vigorously than the city of Rome itself, and that along with it all the sciences will be on the way to renewal.[31]

Alberti's attitude toward the past is more radical than that of Bruni, Niccoli, and Valla, at least in his letter to Brunelleschi. Whereas these other Humanists assumed, as a general premise for the renewal of culture, the mastery of Classical knowledge (although there might be some exceptions to this), Alberti boldly asserts that this mastery is quite unnecessary: "Our fame should be all the greater if without preceptors and without any model to imitate we discover arts and sciences hitherto unheard of and unseen." By positing the absolute originality of an achievement as the measure of its worth, he sweeps away the authority of the Classical past. Indeed, the availability of the whole edifice of Classical learning and precedent diminished the merit of what the Ancients accomplished: "For the Ancients, who had many precedents to learn from and imitate, it was less difficult to master those noble arts which for us today prove arduous." By extension, the value of Brunelleschi's achievement is enhanced, serving as proof of cultural progress: it was "a feat of engineering... that people did not believe possible these days and was probably equally unknown and unimaginable among the Ancients."

Alberti's claim that a work for which there is no model or prototype is better than one that belongs to a tradition is astonishing. He announces a revolutionary approach to the assessment of architectural (and cultural) value by suggesting that the quality of the work resides in its absolute originality. I know of no precedent for such a view within the literature of Classical and Western medieval architectural description or criticism.[32] Nonetheless, this new criterion was not Alberti's invention. Matteo Palmieri may have been the first to interpret the decline in the arts as due to the weight of tradition: whenever man breaks the chains of tradition and has the courage not to bind himself to it, but instead has the will to advance, a golden age of culture arises.[33] For him, this originality was what his own age had in common with Antiquity. Nicholas of Cusa argued that human originality was like God's own creativity and that making something for which there was no model was superior to imitating that which already exists.[34] The criterion of originality would seem to have been applied to architecture first by the Byzantine scholar Manuel Chrysoloras, whose *Comparison of Old and New Rome*, written in 1411, had been circulated among the Humanists.[35] Describing the dome of Hagia Sophia, Chrysoloras says we should admire it not only for its size, but also for the originality of the architects who conceived this great work without any precedent or model.[36]

In applying this new criterion of originality to architecture, Chrysoloras had no direct model in earlier descriptions of Hagia Sophia, although, as we will see, Byzantine attitudes toward the architect's training and the nature of architectural activity did provide the foundations on which the Greek scholar built. His new approach can best be understood as the result of applying Aristotle's standards for assessing human achievement, found in his ethical and rhetorical works, to an architectural subject. In his discussion of the topos αὔξησις (amplification), Aristotle defines the kinds of circumstances by which superior beauty and grandeur—and therefore virtue—may be attributed to an action (*Rhetoric* I.ix.38–39). The first of these criteria is originality: if someone has done something alone or first or with a few, or has been chiefly responsible for it. The excellence of an achievement, we are told in the *Nicomachean Ethics*, lies in its greatness—ἐν μεγέθει (IV.ii.10–11). Aristotle's word means not only "greatness," or "grandeur," but also "great size." For Chrysoloras, who of course learned his Aristotle in Greek, this double meaning (and, therefore, potential application to architecture) would have been obvious. As I will show in chapter 7, the kind of education Chrysoloras had received facilitated the application of general principles to diverse kinds of subject matter. The ease with which Chrysoloras wove connections between different areas of human activity and thought is in sharp contrast to the tendency in the West that, by pressing for logical definition, emphasized their divisions. But if it is difficult to imagine that Alberti's training would have enabled him to make such connections without the model of Chrysoloras's description, it is also true that he recognized in it an appropriate formulation for the reality of Quattrocento Florence.

If we wish to understand how Alberti's experience led him to propose a work of architecture—traditionally, one of the lowly mechanical arts—as paradigm of human intellectual and creative power, we must seek to disentangle a complex web of associations among philology, scientific learning, and the recovery of the Classical past. We will find that this investigation concerns only one aspect of architecture, mechanical engineering; Alberti's claims for the cultural significance of architectural style are of a different and perhaps less ambitious nature. Indeed, in his letter, he praises Brunelleschi's dome for its surpassing great size and for its difficulty, but not for its aesthetic qualities. Salmi was troubled by this, attributing it to a kind of embarrassment on Alberti's part. The dome was, after all, "the logical Gothic completion of a great Gothic edifice."[37] But why would Alberti have been ashamed of its style, when he eulogized the Gothic nave of the cathedral six years later, in *Profugiorum*? Indeed, evidence drawn from this moral treatise supports the view that the question of style was irrelevant to what made the dome an original achievement—that is, its engineering.

In two passages in *Profugiorum*, the process of architectural design is described as essentially eclectic and tradition-bound.[38] In the second of these, the main protagonist of the dialogue, Agnolo Pandolfini, relates that when he designs buildings alone at night he combines various Orders and numbers of columns with diverse kinds of capitals and bases. Thus piecing together elements from an available vocabulary of forms, he invents little. Nor does he hope to achieve more than (in his words) to infuse into his works a new convenience or grace. Yet Agnolo's deference to the authority of Classical Antiquity in matters of design is in the strongest contrast to his claims of originality in the invention of "unheard-of machines for moving and lifting" weights—that is, in mechanical engineering.

Not only Alberti (and his dramatis personae) but all Humanist commentators who were Brunelleschi's contemporaries uniformly praised his genius and originality in engineering while remaining entirely silent about his style. That their view might provide a poor basis for a history of earlier Quattrocento architecture in Florence is unimportant here, since our concern is not with Brunelleschi's actual contribution but with the Humanists' (and particularly Alberti's) assessment of his work. Brunelleschi's tombstone, for instance, eulogizes his intellect, his *ingegno*: "corpus magni ingenii viri Philippi Brunelleschi Florentini."[39] And Carlo Marsuppini's epitaph for the architect, composed in 1446, begins: "How Filippo the architect excelled in the art of Daedalus can be shown not only by the admirable dome of this most famous temple [Florence Cathedral] but also by many machines which he invented with his divine genius."[40] Fra Domenico da Corella praised Brunelleschi for the size of the dome without mentioning its style, just as Alberti does in his letter.[41] Biondo Flavio refers, in two of his works, to Brunelleschi's *ingenio* in constructing the dome.[42] And for Giannozzo Manetti, the dome serves as an example, in his treatise on the dignity of man, of human intelligence, memory, and will, shown in the "many, great and remarkable instruments or machines marvellously invented and comprehended" ("plera magna et ingentia vel facinora vel machinamenta admirabiliter inventa et intellecta".)[43] Sozomeno of Pistoia, recording his impression of the consecration ceremony in 1436, says that everyone was amazed by the fact that the dome had been erected without a wooden armature and that "etiam miribilius" was Brunelleschi's machinery for lifting weights.[44] Landino celebrated the architect's *ingenio* in a poem.[45] Brunelleschi himself seems to have been aware of possessing the quality of *ingegno* and to have valued it highly, equating it with the capacity for original thought.[46]

This current of thought continued in the later Quattrocento, as in Alamanno Rinuccini's summation of the achievements of the moderns, addressed to Federico da Montefeltro in 1473.[47] And Brunelleschi's *ingegno* is a constant theme of Antonio di Tuccio Manetti's biography of the architect, although there the term is extended to include style.[48] I would suggest that appreciation of Brunelleschi's style is difficult to document before the 1460s (in Filarete's treatise on architecture and Giovanni Rucellai's *Zibaldone*) and that Manetti's awareness of this neglect is implied by his words "since you have read the epitaph [which, of course, dwelt exclusively on Brunelleschi's originality and skill in engineering], I will concentrate more on those facts not recorded there."[49]

Alberti's letter to Brunelleschi is one of many Humanist appreciations of the architect's *ingegno*. But it is original in associating the engineering achievement of the dome with the notion of historical progress, measured by the capacity of the individual to invent. This association, and the place of the letter in contemporary Humanist discussion, are precious evidence for the reconstruction of a cultural ambient about which much has yet to be discovered. Although Alberti does not share with his readers the experiences that brought him to write the letter, external evidence permits a tentative reconstruction of this process. Such an endeavor, although of necessity hypothetical, is useful for the interpretation of the significance of Brunelleschi's dome within the culture of early Humanism.

Historical and Intellectual Background

That the art of architecture is intimately associated with the development of human civilization was universally admitted by Classical and medieval authors who, however, assign it a low place in the hierarchy of human achievement. This is, in part, because Aristotle (as, later, the Scholastics) valued knowledge according to the nature of the objects of knowledge, as we saw in chapter 1. In addition, a Classical and, to some extent, medieval bias against manual labor as unsuitable for an educated man contributed to architecture's low status.

Cicero considered architecture to be one of the manual arts, together with agriculture, tailoring, and metalworking, which served human necessity and convenience (*De Natura deorum* II.9). Seneca placed architecture in the lowest of the four classes of arts, those which are "volgares" and "sordidae." A form of mechanical engineering, the making of stage machinery, belonged to the next higher class, the arts of amusement (*Letter* LXXXVIII.21–22). For Vitruvius, the manual dexterity that distinguishes man from animals enabled him to construct shelters (II.i.1–3). Nemesius of Emesa and Philo Judaeus also saw construction as one of the means by which man orders nature and renders it civilized.[50]

Such perceptions, inherited by the medieval Scholastics, were elaborated into a systematic division of the arts and sciences. These systems differ among themselves but consistently place architecture within the mechanical arts that serve human physical needs.[51] We may take as an example one of the most widely influential, Hugh of St. Victor's *Didascalicon* (1133–1141).[52] The mechanical arts comprise woolworking, armament, navigation, agriculture, hunting, medicine, and theater. Armament has two parts, the making of arms and construction (XXII). Probably Alberti's introduction at university to this or a similar system accounts for his definition of the architect (in *De Re aedificatoria*) as constructing for human necessity.[53] As the title suggests, his treatise on architecture is concerned with only one part of armament—that is, construction. Practitioners of a manual art serving necessity are unlikely to be considered paradigms for the possible achievement of the human mind. The case is different, however, for those who make machines, and it is in this art that Brunelleschi was lauded by Alberti and other Humanists. I do not know whether Alberti was familiar with Remigio dei Girolami's *Divisio scientie*, written about 1290 in Florence.[54] Remigio substitutes for the art of armament two arts, architecture and the making of arms, thus anticipating Alberti's division. For Remigio, the art of making machines ("scientia de faciendis machinis et ingeniis"), or "stereometria machinativa," is part of the science of geometry, as is perspective because both are concerned with the measurement of immobile magnitudes. If, as I have suggested, the letter to Brunelleschi and the treatise on painting are connected by their mutual relation to the science of geometry (following Remigio) and if this science does not include architecture (which is, instead, one of the mechanical arts), this distinction may account for the different levels of achievement that Alberti attributes, respectively, to Brunelleschi's dome in the letter and to the art of architecture in his architectural treatise. However, as will be seen, Alberti's notion of engineering was far more complex than Remigio's and could not be encompassed by the science of geometry alone.

The Humanists' term *ingegno*, as it is applied to Brunelleschi, has two closely related, if slightly different, aspects: the first designates a natural capacity for original invention that is not dependent on learning; the second, instead, implies the ability to apply scientific knowledge to a practical problem.[55] The former, akin to brilliance or genius, has no rational explanation and is a mysterious creative force.[56] The latter, instead, is that capacity which transposes speculative thought, or knowledge of unchanging truths, into practical actions; the fruit of this *ingegno* is a demonstration of scientific principles.[57] These two aspects, the creative and the intellectual, may coincide, as in the divine creation of a world ordered by rational principles. But that they may also exist separately is important for Alberti's assessment of Brunelleschi.

Philological study of the term *ingegnio* in reference to architecture reveals a variety of meanings—such as "cleverness," "inventiveness," and "originality"—that do not imply the possession of much scientific knowledge.[58] The term *ingeniator* (engineer) is first documented in 1058, designating men in charge of military machines; by the end of the twelfth century, the term and its variants were widely used, almost always in reference to military machines and fortifications.[59] The place of the engineer within the system of sciences was explained, about 1150, by Domenico Gundisalvo in his *De Divisione philosophiae*.[60] There are, he says, seven parts of mathematics: arithmetic, geometry, music, astrology, perspective, statics (*scientia de ponderibus*), and, subordinate to all these, the science of devices (*scientia de ingeniis*). The task of this last is to provide the mechanical means "for accomplishing all of those things whose modes were declared and demonstrated in the theoretical sciences."[61] His explanation suggests that while it is not impossible that an *ingeniator* might understand something of the higher sciences, it cannot be assumed that he is a master of them. Indeed, as White has shown, there is no clear evidence that any medieval engineer studied at a university, and no instance of a professor lecturing on engineering or writing a commentary on a technological text.[62] He concludes, as does Shelby in his study of medieval architects' knowledge of geometry, that their skills were almost exclusively empirical rather than theoretical.[63] Shelby points out, quite sensibly, that the master masons could not have learned geometry in school, since it was a university-level subject. Thus what they knew were rule-of-thumb procedures transmitted orally from mason to mason. J. Ackerman's examination of the scientific knowledge of the masters at Milan Cathedral around 1400 also showed that it was quite limited.[64] That there is still much to be clarified in our understanding of the question is, however, suggested by texts such as this one of about 1200: "Master Simon the trench-man, so learned in tasks of measurement, advancing in a master's manner with his rod, measured here and there the task already conceived in his mind, not so much with his rod, as with the yardstick of his eyes."[65] Brunelleschi himself, according to Sanpaolesi, had very little theoretical preparation in mathematics, statics, and mechanics, nor could he read Latin or Greek, according to Vasari.[66] All this suggests that the terms *ingenio* and *ingegno* applied to medieval engineers refers to their power to invent without the possession of much theoretical knowledge. The terms would correspond to Vitruvius's definition of the faculty of invention, which is "the solving of intricate problems and the discovery of new principles by means of brilliancy and versatility" (I.ii.2).

Although the mechanical arts were logically separate from the theoretical sciences, it was recognized that, in practice, there was some relation between them. In his *De Ortu scientiarum* of about 1250, the last great Scholastic treatment of the subject, Robert Kilwardby asked, rhetorically, whether the carpenter or the stonecutter could work without the theoretical science of geometry (XLII.393.12–13).[67] Nonetheless, he agreed with Hugh that the mechanical arts (now farming, cooking, medicine, tailoring, armament, architecture, and commerce) served human necessity and were practiced by only the uneducated (XXXVIII). Ovitt rightly points to the difficulty of evaluating views of the *artes mechanicae* in the Middle Ages.[68] The technical accounts of these sciences, written by practitioners, show no interest in the relation of technique to theory, whereas theoretical accounts, concerned with only the metaphysical effects of the sciences, are not concerned with their methods or products. Thus although Kilwardby's statement—"we see, therefore, that the speculative sciences are practical and the practical sciences are speculative"—is indeed interesting, the Scholastic did not, in fact, revise his low opinion of the practical sciences.

Even while Kilwardby wrote, however, events were taking place that would elevate the *scientia de ingeniis* above the mechanical arts and transfer it from the sphere of mathematics to that of physics. These events, which ocurred simultaneously but in isolation from one another, took place in the highest intellectual circles and on the battlefield. A new surge of interest in the science of mechanics—that is, dynamics and kinematics—is manifest in the work of Thomas Bradwardine and the school at Merton College, Oxford; of Nicole Oresme; and of Albert of Saxony at the University of Paris.[69] The new physics was eagerly received and developed at northern Italian universities during the fourteenth century, reaching a high point at Padua and Pavia in the early fifteenth century under Paolo Veneto and others.[70] And it was there, and then, that Alberti must have learned it. He tells us in his autobiography that, having been forced by ill health to abandon the study of law, "at the age of twenty-four he turned to physics and the mathematical arts." And he continues, significantly for our study, that "he did not despair of being able to cultivate them sufficiently, because he perceived that in them talent [*ingenium*] rather than memory must be employed."[71] Vasari recounts that Brunelleschi studied geometry with Paolo dal Pozzo Toscanelli, and that although he had no letters, his natural talent and experience enabled him to confound the masters on more than one occasion.[72] Alberti's dedication of the vernacular version of *Della Pittura* acquires added significance in the context of Brunelleschi's efforts to educate himself. If we now recall the connection between Alberti's letter to Brunelleschi and Bruni's dialogue, in which men of *ingegno* were said to be able to rise above the cultural deficiencies of their times because of their natural talent, we obtain a clue to why Alberti chose an engineering feat as his example of human achievement. The very science on which it depended—mechanics, that is—was perceived as not requiring a mastery of Classical precedent.

This new interest in the science of mechanics, understood as part of physics, coincided with remarkable developments in its practical application. Around 1200 a mechanical catapult of great accuracy and power, the trebuchet, revolutionized medieval artillery; the following century saw the development of cannon and gunpowder.[73] The job of the military engineer became less strictly associated with geometry and began to require a better knowledge of mechanics. At roughly the same time, university-trained men started to use their knowledge of the new physics to make practical devices. White has studied one aspect of this, the contribution of fourteenth-century medical astronomers to engineering.[74] The most famous is Giovanni de'Dondi, professor of medicine and astrology at the universities of Padua and Pavia, remembered as Giovanni dall'Orologio because of the clock he made between 1348 and 1361. Other recipients of the doctor's degree are known to have made artillery and stage machinery, or to have written about military engineering and even rockets. Guido da Vigevano made military machines for Philip VI and, in 1389, was making stage machinery for court festivals. Conrad Kyeser, who studied at Padua, wrote a treatise on military engineering, left incomplete at his death in 1405. Giovanni Fontana of Venice studied medicine at Padua and, while still a student in 1420, wrote about rockets.[75] Indeed, the separation between theoretical knowledge and its application had been broken down by about 1400, at least in the field of mechanical engineering. Alberti's career as architectural adviser is evidence that Hugh of St. Victor's disdain for the mechanical arts had been swept away and that they could perfectly well be practiced by the educated. As we saw in chapter 1, an intimate connection between speculative knowledge and practical application was presented as an ethical ideal in the closing pages of Alberti's *Profugiorum*. This ideal found support in Paul's use of technological images to describe the task of Christians as fellow workmen of God (1 Corinthians 3:9).

Brunelleschi was not university-educated, but the Humanists who assessed his achievement had studied at the Italian universities and felt no reticence about using their knowledge to solve practical problems, including problems in mechanics. This provides us with a second reason for Alberti's choice of the dome in his letter: seen as the fruit of scientific knowledge, it demonstrated the vitality of contemporary intellectual endeavor.

These events, the increased interest in mechanics as a university subject and the development of new mechanical devices, some of them by university graduates, formed the background that sensitized Alberti to the importance of Brunelleschi's dome and prepared the ground for viewing it as a cultural achievement. But these new circumstances existed within an intellectual system no longer adequate to contain them—that is, the Scholastic division of the arts and sciences.[76] Contemporary engineering was, in fact, no longer equivalent to the *scientia de ingeniis*, nor could it really be considered in the same category as cooking or farming. But where else could it fit within the traditional scheme of knowledge? No wonder that Alberti, in his letter, speaks of never-before-seen arts and sciences invented in his own time.

Alberti's reading of Classical Latin authors revealed no alternative system within which to locate an achievement like the dome. As we saw, his beloved Cicero considered architecture to be a manual art serving necessity. True, Vitruvius insisted that the architect have theoretical training, and he devoted an entire book of his treatise to the making of machines.[77] But Vitruvius required that he have only a rudimentary scientific knowledge, drawing a clear separation between the level achieved by the architect and that attained by the scientist. The study of geometry is necessary, he said, because it "teaches us the use of the rule and compass," is helpful in drawing up plans, and provides instruction for how one may "rightly apply the square, the level and the plummet" (I.4). However, "as for [other] men upon whom nature has bestowed so much ingenuity, acuteness and memory that they are able to have a thorough knowledge of geometry, music and the other arts, they go beyond the functions of architects and become pure mathematicians" (1.ii.16). Vitruvius conceived of knowledge in a hierarchical fashion, with manual labor and pure speculation as the two extremes; the architect's place in that order was not, after all, very high.[78] His definition may have been important for Alberti's thought in regard to the art of construction, but it fails to account for the cultural importance assigned to engineering feats in Alberti's letter.[79]

An alternative view of the sciences, in which mechanics occupied a higher place, must be sought within another facet of Alberti's experience. This was, I believe, the influx of Greek works on mechanics into Quattrocento Italy and his encounter with the attitudes toward architecture (themselves indirectly fostered by these ancient treatises) of Byzantine scholars like Chrysoloras.

The Greek-Language Catalyst

Interest in Greek treatises on mechanics ran high in Florence, at least within the small circle of artists and Humanists that included Brunelleschi. Most important of the discovered, or rediscovered, texts were pseudo-Aristotle's *Mechanical Problems*, Pappus's *Mathematical Collection*, and the works of Archimedes, all of which were introduced (or reintroduced) into Western culture through texts brought back from Byzantium. Filelfo owned a copy of pseudo-Aristotle's *Mechanical Problems* by 1446. Although it is not certain that Alberti knew the work, if Filelfo had it with him when he returned from Constantinople in 1427, the year of their meeting in Bologna, Alberti may have borrowed it.[80] Other copies were in the possession of Marco Lippomano in Venice, Cardinal Bessarion in Rome, and Giannozzo Manetti.[81] That the architect Bartolomeo Fioravanti was called "Aristotele" because of his vast knowledge of engineering and mechanics shows how widely diffused this text had become by midcentury.[82] Something of a paradox surrounds this work. On the one hand, its dynamic approach to problems of statics (as opposed to the nondynamic, mathematical approach) strongly influenced Jordanus de Nemore's *scientia de ponderibus* (and his *Elementa super demonstrationem ponderum*) as well as the related treatise (not by Jordanus) *De Ratione ponderis*. On the other hand, there is no evidence that a Latin translation of *Mechanical Problems* existed until 1413 (and that evidence is frail), nor was any commentary on it written until the sixteenth century. Therefore, although the work had been of fundamental importance for the Western medieval science of weights, the arrival of the Greek texts in Quattrocento Italy constitutes a true rediscovery of a lost work.[83] Very important for Alberti is the introductory paragraph of the work, which tells us that the science of mechanics enables man, by skill, to overcome nature, and that mechanics belongs to mathematics in its method and to physics in its practical application. The Middle Ages had considered the science of statics (*scientia de ponderibus*) as entirely different from the science of motion, which, based on Aristotle's *Physics*, included almost all of natural philosophy within its scope (since all things change—that is, are in motion).[84] As we saw, it was within the new physics that advances in mechanics had taken place since the thirteenth century, but this fact had not been associated by theorists with the *scientia de ingeniis*: it is precisely this connection that prefaces the pseudo-Aristotle *Mechanical Problems*. It provided the conceptual framework within which Humanists like Alberti could understand the making of machines (such as Brunelleschi's Great Hoist) as a demonstration of the principles of mathematics and physics.

Aurispa brought a copy of Pappus from Byzantium to Florence in 1423.[85] This manuscript, the archetype of all extant Pappus manuscripts, was later acquired by Lorenzo the Magnificent and eventually passed to the Vatican.[86] The manuscript begins with a fragment of *On Marvellous Mechanical Devices* by Anthemios of Tralles and contains fragments of Pappus's *Mathematical Collection*, including the section on mechanics that opens Book VIII. Charles Stinger has suggested that Brunelleschi learned the worm-gear mechanism from Aurispa's Pappus, the text having been made accessible to him by Ambrogio Traversari.[87] Brunelleschi's Great Hoist corresponds to a device described in Heron's *Mechanics*, although it cannot be shown that he knew this work.[88] A copy of a work thought to be by Archimedes was brought from Byzantium in 1423 by Rinuccio da Castiglione: Niccolò Niccoli and Traversari were anxious to see it. Although William of Moerbeke had translated Archimedes' *The Equilibrium of Planes* in 1269, the work had been entirely ignored because it used a mathematical-statical method instead of the dynamic approach adopted by Jordanus and others.[89] Cosimo de'Medici owned Archimedes' *On Floating Bodies* and *On Balances and Levers*.[90] Paolo dal Pozzo Toscanelli owned a copy of Archimedes in Greek, which was used for the Latin translation made under Nicolas V.[91] In Alberti's *Intercoenales*, dedicated to Toscanelli, the Humanist makes Virtue complain that because of bad Fortune she no longer enjoys the company of her old friends in the Elysian Fields, one of whom was Archimedes.[92] Il Taccola, whose treatises on machines drew on Greek works, signed himself "ego autem Sr. Marianus taccole alias Archimedes vochatus."[93] Alberti attached special importance to the figure of Archimedes, using him in *Profugiorum* as heroic exemplar of virtue who reconciles the active and contemplative lives. In all probability, he derived this interpretation from Pappus.

Indeed, it is Pappus, whose work had aided Brunelleschi in the construction of the dome, who also showed Alberti an approach to the systematization of knowledge entirely different from that of the Scholastics. In his approach, manual labor and speculative thought are no longer the two extreme poles of a hierarchy. Rather, they are the complementary aspects of a single subject of knowledge; a master of the science is equally skilled in theory and application. Mechanics, he says, is that science concerned with the nature of the material elements of the universe.

> The science of mechanics, my dear Hermodorus, has many important uses in practical life, and is held by philosophers to be worthy of the highest esteem, and is zealously studied by mathematicians, because it takes almost first place in dealing with the nature of the material elements of the universe.[94]

Its theoretical aspect is divided into geometry, arithmetic, astronomy, and physics; its manual part consists of work in metals, architecture, carpentry and painting. Master of this science is the μηχανικός. Pappus's scheme corresponded to the new reciprocity between learning and application that, as we saw, had characterized Alberti's life experience. The discovery of the manuscript did not change Alberti's thinking; rather, it provided a new logical framework for it. It is his definition of mechanics, queen of natural sciences and single mode for the investigation of the natural world, that may help to account for Alberti's praise of Brunelleschi and for the logical relation between the letter and the treatise on painting that it prefaces. Nonetheless, Pappus's relevance might have been missed had it not been accompanied by other discoveries that underlined its applicability.

At the same moment that new copies of Archimedes' work were being circulated in Florence, the cultural significance of his achievement was being reassessed. At least four views of Archimedes can be distinguished in Classical, Western medieval, and Byzantine literature. All of them emphasize his *ingegno*, understood as both originality and intellectual attainment; they differ in the definition of his scientific activity.[95] The Archimedes of the crown problem is a natural scientist, important as such to Vitruvius and to Alberti in his *De Ludi matematici*.[96] Of far greater interest to the Western Middle Ages is the philosopher and mathematician, master of speculative thought, who wrote *The Squaring of the Circle*.[97] Most of his other works, although translated by William of Moerbeke (including the *Mechanics*), were ignored by medieval scientists and biographers. Another view of Archimedes, popular in Antiquity and revived in the Quattrocento, saw him as astrologer and emulator of the Divine Mind, whose great achievement was the creation of a sort of orrery.[98] Manetti emphasizes this aspect of Archimedes in his 1452 treatise on the dignity of man, using Lactantius and Cicero as his sources.[99] Finally, there is Archimedes the *ingeniator*, maker of machines and military artillery, whose activity exemplifies the intimate connection between speculative knowledge and practical application.[100] It was in this last aspect that he served Alberti in *Profugiorum*. The tendency to interpret Archimedes as reconciling the active and contemplative lives was already apparent in Petrarch, who defined him as "astrologus ingens et mechanicus.'"¹⁰¹ But it was Pappus who made the significance of his activity explicit, claiming that Archimedes alone had the distinction of applying inventive genius to ordinary needs.¹⁰² That distinction, as Alberti's letter to Brunelleschi shows, was no longer unique.¹⁰³ Indeed, the reinterpretation of Archimedes suggests the emergence of a new cultural ideal in early Humanist thought, grounded in a reconsideration of ethical requirements. Valla seems to share this ideal in a passage in *De Voluptate*, where he attacks Aristotle's hierarchy of knowledge. Aristotle

> does not understand that contemplation is nothing but a progressive process of learning, which we sometimes call interpretive reflection and sometimes invention, which is proper to men and not to gods. Whence it happens that the arts also have been brought to their highest development by the most vigorous human contemplation.¹⁰⁴

That theoretical knowledge *should* have utilitarian application is clearly implied in a letter from Aeneas Silvius Piccolomini to Ladislas of Hungary (1450): "And if anyone should say that geometry be rejected, he may be refuted by the words of the Syracusans who, when Marcellus was moving his engines against the city, were able to prolong the siege by the genius of Archimedes and his knowledge of geometry."¹⁰⁵

The formulation of this cultural model received additional impetus from Chrysoloras's description of the dome of Hagia Sophia. As we saw, Alberti's quite extraordinary claim for Brunelleschi's dome was anticipated by Chrysoloras, who attributed the feat of vaulting Hagia Sophia not only to the originality of the architects, but also to their scientific knowledge. They must have been very skilled in the science of geometry or mechanics, he says.¹⁰⁶ This conclusion was obvious to Chrysoloras, familiar as he was with the Byzantine ekphrastic tradition. Agathias had praise for Anthemios of Tralles for his surpassing wisdom as a μηχανικός (*History* 6.4). "Anthemios," we are told, "worked in the art of discovering mechanical devices; he was one who applied geometrical speculation to material objects."¹⁰⁷ The fragment of Anthemios's περὶ παραδόξων μηχανημάτων (*On Marvellous Mechanical Devices*) that prefaced Filelfo's copy of Pappus was proof that Chrysoloras's claim was true.¹⁰⁸ By and large, Byzantine descriptions of architecture accept Pappus's definition of the science of mechanics, assuming that a distinguished architect manifests his theoretical knowledge above all in the planning and execution of vaulted structures.¹⁰⁹ Moreover (as we will see in chapter 3), in the Greek, but not the Latin, literary tradition, works of great size and difficulty—such as the Seven Wonders—were interpreted as manifestations of human capacity to invent. In *Hippias, or the Bath*, Lucian wrote that just as with doctors and musicians the best are those who not only have knowledge, but apply it, so with engineers (τῶν μηχανικῶν) we admire those who, though famous for knowledge, leave for later generations reminders and proofs of their practical skill. Such were Archimedes and Sostratus of Cnidus, who took Memphis for Ptolemy without seige by turning a river aside and dividing it.[110]

One wonders whether Brunelleschi's ill-fated attempt to take Lucca by converting the city into an island (in 1430) was inspired by the account of Sostratus of Cnidus's feat.[111] His attempt would have reversed the miracle by which San Frediano converted the Lucchese to Christianity, turning aside the Serchio, which had until then endangered the city and flooded the countryside.[112] In any event, we can be sure that his stage machinary for the *sacre rappresentazioni* performed at Santissima Annunziata and the Church of the Carmine in 1439 as part of the entertainments for the Ferrara/Florence Council had an appreciative audience in the Byzantine delegates.[113]

Alberti knew Lucian's works; there is every likelihood that he knew Chrysoloras's *Comparison*; and he may well have seen the fragmentary mechanical treatise by Anthemios. These Greek models showed the Humanist how Pappus's definition of mechanics could be applied to architecture. But, more than that, they demonstrated how an architectural achievement could serve as exemplar of the entire cultural status of a society. The dome of Hagia Sophia, wrote Chrysoloras, is proof of the ability of the human race to achieve great deeds through intelligence and inventiveness; admiring the dome, one recognizes inventiveness, elevated thought, value, and power such as had never before been seen.[114] Chrysoloras's description bridges the separation between science and the art of architecture and, applying Aristotelian ethical criteria to the work, inserts the dome into the context of human achievement and cultural progress. We will see in chapter 7 that the Byzantine scholar's cultural experience, so different from Alberti's, enabled him to perceive and formulate interrelationships between different areas of human endeavor in a way quite new to Western thought. His description served to clarify and focus Alberti's own experience and provided a new framework within which it could be expressed. His account was, I submit, a direct model for Alberti's letter to Brunelleschi.

Conclusion

In writing his letter, Alberti wished to assess the achievement of his own time and define its relationship to the Classical past. In it, he asserts a belief in historical progress derived partly from his awareness of new developments in both the intellectual and practical spheres and partly from his religious and, more precisely, ethical conviction in human ability to perfect the inner self as well as nature. He was disturbed by Lucretius's (and others') degenerative concept of history, as by the pessimistic assessments of the potential creativity of modern culture expressed by some of his contemporaries. Scholastic logic could no longer encompass the reality in which he lived; its framework could not account for some of the most remarkable of his experiences. Of these, witnessing the completion of the great dome in Florence was one of the most impressive and thought-provoking. The visual impact of the dome, and Alberti's investigation of Brunelleschi's mechanical devices for its construction, intersected with his encounters with the Florentine Humanists and with the new texts that they were discussing and the new ideas about man's place in the world that they were considering. This constellation of problems, doubts, and hopes came together in Alberti's mind while he was translating his treatise on painting, itself a manifestation of the dissolution of barriers between the speculative sciences and the mechanical arts, and symptomatic of a redefinition of the kinds of knowledge and activity appropriate for a man of education. As the dome completed the Gothic body of the cathedral, defined its characteristic form, and gave visual focus to the entire architectural perspective of the city of Florence, so it served to clarify Alberti's thoughts about the cultural position of his own time; it made explicit what was specific to that particular moment in history, and revealed its most positive aspects. The treatise on painting had been composed before these Florentine experiences revealed its full significance; now Alberti added a prefatory letter whose cultural import is more profound than that of the treatise itself, and which served to orient the implications of the treatise toward those larger issues that, Alberti only now realized, had underlain its original composition.

THREE

Foul Enormity or Grandiose Achievement? The Moral Problem of Size

IN BOOK II of *Africa*, Petrarch laments that both Scipio and Hannibal will have equal praise from posterity, since "the vulgar multitude cannot discern the gap that yawns between magnificence and deeds of foul enormity."[1] What distinguishes these acts is, of course, their moral value. Within the literature of architectural description, great size is one of the most constant criteria by which the moral value of a building is positively or negatively judged. Saint Bernard's condemnation of "the vast height of your churches, their immoderate length, their superfluous breadth" and Alberti's praise of the dome of Florence as "an enormous construction towering above the skies" represent the two poles of moral judgment possible on the same criterion.[2]

Onians has defined a strong current of opposition to great size in architecture within a medieval, Christian context; and he has explored the sources of this opposition in Christian ethics and Scripture.[3] But his study failed to recognize that the problem of size required discussion within a much broader ethical context. Condemnation of excessive size and cost in building is not restricted to Christian writers or to the Middle Ages; it is common as well in Classical, pagan authors.[4]

Plato's condemnation of the Temple of Poseidon, which measured 164 by 95 meters and was covered with gold and silver, as βαρβαρικόν, may be the first example.[5] Luxurious architecture is a frequent image of moral corruption in Horace's odes: the man who cherishes the golden mean avoids both the foulness of an ill-kept house and a hall exciting envy (II.x.5–8); if you seek a clear conscience, "abandon luxury and the pile that towers to the lofty clouds" (III.xxix.13–15).[6] Sallust associated the decline of Roman morals with excessive riches and the desire for luxury manifested in huge private buildings: "It is worth your while, when you look upon houses and villas reared to the size of cities, to pay a visit to the [simple] temples of the gods built by our forefathers, most reverent of men."[7] Suetonius records that Augustus read Rutilius's speech "On the Height of Buildings" to the Roman Senate to show them that their forefathers had been concerned with this problem.[8] Chastened, Vedius Pollio gave his palace in the Subura to Augustus, asking him to build something for the Roman people on the site. The Portico of Livia, which was erected there, was celebrated by Ovid:

> Where Livia's colonnade now stands, there once stood a palace huge.
> The single house was like the fabric of a city; it occupied a space
> Larger than that occupied by the walls of many a town.
> It was levelled to the ground, not on a charge of treason,
> But because its luxury was deemed harmful.
> Caesar brooked to overthrow so vast a structure,
> And to destroy so much wealth, to which he himself was heir.
> That is the way to exercise the censorship;
> That is the way to set an example, when the judge does himself what
>   he warns others to do.[9]

Domitian's building projects, praised by Martial and Statius, were condemned by Plutarch:

> Should any one who wonders at the costliness of the Capitol visit any one gallery in Domitian's palace, or hall, or bath, or the apartments of his concubines, he would think that Epicharmus' remark about the prodigal, that
>
> > 'Tis not beneficence, but truth to say,
> > a mere disease of giving things away
>
> would be in his mouth in application to Domitian. It is neither piety, he would say, nor magnificence, but, indeed, a mere disease of building, and a desire, like Midas, of converting everything into gold or stone.[10]

Yet vast size is not always a sign of vice. One of the principle reasons for which the cosmos was considered beautiful by biblical Jewish and Classical writers was its size. The Stoic Posidonius, for example, says that the beauty of the cosmos is shown by its spherical shape, color, multitude of stars, and great size.[11] Aristotle considered large size an essential aspect of anything beautiful: "to be great-souled involves greatness just as handsomeness involves size: small people may be neat and well-made, but not handsome."[12] And large size is also a sign of prospering fortune, as Seneca implies: "Oftentimes a reverse has but made room for more prosperous fortune. Many structures have fallen only to rise to a greater height."[13] Enormous constructions may be the visible expression of Christian faith. In Prudentius's *Psychomachia*, "the crane was creaking with the weight on its chains as it whirled the vast gems up to the heights" of the temple that Faith and Concord were building for Christ.[14]

Alberti both condemns and praises great size. In the letter to Brunelleschi, he praises the dome of Florence Cathedral (Figure 1) for its difficulty and its vast size. Yet in Book VII of *De Re aedificatoria*, we read that

> the Milesians built a temple, says Strabo, which on account of its great size remained without a roof. I do not approve of that. The Samians boasted that their temple was the largest of all. I do not deny that a temple should be as large as possible, but it must be possible to adorn it.... In any case, I approve of temples which, in view of the size of the city, you would not wish larger; I am offended by a vast expanse of roof.[15]

Alberti condemns the kings of Asia because "thinking that their works were praised on account of their vast size... they took delight in the colossal scale of their works and competed amongst themselves until they arrived at the insane idea of constructing pyramids" (VI.iii, p. 451). The pyramids are criticized again in Book VIII: "certainly I detest those monstrous works that the Egyptians built for themselves" (VIII.iii, p. 681). Given Alberti's own opposition to excessive size, and the extensive literature—pagan and Christian, Classical and medieval—that associated excessive size with vice, his approval of the dome on this criterion is difficult to account for. Since he praises Brunelleschi's dome by comparing it, precisely, with the hated pyramids, his attitude toward the problem of size would seem to be inconsistent. Alberti does not defend the dome's size in the letter; indeed, he brushes away possible objections, saying that no one could be so hard or so envious as to withhold praise from the architect, seeing a dome so large and so difficult to construct. Yet it would be surprising if none of Alberti's contemporaries criticized the dome. In his influential *Summa* of around 1420, the year in which construction on the dome began, Saint Antoninus warned that "excessive buildings, beyond what suits their status and made for ostentation, are not pleasing to God."[16] Giovanni da Prato, who was consulted on the design of the cupola in its early stages, denounced excessively large structures in his *Paradiso degli Alberti* of around 1425.[17] What is the sense of raising proud towers to the heavens or building magnificent, large palaces when only virtue is eternal?[18] Giovanni's argument repeats that of Salutati, who had presented the cathedral (before the dome was built) as evidence of the vanity of earthly efforts.[19] There is no evidence that Giovanni criticized the size of the dome, but he certainly understood, and endorsed, the traditional grounds on which it might be condemned.

Alberti's praise of the dome for its large size has three justifications:

1. As evidence of man's God-given power to invent, the dome could be associated with the theme of the dignity of man.
2. As evidence of the excellence of Florentine culture and the greatness of its citizens, the dome served a useful political purpose.
3. As evidence of the immense wealth of the Florentines, it revealed the flourishing condition of the city.

But before coming to these justifications, it is necessary to examine more closely just what Alberti says.

The dome is said to be "towering above the skies" and "vast enough to cover the entire Tuscan population with its shadow." Alberti's remarks were not intended as factual description: this is a panegyrical ekphrasis based on the topos of "outdoing" (as Curtius defines it),[20] which requires that the thing praised be said to surpass anything else of its kind. This topos is discussed in Aristotle's *Rhetoric* as amplification (αὔξησις), which, as we saw, permitted superior grandeur and beauty to be attributed to an action on the basis of its orginality or its superiority to comparable feats.[21] The topos, then, calls for comparison with things of the same class and for exaggeration.

Alberti's rhetoric serves not only to praise the dome, but also, through the device of amplification, to create a mental image. As Cicero recognized, exaggeration is more effective than clarity as an aid to visualization: "The one [clarity] helps us understand what is said, but the other [exaggeration] makes us feel that we actually see it before our eyes."[22] Thus Alberti's panegyrical description of the dome, like his evocation of the cathedral interior in *Profugiorum*, serves to clarify his argument and is further evidence of the early Humanists' desire to understand through sight.

What are the other, similar structures surpassed by Brunelleschi's dome? One might assume that Alberti had the dome of the Pantheon in mind, since the diameters of these two structures are almost identical, approximately 43 meters. Well he might have. However, he could not have formulated his claims of "outdoing" within the Pantheon's literary tradition because, curiously enough, that structure was almost never described by ancient or medieval writers.[23] Ammianus Marcellinus's rather tepid comment that it was "vaulted over in lofty beauty" is one of the few texts that can be cited in this connection.[24] His work, which had been rediscovered by Poggio Bracciolini in 1417 and was therefore receiving attention in Humanist circles, was probably known to Alberti but does not seem to have been his source.[25] Appreciation of the Pantheon, or at least verbal expression of appreciation, is a phenomenon of the Renaissance itself, following rather than preceding Alberti's praise of Brunelleschi's dome and Chrysoloras's praise of Hagia Sophia's. Biondo Flavio was the first to speak of the Pantheon as Rome's most beautiful building (in 1446); not until Serlio was it judged to be the superlative example of architecture (in 1540).[26]

The first comparison between the dome of Florence Cathedral and that of the Pantheon is in Pius II's *Commentari*: "But among the buildings [in Florence] none is more deserving of mention than the church of the Reparata, the dome of which is nearly as large as that which we admire at Rome in the temple of Agrippa called the Pantheon."[27] The first dome said to exceed that of the Pantheon in size was that which Bramante planned for new St. Peter's.[28] Whether or not Alberti thought Florence's dome surpassed that of the Pantheon, his topos must derive from the literary tradition of some other structure or structures.

Domes are rarely mentioned by Classical authors, partly because the first fully developed cupola (at the temple of Mercury at Baiae) dates from only the late first century B.C., and the dome was not a widely used form until the second century A.D.[29] This accounts for the silence on this subject of some of Alberti's favorite authors, such as Cicero and Vitruvius. But it leaves unexplained the silence of later writers, whose works were also read by Alberti. Very few Classical literary models were available to the Humanist. There is a description of a vault, probably not domical, in Domitian's palace on the Palatine (which Plutarch had condemned) in Statius's *Silvae*, another work that Poggio Bracciolini had discovered in Constance in 1417: "Far upward travels the view; scarce does the tired vision reach the summit, and you would deem it the golden ceiling of the sky."[30] Like Alberti, Statius measured the height of the dome in relation to the ceiling of heaven. One of Martial's epigrams, which were just becoming known to the Humanists, claims that this vault was a greater wonder than the pyramids:

> laugh, Caesar, at the regal wonders of the Pyramids:
> now barbaric Memphis speaks not of her Eastern work.
> How small a part of the Palatine hall would Egypt's toil achieve
> . . .
> Heaven it so pierces that, hidden amid the lustrous stars,
> its peak echoes sunlit to the thunder in the cloud below.[31]

Another possible source is an epigram in the *Greek Anthology* in which the author asserts that he has seen the walls of Babylon, the Hanging Gardens, the pyramids, and the Mausoleum of Halicarnassus, but that they are nothing compared with the Temple of Artemis at Ephesus, which "mounted to the clouds."[32] The *Anthology* was also a new work to the Humanists: Aurispa brought the first copy to Italy in 1423.[33] In order to surpass a vault that touches the clouds, it is necessary to claim, as Alberti did, that Brunelleschi's dome towers above the heavens themselves. Thus Alberti claimed that Brunelleschi's dome outdoes the Temple of Artemis at Ephesus, one of the Seven Wonders of the ancient world, and Domitian's palace on the Palatine. If epigrams from

Martial and the *Greek Anthology* are the sources of Alberti's topos, then his formula of praise derived from literary models, not his own observations and visual comparisons. Alberti was not making an art-historical statement about his perception of Brunelleschi's dome, or even a statement that he believed to be true; he claimed superiority for the dome solely within the context of a literary tradition. The significance of his statement depended, and depends, on the reader's knowledge of this literary context.

Having defined its height, Alberti described the breadth (or, more precisely, the total volume) of the dome: it is "vast enough to cover the entire Tuscan population with its shadow." This unusual criterion for measurement, the length of a cast shadow, can have but one, very well-known source. The pyramids were said to cast shadows that, from sunrise to sunset, were as long as several days' journey. Since no other monuments than these were ever measured by their shadows, Alberti's referent for his second topos of "outdoing" is certain. Probably, his source for this information was Manuel Chrysoloras's *Comparison of Old and New Rome*.[34] One must suppose that Alberti calculated the boundaries of the Tuscan state to be several days' journey distant from the dome, if it does indeed equal or surpass the pyramids in size.[35] The logic of his claim, however, is less important than his intention, placing the dome of Florence Cathedral on a par with another of the Seven Wonders of the World and with the most colossal structures of antiquity. Giannozzo Manetti's almost exactly contemporary description of the dome explicitly compared it with the Seven Wonders: "I should think it must, not undeservedly, be counted as one of the Seven Wonders of the World on account of its incredible magnificence."[36] He claimed with pride that its height exceeds that of any other building in the world. Both texts, written in the year of the cathedral's consecration, reflect a new, positive, awareness of the Seven Wonders. In the same year, Cyriacus of Ancona translated a poem describing the Seven Wonders, wrongly attributed to Gregory Nazianzus.[37] Iacopo Zeno probably used that translation for his *Vitae summorum Pontificum*, of around 1441 to 1442.[38]

It might seem that the Wonders had always been admired and that the Humanists were repeating old clichés, but this is not the case. The literature on the Wonders reveals two traditions of criticism, of *laudatio* and *vituperatio*. Viewed in terms of their utility and patronage, the Wonders were condemned; seen as manifestations of the architect's power to create, they were extolled. In Classical literature, these opposed positions were represented by Latin authors, on the one hand, and Greek, on the other. This dichotomy was not apparent to Alberti and his contemporaries who, having inherited the Latin insistence on utility, welcomed the additional criterion of creativity. In actuality, only Brunelleschi's dome was defended on these criteria, not only because of its unique status as symbol of Florence and result of one of the greatest engineering feats in history, but also because other criteria came to the fore after midcentury.

The Seven Wonders and Human Capacity to Create

The early Humanists' positive attitude toward the Seven Wonders is a departure from that of Classical Latin authors, whose concern for utility and the public good shaped their negative evaluations of these works. The Wonders' most vehement critic was Pliny, who found them positively enraging. The pyramids "rank as a superfluous and foolish display of wealth on the part of kings; and chance, with the greatest justice, has caused those who inspired such a mighty display of vanity to be forgotten."[39] The labyrinths are "quite the most abnormal achievement on which man has spent his resources."[40] Instead, a work worthy of genuine admiration was the aqueduct system that supplied Rome with water, and the greatest achievement of Roman architecture was Rome's sewage system.[41] Frontinus, whose treatise on the aqueducts of Rome had been found by Poggio and brought to Rome in 1429 (where Alberti could have consulted it), denounced the pyramids as superfluous and useless in comparison with these.[42] These criticisms focus on the uselessness of the Seven Wonders; the authors refuse to recognize the positive value of human intellect and imagination in their creation.

Alberti was much influenced by this Latin heritage. We have seen how he condemned the folly of the pyramids, connecting them with barbaric taste and personal ambition. His negative attitude toward colossal tombs is connected with their uselessness in his assessment of Porsenna's tomb at Chiusi: "I will never entirely approve of such works, marvellous but useless" (VII.iii, p. 683). Elsewhere in *De Re aedificatoria*, he revealed his awareness of Pliny's praise of the sewers of Rome: "The Ancients attached such importance to the use of drains that in no other work did they invest such expense and care. Indeed, they count the sewers among the first marvellous works of the city of Rome" (IV.vii, p. 323).

Despite his acceptance of the value of utility, Alberti also advocated other values. In *De Re aedificatoria*, he urged the primary importance of beauty: "gazing at the heavens and their wonderful works, we admire the work of the gods more for the beauty we see, than the utility that we recognize"; and "to have satisfied necessity is of little account and a small matter, to have satisfied covenience will receive no thanks, where you are offended by the inelegance of the work" (VI.ii, pp. 445, 447). The sources and implications of Alberti's view, that both beauty and utility were necessary, will be examined at greater length in chapter 5.

Appreciation for architecture as a manifestation of human intelligence and skill occupied a much more prominent place, and is more frequent, in works by Greek and Byzantine rather than Latin authors. It is not surprising, therefore, that the earliest extant list of the Seven Wonders was drawn up by a Greek, Antipatros of Sidon; of these, five are architectural achievements: the walls of Babylon, the Hanging Gardens of Babylon, the pyramids, the Mausoleum of Halicarnassus, and the Temple of Artemis at Ephesus. This text, by the way, is the one that Alberti knew from the *Greek Anthology*. All the Wonders are considered marvels because of either their great size or their difficulty of execution. The earliest treatise on the Wonders is also by a Greek, Philo of Byzantium.[43] Greek historians, beginning with Herodotus, saw the Seven Wonders as achievements of the cultures that produced them.[44] Diodorus Siculus was enchanted by the Seven Wonders: the pyramids surpassed all other constructions in Egypt, "not only in their massiveness and cost but also in the skill displayed by their builders." "And they say," he continued, "that the architects of the monuments are more deserving of admiration than the kings who furnished the means for their execution."[45]

This predominantly Greek-language tradition of admiration for the Seven Wonders and, in general, for huge constructions, as manifestations of the greatness of the architect's mind and of the culture that produced him, was brought into the literature of early Humanist description by Manuel Chrysoloras. His *Comparison* opens with a review of six of the Seven Wonders, all of which are said to be surpassed by the city of Rome; Constantinople is compared with the seventh (chapter 8). Having praised the dome of Hagia Sophia for its immense size, he concludes:

> One should not only admire its great mass and skillful construction, but also those men who mentally conceived of such a great work without the help of any similar model and who, having formed a mental image of the work, believed themselves capable of actually realizing it and bringing it to completion.[46]

Chrysoloras's words are close to a passage in *De Re aedificatoria* where Alberti recommended that the architect study, above all, the biggest and most important buildings: "Yet he should not be so affected by vast size of the work as to be satisfied with that alone... but above all search out what is rare and admirable in each on account of considered and hidden artifice or of invention" (IX.x, p. 856).

The similarity between Chrysoloras's thought and that of Alberti, in both his letter to Brunelleschi and his treatise, is striking. Both ignore the motive of the patron and the social benefit brought by the work (criteria of excellence common in Classical Latin criticism) and focus attention on the mind of the maker.

As we saw in chapter 2, Brunelleschi's contemporaries praised the dome as the achievement of his *ingenio*. Thus they interpreted its great size according to the criteria used by Greek, but not Latin, authors for the Seven Wonders: as a manifestation of the mind of the architect and evidence of human power to create. This does not imply the ascendance of Greek over Latin literature, but the emergence of new values within early Humanist culture itself. The early Humanists' new emphasis on the power of the human mind was intimately related to their conception of man's God-given role as second creator. Petrarch had argued that human dignity derives from theological and philosophical reasons, but also has historical and existential explanations.[47] Dignity derives from individual and collective actions and creations in this world, from which come earthly fame and greatness, tokens of the contributions to the cultures and civilizations that mankind invented and constructed. This current of thought was especially nourished by revived interest in Patristic authors, both Greek and Latin. Chrysostom explained that man was made in God's image not in dignity of substance, but in similarity of power; what God is in heaven, man is on earth.[48] Particularly important was Nemesius of Emesa's *De Natura hominis*, written in the late fourth century.[49] Giannozzo Manetti owned a copy and drew heavily on it for his own *De Dignitate et excellentia hominis*. Nemesius argued that man, through his skill in the arts and scientific learning, gave order to creation.[50] Manetti saw man's second creation as the operation of human inventiveness on nature.[51] Book II of *De Dignitate* praises human operative and inventive faculties, whose powers are shown by the "many and huge actions and mechanisms admirably invented and known" by man.[52] Great size is that which characterizes human works, as he tells us in his account of the consecration of Florence Cathedral, describing Brunelleschi's dome:

> The structure of so amazing a building seems to have been divinely erected from its bottom to its top out of huge stones dressed with precision and highly polished. And yet, because of the unbelievable size of the building, the eyes of the onlookers surmise, not unjustly, that it has been made by man.[53]

Manetti's observation is provocative, associating beauty of material and execution with divine, and size with human, work. Perhaps he had in mind the Seven Wonders, the largest constructions of antiquity, none of which were thought to have been divinely inspired. Like the Tower of Babel, the moral status of some of them was not high, at least among Latin authors. If Manetti intended to praise the human capacity to create enormous structures such as these, or to see this capacity as characteristically human, it must have formed part of a larger reinterpretation of moral action. Unfortunately, Manetti's theological and ethical thought has not yet been sufficiently studied to explain this altogether suprising view. Although Chrysoloras had argued that the vast dome of Hagia Sophia was proof of human capacity for achievement and Alberti considered the Florentine dome to be evidence for a view of history as culturally and spiritually progressive, neither went so far as Manetti in endorsing large size.

Alberti's justifications of Brunelleschi's dome on the grounds we have examined were not further developed in the second half of the century. Only in the *Hypnerotomachia Poliphili*, a work much influenced by Alberti, do we find enormous size praised as revealing the mind of the architect:

> In itself it [the obelisk] contained such a store of wonders that I stood and contemplated it in an insensible stupor. And much more on account of the immensity of the work and the surpassing subtlety of its opulence and the rare skill, great care and exquisite diligence of the architect. What boldness of inventive power! What human strength, art, device and incredible enterprise was needed to raise up into the air such a weight, rivalling the sky.[54]

Rossellino was not principally praised for his engineering at the difficult site of Pienza, or Francesco di Giorgio for his remarkable achievement with the site at Urbino. Increasingly, in the second half of the century, the magnificence of the patron and the institutional authority acquired through enormous buildings, rather than the *ingenio* of the architect, justified excessive size.[55]

Utility

Alberti and Manetti also drew on a second line of defense. By interpreting the dome as proof of the status of the whole city of Florence, they associated its great size with civic welfare.[56] There is some Classical precedent for the association of architecture and the authority of the state in Cicero, who saw "the great size and surpassing beauty of a city" (although not its individual buildings) as contributing to the safety and security and also to the importance and power of the state.[57] But this argument owes less to Classical models than to developments in historiography in early Quattrocento Florence. Leonardo Bruni had claimed, in *Laudatio florentinae urbis*, that the city's splendid architectural monuments were proof of the greatness of its citizens.[58] When he wrote, the dome had not been begun. In chapter 2, Alberti's letter to Brunelleschi was interpreted as a continuation of discussions of Florence's status from the first decade of the century; his claims for the dome suggest not only the grandeur and piety of Florence, but also its capacity to protect its territory. Bruni's *Laudatio* had been republished shortly before Alberti wrote his letter. Both works furthered Florence's political ambitions of that moment, as we will see in chapter 9. Nor were these isolated instances. Matteo Palmieri emphasized the advantage to the community of splendid public buildings in an appendix to his *Della Vita civile*, almost contemporary with Alberti's letter.[59]

In *De Re aedificatoria*, Alberti again affirmed the civic importance of impressive buildings, declaring that the temple is the most important of architectural structures since it is the greatest ornament of the city (VII.iii, p. 543). And there is some indication of the same view in the preface to the same work, where he writes that

> Delos was visited not so much on account of Apollo's oracle as because of the beauty of its city and majesty of its temple. As for how much architecture has contributed to the fame of the Latins and their imperial authority, I need only mention that the tombs and other remains of ancient magnificence which we see everywhere have led us to believe as true many things written by the ancient historians which would otherwise have probably seemed much less convincing. Thucydides was right to approve the prudence of the Ancients who had adorned their cities with every kind of building so as to seem much more powerful than they were. (Prologue, p. 13)

It was Giannozzo Manetti who most explicitly associated the glory of the Florentine state with the incredible achievement of the dome. In his account of the consecration of the cathedral, Manetti argued that

> if the Athenians take great pride in their arsenal for this reason, namely that the work will have had to be seen for its cost and elegance, and are also proud that those who wrote to preserve the memory of their deeds have transmitted the name of Philo, its architect, with great praise; ought not the Florentines, too, for the same reasons take the greatest pride that they have built an admirable work, so sumptuous that it will be very difficult to reckon the unbelievable cost thus incurred by its amazing construction, and so extraordinarily elegant that it will draw from various places many foreigners to gaze on its astonishing elegance and, once drawn, it will hold their gaze more often and for longer periods of time and stun their minds.[60]

Seen in this context, Alberti's letter to Brunelleschi belongs to a trend in early Humanist historiography in which the built environment was used as historical evidence. As he wrote, the controversy over whether panegyrical amplification was permissible in this genre was imminent; Bruni's *Laudatio* would be its principal target.[61]

Just as panegyric, seen as a tool for political propaganda, won acceptance in the later Quattrocento, so did the use of the built environment as evidence for the authority of the state. Although Poggio Bracciolini and Piero da Noceto urged Nicholas V to stop his grandiose construction projects, which everyone hated, Nicholas was still defending their utility on his deathbed:

> Not for ambition, nor pomp, nor vainglory, nor fame, nor the eternal perpetuation of my name, but for the greater authority of the Roman Church and the greater dignity of the Apostolic Seat among all the peoples and the more certain avoidance of the usual persecutions we conceived such buildings.[62]

Michele Canensi dismissed these criticisms with words almost identical to Alberti's: "Who could be so depraved of mind or hard of heart" as not to recognize Nicholas's piety in these works?[63] The association between personal piety and the procurement of institutional prestige would be a fruitful one, as we will see.

Wealth

Large structures entail great expenditure. That the Florentines had been able to sustain the cost of the dome was proof positive of the exceptional wealth of the city. But the question of wealth was a particularly sensitive one in the merchant republics, and had been at least since the time of Saint Francis. Salutati, advocating the ideal of apostolic poverty, argued that Constantine had corrupted the Church with wealth and that this had resulted in its spiritual decline.[64] In both the Classical and the Christian tradition, intrinsic values such as virtue and goodness are always set over extrinsic ones like fortune and riches.[65] Baron has examined the reevaluation of the moral status of wealth in early Renaissance Florence, showing how riches were praised as essential to civic culture. In the preface of his translation of pseudo-Aristotle's *Economics* (1420–1421), Bruni denied the value of poverty and urged the connection Aristotle had made between ethics and the conduct of the citizen; material goods are a necessary tool for the exercise of virtue.[66] In Rome, Stefano Porcari, *capitano del popolo* in 1427, asked rhetorically: From whence come our houses and palaces...? From riches!... These consecrated churches with their decorations, your walls, towers, defenses... your palaces and dwellings, your most noble buildings, bridges and streets: with what have you built them, and where do you obtain the means of preserving them, if not from riches?[67] Poggio argued in favor of avarice, on the grounds that the poor man satisfies only his own needs, whereas riches contribute to the creation of a common civilization. Without wealth "all the splendor, beauty and adornment of cities would disappear; no more temples, no monuments and no arts... our whole life and that of the state would be undermined if everyone procured for himself only what was necessary."[68]

By midcentury the new materialism was being defended as a historical necessity. Lapo da Castiglionchio argued that while poverty had been appropriate for Christ and the early Church, now the Church needed wealth since the present age admired riches and depised poverty.[69] Benedetto Accolti shared this view, adding that in the modern age of materialism the Church could maintain its authority only through external splendor and magnificence.[70] This line, in full contradiction of the pagan Classical and Christian tradition of condemnation of extrinsic values, would seem to be one of the great innovations bequeathed to the modern world by the Renaissance.

Nowhere is this clearer than in Julius II's project to erect the most immense church in all of Christendom, new St. Peter's. All observers of Julius's project commented on its enormous size; none had any moral objection to offer. S. De'Conti (1506) related that the church was to surpass all the works of antiquity in size and beauty and that its dome was to be wider and taller than that of the Pantheon.[71] Giles of Viterbo (1507) compared Julius with Solomon, Zerubbabel, Sylvester, and Constantine and urged him to raise a building as high as the heavens.[72] Although, like Salutati, Giles believed that Constantine had corrupted the Church by granting it money and buildings, he seems to have been convinced that the size of new St. Peter's was evidence for Julius's great religious devotion.[73] Francesco Albertini (ca. 1510) claimed that the new church would exceed the Temple of Diana in Ephesus in size, using as proof the measurements given in Pliny.[74] It was true, he said, that the Cathedral of Florence had already surpassed the Temple at Ephesus, but now the papal basilica would outdo Brunelleschi's dome, rising even higher.[75] Those who disapproved of the project, like Maffei, complained that its cost would be a source of grief to Julius's successors and that such a work was not necessary, but no one suggested that its size was in itself objectionable.[76]

After the death of Julius, Giles of Viterbo urged his successor to build an even larger structure. "Since the very great mind of Julius II . . . began to raise the mass of the temple up to heaven, may it be your concern [Leo X] first to finish the temple of your immediate predecessor, then raise its tower higher than the highest temple."[77] Giles went on to imply comparisons between Asia (the Temple of Diana?), Etruria (Brunelleschi's dome?), and the Tower of David, to be realized by Leo X. His theme seems, at least in part, to be borrowed from the Florentine Albertini's slightly earlier comparison of the Wonder at Ephesus, Brunelleschi's dome, and new St. Peter's. And behind both is, of course, Alberti's comparisons between Brunelleschi's dome and the Seven Wonders.

Despite this evident literary continuity, great differences separate the early Humanist's intentions and those of the papal panegyrist. Alberti struggled between his admiration for the mind of the architect and civic pride, on the one hand, and the tradition of condemnation of great size, on the other. Giles was unconcerned with Bramante, evidently the mere instrument of papal will, and with the city of Rome: the personal fame of Leo X and the authority of the Church are all that matter. Yet Giles was also a reformer, who had claimed that the Church declined from the moment when Constantine endowed it with earthly riches. His views are clearly contradictory, and cannot be reconciled.[78] Like Alberti, he opposed excessively large structures but felt obliged to make an exception for a project that he believed would further just causes.

Both the shifting grounds on which large size was justified and the self-contradiction of its defenders suggest that the moral problem of size could not be resolved by Renaissance thinkers. Despite Alberti's justifications of the dome as useful to society, as manifesting an intellectual achievement, and as contributing to the role assigned to man by God, his attitude toward the question of large size remained ambivalent and his praise of Brunelleschi's dome, although it can be associated with ideas in other of his works and in those by contemporary Humanists, remained exceptional. The problem was no closer to resolution seventy years later, when Giles of Viterbo, like Alberti, could justify Bramante's even larger dome only by ignoring his own fundamental distrust of wealth and worldly possessions. Evidently, size posed a moral problem for which there was no compelling justification in the Renaissance. Indeed, the grounds on which excessive size can be justified are few, and have but a weak tradition in the West, whereas the moral grounds for condemnation are many and find support in the most authoritative Classical pagan and Christian writers. Both Alberti and Giles, profoundly concerned with moral virtue, knew that the issue centered on the danger of loving earthly things instead of loving God. Neither could have answered Augustine's question: Why should the soul devote itself to raising up vast monuments in this visible world, when all they amount to in the end is an array of mud huts, slapped together from the soul's own ruins?[79]

PART II

Aesthetics

FOUR

Norm Versus Innovation: The Problem of the Middle Ages

LEON BATTISTA ALBERTI'S admiration for Classical architecture is common knowledge. He assumes, in both *Profugiorum* and *De Re aedificatoria*, that the Classical Orders are the basic vocabulary of architectural design; indeed, no other vocabulary is mentioned in his writings. Yet his only extended description of a building is a panegyric of Florence Cathedral, the most fully Gothic building in the city. This paradox demands resolution. It cannot be solved with the evidence of his writings on architecture alone, not only because they nowhere address our problem directly, but because the issues involved were most fruitfully explored by the early Humanists in regard to language rather than building. This chapter examines those issues—the problem of the authority of the Classical past, of imitation versus innovation in matters of style, and the status and character of the Middle Ages—in the light of early Humanist writings, and shows their application to architectural theory. It will be seen that Alberti's thought was consistent, and that his acceptance of Gothic buildings and motifs finds its analogue in his acceptance of the vernacular and of neologisms. At the end, a new paradox will emerge: Alberti's acceptance of the Gothic reveals the profound classicism of his thought.

The architectural tradition that concerned Alberti was the Roman one. Although he distinguished four types (or phases, or styles) of architecture—Egyptian, Greek, Roman, and Etruscan[1]—he did not identify those later, and non-Classical, developments that we know as Gothic and Byzantine. Therefore, he did not recognize the invention of any post-Classical formal vocabulary or structural system. This led Krautheimer to suppose that, for Alberti, Roman architecture represented the perfect culmination of a progressive development, and that "that which was, becomes the yardstick for what should be."[2] My study takes issue with this conclusion. If the Romans had achieved perfection, then no further development would be possible. But this is not Alberti's view since, within the Classical tradition, he allows place for innovation. We may consider as evidence a passage from *De Re aedificatoria* where Alberti discusses what kind and degree of variety is permissible in modern design: in other words, the problem of tradition versus innovation.

> In other respects [such as introducing variety], these rules will have to be followed as usage, appropriateness and even the respected practice of the experts would demand. For indeed, in a great many instances to resist custom takes away beauty, while to assent to it is a source of benefit and profits extraordinarily since the work of the best architects seemed to show that the division into Doric, Ionic, Corinthian and Tuscan is the most convenient. Not that we must cling to transferring their descriptions into our work as though we were restricted by laws, but that, advised by them, we may produce new discoveries and strive to attain a reward of praise equal to them or greater, if that is possible. (I. ix, p. 69)

This passage affirms the value of tradition and cultural continuity; yet it also expresses Alberti's belief in progress, and therefore in cultural change over time. Does his acceptance of innovation within a tradition include or exclude modern—that is, Gothic—architecture?

Alberti's Understanding of the Gothic

It is usually thought that Alberti condemned the Gothic, even though he never specifically mentions it in his works. Frankl found his very silence eloquent, remarking that "in his writings he is far too dignified to combat Gothic architecture or even recognize Gothic architects as rivals, and much too tactful to curse an already defeated opponent."[3] Such an interpretation proceeds from questionable assumptions. First, it assumes that early Humanists, such as Alberti, rejected the Middle Ages. I will present evidence that shows that this was not always the case. Second, it assumes that the disdain expressed for the Gothic by later architectural theorists, from Filarete to Palladio, was shared by Alberti. This assumption would consider the period 1400 to 1600, corresponding to the Renaissance in Italian architecture, as a single cultural unit; I will argue that the values of the early Renaissance were quite different from those of the high and late Renaissance. To interpret Alberti's thought in the light of views expounded by later theorists is to distort it.

Our first task is to examine those passages in *De Re aedificatoria* where

Alberti seems to condemn the Gothic. The most important of these is the following: "Those who build today have been charmed by tasteless and absurd novelties rather than the best rules of the most praised works" (VI.i, p. 443). Read together with the passage already examined, it seems clear that Alberti does not intend to condemn novelty as such, since that would eliminate the possibility of progress. Instead, he compares the best works with tasteless and absurd ones. Evidently, Alberti felt that modern architecture had lost touch with those criteria on which good architecture is based. His criticism concerns quality, but not a specific style; he rejects some unspecified manifestations of modern architecture, not the Gothic per se.

In another passage, Alberti criticizes modern architects for leaving large holes in the wall for scaffolding. The use of the "Ancients," in which scaffolding beams were rested on stone mensoles projecting from the walls, was better (III.xii, p. 229). Again, this is not a criticism of the Gothic style, but of deviations from good practice by modern architects. Another modern practice also comes in for criticism: the immediately precedent generations placed steps rising to the portal of a church and then coming back down on the other side of the threshold to pavement level: Alberti comments that without calling this absurd, he cannot see the point in it (VII.v, p. 559). I do not know what works he could have had in mind, but this practice is not a typical characteristic of Gothic design. Alberti also condemns the tendency since 1200 to build towers everywhere, even in small centers; he contrasts this with the Romanesque, when people did nothing but build churches (VIII.v, p. 699). Once again, this is a criticism of modern practice but not of Gothic style. There is no rejection of the Gothic in his chapter on building restoration, either. He suggests that it is sometimes necessary to correct the style of a preexisting building for the sake of "elegantia." Thus if a wall is too tall, horizontal cornices should be added, and if too long, columns, so that the eye can find repose and not be offended by the vastness of the ambient (X.xvii, p. 1001). Here, Alberti is objecting to the insufficient articulation of planes, not to the proportions of any particular style. He is not totally disapproving of modern practice: patrons looking for solid ground on which to construct should consult the local architects, whose opinion is based on the example of old buildings and on modern usage (III.ii, p. 181).

It may be supposed that no Gothic structure could meet Alberti's criteria for good architecture, but his description of Florence Cathedral proves that this is not the case. In his treatise, good architecture is defined, above all, in terms of its desired emotional and visual effect on the viewer; style is what helps to bring this about. Although Westfall suggests that Alberti's conception of architectural beauty as didactic, aiming to move men to forsake vice for virtue, derives from poetic theory and may be connected with Neoplatonic ideas, I would prefer to connect it with rhetorical theory, particularly with that of Augustine in Book IV of *De Doctrina christiana*, where he discusses the need for a Christian orator to move the spectator and give him pleasure.[4] Alberti's first requirement for a religious building is that it should fill the soul with joyous marveling, and the spectator should be struck with amazement at such a wonderful sight (V.vii, p. 361; VII.iii, pp. 543–45). Its style should be such that someone who has never seen it should be attracted to visit it, and those who see it should be charmed by the marvelousness and rarity of the work. In *Profugiorum*, Agnolo calls the cathedral a "nest of delights," and speaks of the deep pleasure he feels as he experiences it; his warm response is proof, therefore, of its worth.[5] It would be wrong to mistake Agnolo's (or Alberti's) feeling for an effulgence of uncritical pious sentiment, since he identifies those specific stylistic qualities that have affected him. These, as we saw in chapter 1, were pairs of opposite qualities—grace and majesty, charm and robust solidity, ornamentation and solid construction—in an ideal equilibrium. The union of firmness and delight was what Alberti considered to be the great achievement of Roman architecture, and an essential principle of all good architecture.[6] Thus, unlikely though it may seem, Alberti credited the Trecento cathedral of Florence with the virtues of Classical design.

This was facilitated by the terms of his description in *Profugiorum*, which, being purely aesthetic, could be applied to any historical style. Although we might feel, for example, that the Pantheon possesses rather more majesty than grace, and the Sainte-Chapelle more grace than majesty, either term might conceivably be applied to each of these buildings. Such critical tools are useful in determining the quality of a building and its aesthetic effect, but not its historical category. Alberti's descriptive vocabulary is style-neutral; in characterizing the supports of the cathedral as slender, he does not tell us that they are Gothic piers instead of Classical columns. The cathedral meets Alberti's criteria for good architecture through its effect on the spectator and the principles of its design. There is no hint that its formal vocabulary is other than Classical.

We may wonder to what extent Alberti was able to define the Gothic style. His vocabulary permitted him to describe variant forms of columns, but he had no word for piers. He suggests, in *De Re aedificatoria*, that columns have undergone a natural evolution from being round to being, at least on occasion, square (I.x, p. 73). Alberti may have been misled by Pliny, who relates that "Atticae columnae" have "quaternis angulis pari laterum intervallo" (*Natural History* XXXVI.lvi), or Isidore of Seville, who follows Pliny on this point (*Etymologiarum* XV.8). In two passages in *De Re aedificatoria*, he endorses the use of these "square columns": for a colonnade of arches (as opposed to a trabeation), "columnae quadrangulae" should be used (VII.xv, p. 643); if you wish to provide robust supports for a private house, a square column with two adossed half-columns is a good solution (IX.i, p. 787). This last recommendation describes a form very nearly like the compound pier used in Romanesque and Gothic architecture. If Alberti could not distinguish any difference in genus between a column and a pier, considering the latter to be a variant of the former, on what basis could he have considered the new vocabulary of the Gothic to be un-Classical? Although he might have observed differences in proportion and handling between ancient and modern architecture, his vocabulary did not recognize differences in morphology between the two.[7] A similar conclusion is suggested by his discussion of the arch. There are three kinds of arches: semicircular, depressed, and pointed. The pointed arch, called "compositus" or "acutus," is obtained by the conjunction of two arch-segments (III.xiii, p. 235). This type of arch was not used by the Ancients, Alberti says; it may have been introduced for tower openings in order to divide the load on the arch, since the pointed arch is strengthened rather than weakened by weight load (III.xiii, p. 237). There is no sign of condemnation here: the semicircular arch is said to be the most stable, but the pointed arch is stronger under pressure. These are alternatives between which the architect can choose, depending on his specific need. Once again, the morphology of Gothic architecture is considered as an extension of the Classical and as a response to new needs. Alberti notes that the Ancients made all doors and windows quadrangular (VII.xii, p. 619), but not for this does he proceed to condemn the round-headed windows and portals traditional in, for example, Florentine domestic architecture.

Specific terminology for the description of Gothic architecture is absent from Alberti's treatise. This may be contrasted with the documents for Milan Cathedral in which, perhaps under the influence of transalpine architects, piers are "piloni," and thus clearly differentiated from columns; pointed arches are "voltae acutae"—vaults, that is, not arches.[8] Unlike Alberti's, this terminology recognizes a morphological and static system different from the Classical. This style-specific terminology may have heightened Filarete's perception of the differences between ancient and modern architecture; his treatise, written about 1460 in Milan, is one of the first to criticize the Gothic. Door and window openings must always be quadrangular; "colonnette piccole" are made by people who do not know the proportions proper to columns.[9] It is Filarete, not Alberti, who presents modern practice and the "modo antico" as opposites, like day and night (XIII, pp. 380–81); who undertook to teach his patron to distinguish between "le cose antiche e le moderne" (VIII, p. 220); and who insisted that love of ancient architecture must engender hate of the modern (i.e., Gothic) style, which "gente barbara" had brought to Italy (VIII, p. 227). None of these notions is explicit in Alberti's writings.

Despite Filarete's dislike of Gothic, he had no systematic understanding of it; such knowledge was but slowly acquired, and later. Cesare Cesariano saw no contradiction (in 1521) in demonstrating Vitruvius's theory of proportion with a cross section of Milan Cathedral, a work that he himself referred to as "Germanico more."[10] Panofsky has shown that Vasari's rejection of the Gothic led him to write the first real description of the style;[11] yet even Vasari considered Gothic to be a kind of Order, in the Classical usage of that word. That Gothic could not be discussed outside the traditional genres of Classical architecture was still true in 1589: Francesco Terribilia considered San Petronio in Bologna to be in a style of architecture "chiamata Tedesca, imitante l'ordine Corinthio."[12] He explained that the Goths, or Germans, had mixed together elements of Greek and Roman architecture with their own, and although this hybrid was said to be based on rules of its own (he referred to triangulation), he considered it necessary to discuss it "within the framework of our natural and universal rules according to the guidelines laid down by Vitruvius."[13] Terribilia did not condemn the Gothic; indeed, he claimed to know many churches built in "good German manner."[14]

Acceptance of, and even enthusiasm for, the Gothic was even more widespread in the earlier period, which concerns us. Gombrich has emphasized the continuity between Trecento and Quattrocento architecture, as have Klotz, Trachtenberg, and, most recently, Morolli.[15] Gilbert demonstrated a persistent admiration for Gothic buildings even among those exposed to the new Renaissance style.[16] This line of scholarship is supported by the statements of the early Humanists. Leonardo Bruni thought that the Gothic buildings of Florence were worthy of praise in his *Laudatio florentinae urbis*, saying that "if you are looking for contemporary architecture, there is surely nothing more splendind and magnificent than Florence's new buildings."[17] That so respected a Humanist scholar as Aeneas Silvius Piccolomini commissioned a Gothic Hall Church for Pienza is well known. His writings suggest a real admiration for the Gothic style. For example, "In my opinion the Germans are wonderful mathematicians and in architecture they surpass all peoples."[18] Or, in his description of Nuremburg: "What is there more impressive than the church of St. Sebaldus, what is more resplendent than the church of St. Laurentius?" And again, of Strasburg: "The episcopal church, called the minster, built most magnificently of cut stone, rises as a very extensive edifice, adorned with two towers, of which the one that is completed, an admirable work, hides its head in the clouds."[19] Like Alberti, Pius II used Classical terminology to describe Gothic forms. Thus he tells us that in the Cathedral of Pienza "octo columnae spissitudine et altitudine congruentes universam testudinum sustentant molem" ("eight columns, all of the same height and thickness, support the entire weight of the vaulting").[20] For him, neither the morphology nor the proportions or the statics of the building were anything but Classical. At another point he refers to the piers at Pienza as "columnas quatuor habentes facies" ("columns with four semicircular faces") and therefore as variants of the single column.[21] Yet Pius had studied Vitruvius during the Council of Mantua (1459) and must have been aware that the style of the work he had commissioned was not Antique.[22] Such use of Classical terms for Gothic forms was common. Manetti described the piers of Florence Cathedral as "saxeae columnae."[23] For Giovanni Caroli, in the last decades of the Quattrocento, the Gothic piers of Santa Maria Novella were still "columpnarum."[24] That he evaluated the aesthetic merit of the architectural complex using Alberti's own critical criteria (from *De Re aedificatoria*) supports my argument that these were style-neutral.

> The kind and quantity of magnificence and grace these buildings have is clear to everyone. In fact, if they would observe the dignity of the temple from all sides they would see that it has been congruently planned both in the solidity of its structure and in the appropriateness and relation of its single parts, and is so perfect that virtually nothing could be added or taken away which would improve its stability or decorum.[25]

As late as 1515, a poem in praise of one of the piers of Sant'Agostino in Rome referred to it as "augusta columna."[26]

Undoubtedly, the architects of the earlier Quattrocento—Brunelleschi and Alberti, above all—*did* reinstate the Classical Orders as the basic vocabulary of design, and we must therefore assume that they could distinguish between a Gothic and a Classical building (if not, perhaps, between a Romanesque and a Classical structure). But there is little evidence that they possessed any systematic understanding of Gothic architecture, and even less that they were able to define non-Classical styles. Earlier assumptions that the differences between Classical and Gothic were clearly known, that these styles were regarded as antithetical, and that the Classical was believed to be superior have been weakened by recent research. According to Burns, the correct use of the Orders was not an important concern in the Quattrocento, or considered essential to *all'antica* design.[27] Günther would see the importance of the Orders for architectural theory as beginning with Giuliano da Sangallo and Bramante.[28] Indeed, recent opinion has tended to reinforce Fabriczy's intuition that Quattrocento architects were unaware of a break with the forms and ideas of the past, were unable to define their relation to Ancient art, and therefore had no clear program to revive it.[29] Their knowledge of Roman architecture was so fragmentary that they could form no general notions about it, he claimed, and their chief interest in Classical architecture was in learning techniques they could apply for their own purposes.

These views go directly against Mommsen's famous definition of the Renaissance, according to which individuals, beginning with Petrarch, viewed the preceding age as one of obscurity, from which they wanted to break away.[30] Yet it was Petrarch who praised Cologne Cathedral: "In the middle of the city I saw an uncommonly beautiful temple, which, though still incomplete, they call with reason the most magnificent."[31] What was really obscure to Alberti and his contemporaries about the Middle Ages was its duration and character. For it was by no means clear when Antiquity had ended, or rebirth begun, and the Middle Ages was perceived as different from Antiquity and the Renaissance in quality, but not in kind. In the absence of firm historical definition, it was difficult—perhaps impossible—to define stylistic categories for architecture.

Nowhere in his writings does Alberti address the problem of what is Classical and what is non-Classical, or medieval. We know that these historical distinctions were no easy matter for Alberti's contemporaries.[32] People in the fifteenth century believed that they were living in the sixth age, which had begun with the birth of Christ, and in the time of the fourth (Roman) empire, which still existed.[33] This historical scheme was quite sufficient, for example, for Saint Antoninus when he wrote his *Opus excellentissimum historiarum seu chronicarum* in 1458, and for Antonio Pierozzi in his *Chronicon universale* of 1459.[34] Nonetheless, if from a spiritual and, to a lesser extent, political point of view the preceding 1,500 years or so constituted a single age, it also appeared to be divisible into phases, even if no consensus existed about where these historical divisions should be placed. The chronological duration of the Middle Ages was uncertain, because there was no agreement on when Classical Antiquity had died or when the modern rebirth had begun. Moreover, as Ferguson observed, "as long as historical thought remained within this framework of supernatural teleology there could be no idea of a distinction between ancient Roman civilization and that of the age following the breakup of the empire nor of the rising civilization after the darkest period was passed."[35]

The development of the concept of the Middle Ages was intimately connected with Humanist historiography and, in particular, with the desire to record the rise of the city-states in the communal period. It is no accident that it was Leonardo Bruni who coined the term "medium tempus" while describing the period after the destruction of Florence in 450 and its rebuilding under Charlemagne around 800.[36] Bruni's nationalism helped shape his chronological definitions. He felt that "the decline of the Roman Empire must begin from that time in which, giving up its liberty, Rome began to serve the emperors."[37] For him, loss of civil liberty brought about the decline of civic virtue; the empty shell of empire collapsed when the capital was moved to Constantinople, leaving Italy prey to tyrants and barbarian invasions. Thus Bruni's Middle Ages began in the first century B.C., were fully established by the fifth century A.D., and lasted until the rise of the Italian city-states beginning in the ninth, and fully accomplished in the thirteenth, century. For Biondo Flavio, instead, the Middle Ages did not begin until the sack of Rome in the fifth century; the Roman Empire died in the sixth century, and a new era had begun only in 1410, the millennium of the sack of Rome.[38] Like Bruni, Biondo felt that things had been steadily improving ever since the time of Charlemagne, but he did not feel that the Dark Ages had yet come to an end at the time he wrote (1417–1442).[39] Bartolomeo Scala was not convinced that the period of darkness was over even in the 1490s.[40] Many estimates regarded only the fifth through seventh centuries as really dark; this was the opinion of Filippo de'Medici, Domenico Silvestri, and Donato Acciaioli as well as of Leonardo Bruni and Biondo Flavio.[41] But more radical views, as we saw, extended the period from the birth of Christ into the Quattrocento. These views, had they been applied to the history of architecture (which they were not), would have produced a most startling categorization. Petrarch, having ceased his praise of Rome with the reign of Titus, unwittingly made the Pantheon a medieval monument.[42] Within Bruni's chronology, all works after the time of Vitruvius were works of decadence, yet the cathedrals of Siena and Florence would be renascence structures. Indeed, for political historians with nationalist leanings, all of what we consider Gothic was the art that not only accompanied but also expressed the revival of the cities and the rise to power of the new city-states. It is unlikely that a historian like Bruni would have condemned the structures built in thirteenth- and fourteenth-century Florence. National pride required political historians to applaud the Gothic.

Of course, these implications were not intended by the political historians, and historians of culture used different frameworks and different criteria to define the Dark Ages. Ferguson pointed out that whereas the political history of the medieval communes was of interest to most citizens, the history of medieval culture was neither interesting to the educated layman nor related to local patriotism; therefore, it was less discussed.[43] In general, the period of darkness was said to have lasted longer in cultural history than in political history; the revival was seen as more recent, more sudden, and more connected with individual men.[44] Ullman has argued, persuasively, I think, that the notion of cultural renascence began in the field of poetry: Boccaccio, Salutati, Polenton, and F. Villani all speak of Dante as restoring good poetry.[45] From poetry, the concept was extended to painting, where Boccaccio and Villani applied it to Giotto.[46] Some areas of cultural achievement seem to have lagged: Alamanno Rinuccini, Lorenzo Valla, Guarino Veronese, and Ambrogio Traversari (among others) would have seen oratory and Latin style as being revived shortly before the mid-fifteenth century.[47] Others, such as Matteo Palmieri and Biondo Flavio, thought that Latin letters had been revived around 1400.[48]

The date at which decline began, for both architecture and eloquence, was particularly difficult to establish. For the latter, dates ranged from the second century (after Juvenal),[49] the fourth (after Lactantius and Jerome),[50] the fifth (after Orosius and Claudian),[51] the sixth (after Cassiodorus),[52] the early seventh (hastened by Gregory the Great's burning of the Palatine library),[53] to even as late as 800.[54] The question of the decline of eloquence was decided on the criterion of the style of individual writers; it was hardly possible to assess architectural merit in the same way. That decline had somehow occurred was clear to all, but few early Humanist writers wished to hazard an explanation about why or how, or even when, it had taken place.

Biondo Flavio thought the decline of architecture was in large part the fault of the Goths, who, unlike the Romans, did not know how to build solid structures.[55] Fazio degli Uberti blamed Gregory the Great for the beginning of the destruction of ancient architectural monuments and therefore, by implication, for the loss of the Classical tradition of building.[56] Matteo Palmieri, who could not establish a firm date, thought that architecture had been in decline "da noi indietro lunghissimo tempo" but had returned to perfection by the 1430s.[57] Valla agreed that both eloquence and architecture had been restored by the second quarter of the century.[58] Filarete tells us that the renewal of architecture began in the 1420s or 1430s; he does not give a date for its previous decline.[59] In the last decades of the Quattrocento, Francesco di Giorgio disagreed with Filarete; although many think that good architecture has recently been restored to light, "si può dire che tutti li edifici moderni sieno pieni di errori e di parti senza la debita proporzione o simmetria."[60] Antonio Manetti saw the decline as beginning with the abdication of the emperor in the West; the renewal began, of course, with Brunelleschi, the subject of his biography.[61] Yet at almost the same moment, Francesco Colonna lamented that in regard to architecture, he and his contemporaries were still living in the Dark Ages. "O infoelici tempi et aetate nostra, come dagli moderni (usando conveniente vocabulo) sì bella et dignifica inventione [the Classical temple of Venus] è ignorata."[62] Sixteenth-century writers saw the revival as even more recent, having been brought about by Bramante.[63] It would seem, then, that there was some uncertainty about whether architecture had revived together with eloquence in the course of the fifteenth century, about 100 years after painting had revived together with poetry.

Alberti was a student of history, literature, painting, and sculpture as well as of architecture. Did he define decline and rebirth differently in each of these spheres of activity, or did he adopt a single chronological scheme for the duration of the Middle Ages? Was Florence Cathedral a medieval monument in his eyes? Judged by analogy to contemporaneous developments in poetry (Dante), painting (Giotto), and, most important, politics (the rise of the city-state), it was not, whereas by analogy to letters and eloquence, it was. My discussion suggests how unlikely it is that Alberti had any firm idea about the chronological extension of the Middle Ages and that, as a consequence, his conception of both Antiquity and the modern period was blurred at the edges.

Alberti surely realized that even if by some definitions the Cathedral of Florence could be seen as a renascence structure, it hardly accorded with Vitruvius's recommendations for architecture, nor did it resemble the monuments of ancient Rome. Yet it was also true that many Roman monuments did not correspond to what Vitruvius described, since his treatise was written before they were built. Vitruvius's text by no means accounted for the gamut of architectural types and styles that Alberti could have considered to be Classical. Omission of the Colosseum and the Pantheon is but one dramatic instance of the fact that Vitruvius did not discuss those arched and vaulted constructions, which are the great achievement of Roman imperial architecture.[64] Vitruvius considered the Orders to be the basis of both stylistic and constructional vocabulary, because they were either the only systems he knew or the only ones of which he approved. But although the Orders might constitute norms for Alberti, they could not define the boundaries of Classical practice. The utility of Vitruvius's text was, therefore, prescriptive rather than definitive.

Alberti's troubles in defining the boundaries of the Classical were compounded by the lack of chronological information available for many monuments. Had he thought about it, he would have been puzzled by the stylistic similarities between the Baptistery of Florence, thought to be an ancient temple to Mars, and San Miniato al Monte, known to date from the eleventh century.[65] How could Alberti define the specific character of medieval architecture when it was, as in this case, virtually identical with what was believed to be the Classical? Guenée is probably right that the enthusiasm of both medieval and Renaissance historians for funerary monuments, and for monuments with inscriptions in general, was stimulated by the fact that they provide dates.[66] Many Classical and medieval buildings have no such helpful inscriptions; how could Alberti have dated them? Certainly not with the aid of the kind of critical vocabulary he used in *Profugiorum*. And if he could not date them, how could he judge whether they were Classical or medieval?

It would be imprudent to assume that Alberti had any systematic understanding of Classical architecture, and without this, he was hardly in a position to define the non-Classical. Even in high Renaissance Rome, a man as well versed in Classical architecture as Raphael was unable to provide any coherent account of the principles or development of Classical architecture. He asserted that all architecture built under the emperors followed "una ragione." Two hundred years of darkness followed this period, and then the modern era began.[67] If we may guess that the dark centuries were the sixth and seventh (implying that he used the Humanists' political scheme for his history of architecture), then Raphael considered some 500 years of Roman architecture to have followed a single aesthetic principle, and he saw no development in architectural style from about 800 to about 1515.[68]

Sauerländer has argued that the idea of style as a means of periodization was conceived in the eighteenth century, and that the discipline of art history depends on the interconnection of "stilus" and "chronos."[69] What I am suggesting is that Alberti was unable to make this connection in any systematic way, due to the state of historical consciousness and factual knowledge in the fifteenth century. Moreover, unable to conceive of any architectural system other than the Classical Order, he had no means with which to construct a category that would be antithetical to it, such as the Gothic. He could not, in short, have condemned the Gothic because this category—whether formal, historical, or cultural—did not exist.

The Humanists' solution to the problem of periodization, as Black has recognized, was to simplify the differing views of antiquity and modernity in such a way that the Ancients became Greeks and Romans, even if it was not clear when the period of the Ancients ended and that of the Moderns began.[70] Indeed, this is the scheme Alberti adopts in *De Re aedificatoria*: he speaks of usage *apud veteres* and that of *huius aetatis*. There was no middle period for him. This subtle but important clarification has so far not received the attention it deserves. It is not that Alberti ignored the Middle Ages, as is suggested by Frankl's comment about Alberti's silence on Gothic architecture, and the following by Krautheimer: "Into the realm of history he admitted only Roman antiquity and his own times, as long as they yielded to the program set down by him. Everything else was eliminated."[71] These imply that Alberti censored history and eradicated from his mind what did not suit him. Instead, the reality is that Alberti understood only two categories: Ancient and Modern. For him, the latter was not essentially different from the former, representing only a later stage in its evolution. Thus it is most accurate to say that Alberti really conceived of only a single historical category, or period, which had an earlier phase (*apud veteres*) and a later one (*huius aetatis*).

As Black has shown, many Humanists adopted the concept of "shifting antiquity," in which all that is not in the most recent past is considered ancient. This permitted Villani, for example, to refer to Dante's contemporaries as "antiqui."[72] The concept of shifting antiquity is perfectly serviceable if, as I have suggested, Ancient and Modern are thought of as different segments of a continuous culture. Seen in this light, Alberti's term *veteres* is not a historical term in the modern sense, since it is not inextricably tied to any limited segment of time and simply refers to all those who do not live in *huius aetatis*, including, for example, Arnolfo di Cambio and other "Gothic" architects. Far from defining the Gothic, Alberti failed to recognize the existence of any separate period between Antiquity and the present. Antiquity might have declined or become corrupted, but it had not given way to some other culture. The *columnae quadrangulae* are, therefore, a development within the Classical tradition—whether for good or ill, we will see—and not a non-Classical or post-Classical formal invention.

Despite the limitations of Alberti's perception of Antiquity and the Middle Ages, he, like most of the early Humanists, was concerned to define the relation of his own culture, that of the Moderns, to that of the immediate and more remote past. Ferguson has emphasized how the task of glorifying the history of their city-states forced the Humanists to treat the period after the decline of Rome in a positive light, since it was in the Middle Ages that the city-states rose to greatness.[73] And although, as we have seen, these periods were by no means clearly defined, the Humanists still managed to discuss the relative merits of the Ancients and the Moderns. Such discussion centered on linguistic and literary problems: the so-called quarrel of the Ancients and Moderns, the use of the vernacular, and the permissibility of neologisms. The Humanists did not discuss the problem of architectural style, but the underlying issues are not much different in architecture and language: whether a culture should evolve and change, and whether there is a single "right" style. Indeed, Alberti's writings make clear that general premises about his culture made in reference to literature hold true also for architecture. It is this correspondence that permits him to use architectural metaphors for literary and, in general, cultural problems.

Classical Authority Versus Cultural Evolution

Nicola's description, in Book III of *Profugiorum*, of the process by which the Temple of Artemis at Ephesus acquired its pavement is an allegory for the problem of convention versus innovation in early Quattrocento literary culture.[74] The major features of the temple had been constructed of the finest materials and on the grandest scale: its walls were of huge squared stones taken from marble mountains; its columns were enormously tall; the roof was of bronze and gilded; inside and out, the temple was encrusted with large slabs of porphyry and jasper. Everything about the building was splendid except for the floor, which remained bare and neglected. After some time had passed, an architect undertook the creation of a pavement that, while harmonizing with the rest of the fabric, would introduce variety into the whole. His materials were humble and poor, mere scraps left over from the great construction. Having studied their patterns and shapes, the architect arranged the remnant stones into a series of pictures that were in no way less pleasing than all the rest of the building. In the same way, explains Nicola, the Asians and the Greeks invented all the arts and disciplines: their discoveries erected a temple whose walls were the investigation of truth, whose columns were the observation of nature. Its roof protected the whole through ethical theory and precept. From this noble and public edifice, modern writers remove such small fragments as serve their needs. It is true, he concludes, that "nihil dictum quin prius dictum."[75] Yet in the same passage he also claimed that the needs of the Moderns were new and that their solution was aesthetically valid.

Alberti's assessment of what kind of originality was possible for Quattrocento writers, philosophers, and scientists applies to architecture as well. The cited passage in *Profugiorum* is very close in significance to the discussion of variety in *De Re aedificatoria*, which I quoted at the beginning of this chapter, particularly in the stated goal of introducing variety into a strongly defined tradition in such a way as to harmonize with it. Further application to architecture of these points is in *Profugiorum*, when Agnolo describes the buildings he designs alone at night.[76] Like the designer of the pavement at Ephesus, he invents little. Combining various Orders and numbers of columns with diverse kinds of capitals and bases, his aim is to infuse into his work a new convenience or grace. His work can be called eclectic, but it is not imitative.

This approach, in which Classical Antiquity serves as a great warehouse from which modern men may remove goods in order to fill modern needs, is entirely different from that of Niccolò Niccoli's. Niccoli denounced the culture of his own times as corrupt, denied its capacity for innovation, and called for the restitution of a pure Classicism.[77] Considering all departures from ancient practice degenerative, he thereby asserted that culture does not (or at least should not) evolve. His approach to architecture was, consequently, concerned with attaining correct and precise knowledge of authoritative models. Guarino Veronese described it thus:

> This man, in order to appear also to expound the laws of architecture, bares his arm and probes ancient buildings, surveys the walls, diligently explains the ruins and half-collapsed vaults of destroyed cities, how many steps there were in the ruined theatre, how many columns either lie dispersed in the square or still stand erect, how many feet the basis is wide, how high the point of the obelisk rises.[78]

Niccoli's practice seems almost to illustrate those advances that Vasari claimed were made in the Quattrocento—that is, in rule, order, and proportion.[79] Certainly, Alberti also studied the Orders and measured ancient buildings, but he explicitly tells us that the modern architect—Agnolo, that is—is free to mix the Orders together to create something new. That Alberti practiced what he preached is shown by the innovative capital he designed for the façade of the Tempio Malatestiano.

These divergent attitudes toward the authority of the past and the degree of creativity possible in the present illustrate a dynamic tension in early Humanist thought. Two general positions may be delineated, exemplified by that of Niccoli, on the one hand, and of Alberti, on the other. The "Ciceronians," as Sabbadini called them, advocated a return to pure ancient style, rejected neologisms, urged the imitation of a limited number of authoritative Classical models, and denied the use of the vernacular as a language of literature.[80] Niccoli evidently represents this current. Gombrich, relating Niccoli's "revival of letters" to Brunelleschi's "reform of the arts," attributed to the architect a similar approach to Classical models.[81] I have reservations about this association, given Brunelleschi's eclectic use of sources and the acceptance of the Trecento in some of his works.[82] It has also been challenged by Onians, who preferred to discuss Brunelleschi's style within the context of "Rome versus local tradition" rather than "Latin versus *volgare*."[83] But Gombrich's suggestion would certainly apply to Brunelleschi's biographer Manetti, who tells us that when Brunelleschi first saw the architecture of ancient Rome, it was "just as if God had enlightened him about great matters," and that he "decided to rediscover the fine and highly skilled method of building and the harmonious proportions of the ancients and how they might, without defects, be employed with convenience and economy."[84]

This approach, which is inherently Idealist and aims at standardization and systematization, increasingly dominated architectural theory from around 1470 to 1600. The recuperation of Classical Antiquity and the definition of the ideal church, ideal city, and ideal style are concerns evident in Manetti, Francesco di Giorgio, Raphael, Serlio, Vignola, Vasari, and Palladio.[85] It may be appropriate to connect its ascendance with the popularization of Platonic and Neoplatonic philosophy, and in general with a shift from rhetorical to metaphysical concerns, beginning with the Ferrara/Florence Council (1438–1439), developing under Argyropoulus's teaching at the Florentine Studium (1456–1471),[86] and culminating in the Platonic academy. By around 1480, the main tenets of Platonic Idealism had permeated broad sectors of Italian culture, reinforcing that Idealist view of Classical Antiquity espoused by Niccoli, and providing it with a philosophical—indeed, metaphysical—justification.

What of the second current, the approach of those whom Sabbadini termed "eclectic"? Believing that culture has to grow and change if it is to remain useful, they permitted neologisms. Aiming to rival rather than to re-create the Classical past, they placed more emphasis on creativity and innovation.[87] And they accepted the use of any literary model, whether Ancient or medieval, on the basis of its quality alone. Thus they emphasized the importance of change, growth, progress, innovation, and personal style. Their view was not idealist, but relativist: they believed that conventions varied from author to author, age to age, and culture to culture. Above all, they felt that good style meant effective communication, and therefore the use of a living language (the vernacular), accessible to all and not just to an elite. As Bruni put it, it makes no difference whether one writes in Latin or the vernacular, or in Greek rather than Latin, since "each language has its own perfection, its own music, and its own means of elegant and precise expression."[88]

It is my thesis that Alberti's architectural though and practice are most fruitfully examined within the second of these two trends. This is not to deny his admiration for Classical Antiquity, but to reevaluate the nature of his classicism. Niccoli's beliefs about cultural change posited a single "good" style, the Antique, in opposition to all others. Thus his classicism required the rejection of the modern in favor of the ancient. But Alberti's approach posed the very different problem of selecting the superior and rejecting the inferior, wherever they might be found. His approach, as will be seen, was that advocated by the Classical authors themselves.

Alberti's place in the second of these currents is dramatically illustrated by the *certame coronario* of 1441, a contest in vernacular poetry held in the cathedral.[89] He had promoted this venture, even offering to pay for the prize, a silver laurel wreath. Alberti's most extensive justification of the use of the vernacular as a literary language is in the proem to Book III of *Della Famiglia*.[90]

The first problem to be resolved in the discussion was the origin, and therefore the character, of the vernacular: this depended on whether the Romans had had different spoken and written languages. Alberti's conclusion was negative: the Romans used the same language in literature and life; therefore, the vernacular did not originate as a popular, nonliterary ancient language. It developed, instead, as the result of corruptions introduced into Latin by the barbarians, and therefore represented the historical evolution of Classical speech. He concludes, "I certainly admit that ancient Latin is very rich and ornate. Nonetheless, I don't see why our modern Tuscan should be so despised. If only learned men would seek by study and care to make it refined and polished, it would have as much authority as Latin."[91] Alberti, then, did not recommend purifying modern language of its barbarian elements by returning to ancient Latin usage, but by conferring on the vernacular a polish like that of Classical language.[92] Clearly, he did not believe in the desirability of the restitution of pure Classical form. Moreover, he was more concerned with communicative effectiveness than correctness according to Classical norms.[93] We must use the vernacular, he says, if we wish to be useful to others because it is the language in common use; very few people understand Latin.[94]

An analogy to the situation in architecture is not hard to draw. It, too, had been corrupted by the barbarians, but this did not mean that all modern architecture should therefore be despised. Alberti greatly admired Classical architecture, but if Florence Cathedral succeeded in moving the spectator to religious sentiment—if it could communicate, that is—it was worthy of praise. In architecture as in language, Alberti never lost sight of the importance of social function. He rarely cited modern authors or works in the vernacular; it was difficult for them to attain the level of ancient eloquence. In the same way, he rarely mentioned non-Classical structures. But he did advocate the use of the vernacular for poetry, and his only extended description of any building is a panegyric of Florence Cathedral. Alberti wrote the first grammar of the Tuscan vernacular, showing that it was as "regular" as Latin.[95] The corruption of language, then, did not entail the loss of Classical order and structure. By the same token, Alberti's description of the cathedral showed that the principles of Classical architecture were as evident in this Gothic structure as in an ancient temple. For him, an equilibrium between beauty and sound construction did not depend on the formal vocabulary that embodied it, or what we would call style, in the narrow sense. Alberti's advocacy of the vernacular, and the cultural attitudes that determined it, weaken the assumption that he wished to reinstate a pure Classical style in architecture.

Recent studies have shown that Alberti respected non-Classical monuments, and that he incorporated medieval elements into his own designs.[96] In *De Re aedificatoria*, for example, he advised the architect, when completing a work begun in any style, to try to comprehend the intentions of the original architect even though they may be different from his own (IX.xi, p. 867). In Book X, discussing the art of building restoration, Alberti's assumption is that the structure is to be restored to its original appearance, regardless of what that was (X.xvi–xvii, pp. 991–1001). Alberti drew from medieval sources for the campanile at Ferrara, the Tempio Malatestiano at Rimini, the façade of Santa Maria Novella, and the Rucellai chapel at San Pancrazio.[97] Thus his style was not homogeneous, but eclectic, like that of the designer of the pavement at Ephesus mentioned in *Profugiorum*. I would suggest that Alberti's architectural practice of grafting new (that is, medieval) motifs onto a design that is otherwise classicizing is analogous to the adoption of neologisms in language.

The problem of neologisms is intimately connected to that of the nature of historical process. Some of the Ancients advocated an approach to language and culture that affirmed its evolutionary character. Cicero, for example, warned against using words no longer in common usage; his ideal was to steep oneself in the works of good ancient authors and then express oneself in a current vocabulary.[98] "The worst vice," he said, "is to depart from the ordinary way of speaking ["a vulgare genere orationis"] and from the usage of common sense."[99] Cicero accepted neologisms.[100] Horace was even more strongly in favor of linguistic innovation:

> It has ever been, and will ever be, permitted to issue words stamped with the mint-mark of the day. As forests change their leaves with each year's decline, and the earliest drop off; so with words, the old race dies, and, like the young of human kind, the new-born bloom and thrive.[101]

Particularly persuasive for Alberti, because of his social concern, would have been Tacitus's denial that oratory had declined in his own times: unless oratory changes with the times, it cannot meet modern needs or be effective in the active life.[102] As we saw, Alberti said almost the same thing about the need to use the vernacular.

This evolutionary approach to language was typical of Humanists of the second, or "eclectic," current. Salutati, in a letter to Poggio of 1405, wrote: "Give me one small reason, besides empty glory and the opinion of antiquity, why we should prefer the old and vanquished to the later and more recent."[103] At stake here was the larger issue of whether reason or authority should be followed. Poggio criticized Valla's historical conception of language, saying that he was crazy not to realize that words are established not by reason, but by the authority of ancient writers.[104] Biondo Flavio asserted the necessity of using modern words for modern institutions and tools, although he recognized the problems this caused. When he used ancient words, he himself could not understand what he had written; when he used modern words, it made him sick. On a more positive note, he continued:

> The industry of men of our age in beseiging is no less than that of our ancestors. New names have been invented at once with the invention of new instruments, the unusualness of which many shrink from who, unless they read of battering rams, missiles of burning pitch, slings, and scorpions, think the siege no siege at all or, at best, a very inept one. But if they would read attentively the writings of the present day, they would enjoy the comparison of the intelligence and power of industry of the Moderns and the experience and fortitude of the Ancients.[105]

Valla, especially, argued the importance of neologisms: even if his use of words or spellings was not antique, he said, it did not follow that they were corrupt. It was more important that speech be intelligible than antique; word use should conform to what the reader was accustomed to.[106] His recommendations for modern eloquence did not preclude an interest in ancient usage: the *Elegantiae linguae latinae* aimed at establishing correct usage of ancient Latin in regard to grammar, phraseology, and style.¹⁰⁷ His position, combining admiration for Classical style with the recognition that modern-day eloquence had to take common usage into account, seems particularly close to Alberti's views on architecture. Pius II took a strong stand on the necessary evolution of language: "It would be utterly ridiculous to prefer the language which men have spoken to that which they are now speaking."¹⁰⁸ Pius is quoting Quintilian: "nam pene ridiculum fuerit, malle sermonem quo locuti sunt homines, quam quo loquuntur" (*Institutio oratoria* 1.6.43). But in the same letter to young Ladislas of Hungary, he warns that "to very few, the great creators of a tongue, is it given to coin new words with impunity." His position is almost identical to Alberti's on innovation in architecture: novelty is approved, but with the warning that there is risk involved. At a later moment, we find Poliziano defending the writers of the silver age:

> I do not much admire that argument that objects that the eloquence in the age of these writers was corrupt, since, if we enquire more correctly, we shall see it not so much a corrupt or depraved but simply a changed kind of speech. Nor should we so quickly call worse what is different.¹⁰⁹

Poliziano was especially in favor of neologisms. In *Panepistemon* he writes: "You will be greatly pleased by my sort of strange and erudite variety of novel and different terms" (I.24). This evolutionary approach to language could also be applied to architecture, and was. In his treatise on human happiness (1468–1470), Benedetto Morandi argued that when houses were burned down, struck by lightning, or ruined by earthquake, this provided the occasion for "succession or rather innovation." In place of Fiesole, the Tuscans now had Florence, "the most beautiful and richest city."¹¹⁰ Trinkaus showed that Cicero was Morandi's source for this argument:

> There is nothing made by work or hand which antiquity does not sometime consume and use up. And so it is not to be wondered if homes, cities and other things constructed by art fall and perish. For there would be no business left to posterity if what had been previously constructed perpetually remained.¹¹¹

Similar views were expressed by Biondo Flavio: "In this description of buildings, account must be taken of time and place. One way of building prevailed at the time of the first founding of the city, another way after opulence was born and had grown."¹¹²

Alberti's views on neologisms belong to this evolutionary current. He tells us that in writing *De Re aedificatoria*, he frequently encountered difficulties because he had to invent words (VI.i, p. 441). This may be compared with Lucretius's statement, which was known to Alberti, in *De Rerum natura* (I.136–

39): "Nor do I fail to understand that it is difficult to make clear the dark discoveries of the Greeks in Latin verses, especially since we have often to employ new words because of the poverty of the language and the novelty of the matters."[113] Elsewhere in the treatise, Alberti says that he has tried to use Latin that everyone can understand, and that when no appropriate terminology was available, he had to invent it (VI.xiii, p. 525). And in a third place, he explains that in his discussion of vaults he had to invent new words in order to explain his meaning (III.xiv, p. 241).

All those who defended the vernacular and neologisms did so on the assumption that communicative effectiveness necessitated the use of what was familiar to the audience. Thus they followed Augustine, who told those in his audience that he often used incorrect terms so that they would understand him better and that understanding him through his use of barbarisms was preferable to speaking correctly.[114] This belief helps us understand Alberti's use of medieval elements in some of his designs. For his works in Florence, at Santa Maria Novella and San Pancrazio, he drew on Florentine Romanesque and Gothic sources; at Rimini, it was the Gothic of the Veneto.[115] Thus his eclecticism was not arbitrary; rather, he drew motifs from regional traditions in order to soften and domesticate the Classical vocabulary of his designs as well as to harmonize the new creation with its ambient. In Alberti's latest works—Sant'Andrea and the tribune of Santissima Annunziata—medieval references no longer appear. But by this time, the Classical vocabulary had been reinstated and was part of common usage.

Eclecticism

The maker of the pavement at Ephesus in *Profugiorum* culled scraps from diverse sources and rearranged them into a new design. He was something more than a compiler, but he did not invent the constituent elements with which he worked. Moreover, he took fragments from works of different kinds and dates, without regard to their homogeneity, from the columns, walls, and roof of an edifice constructed over many centuries. There is some evidence in *De Re aedificatoria* that Alberti advocated an eclectic use of the monuments of the past. He recommends studying all the edifices that have merit, and using them in modern design:

> I would wish the architect to behave like those in literary studies. For no one will think he has sufficiently devoted himself to literature unless he has read and understood all the authors, even the bad ones, who have written something on the subject which they pursue. Thus the architect, wherever he finds works approved of by the opinion and consensus of men, will examine them with

the utmost diligence, make drawings of them, take note of their measurements, and deduce their modules and models.[116]

Alberti makes it perfectly clear that these exempla are to be selected only on the basis of their quality, not because they belong to a certain style or period. Commenting on this passage, Nelson suggested that it was dependent on Quintilian and on St. Basil's *Letter to Young Men on Reading Pagan Literature*. He noted its parallelism to Pius II's statement that "in the reading of poets and other writers it will be proper to imitate the bees."[117] Another source for Pius's statement could be Petrarch: "We write as bees make honey, not conserving the individual flowers, but converting them at the honeycomb, so that from many and various things one thing is produced, and that different and better."[118] But I believe that yet other sources, which urge the desirability of eclecticism not only in the study of the past, but also in the creation of new works, are closer to Alberti's meaning. Seneca, who also used the metaphor "like the bees gathering honey" in his letter on how to read and study urged the student to "blend those several flavours into one delicious compound" so that it is "clearly a different thing from that from whence it came."[119] Aristotle, in the prologue to his *Rhetoric*, says that "following the lesson taught us by Nicanor, we have adopted from other writers anything that has been well-expressed by anybody writing on the same subjects."[120] Cicero echoes Aristotle in his *On Invention*:

> When the inclination arose in my mind to write a textbook of rhetoric, I did not set before myself some one model which I thought necessary to reproduce in all details, of whatever sort they might be, but after collecting all the works on the subject I excerpted what seemed the most suitable precepts from each, and so culled the flower of many minds.[121]

Alberti followed exactly this method of composition in *De Re aedificatoria*, as he tells us: "Having to write about architectural design, we will gather and include in this work the best and most refined things written on the subject by the greatest experts, and what we have observed to be operative principles in their works" (I.i, p. 19). Alberti says that he intends to recount what Theophrastus, Aristotle, Varro, Pliny, and Vitruvius have said about materials (II.iv, p. 111).

In such passages, Alberti reiterates that the sole criterion for imitation is the excellence of the model, following Aristotle and Cicero. Elsewhere, Cicero argued that excellence may be "exhibited in a variety of artistic and rhetorical styles: there is a single art and method of painting, and nevertheless there is extreme dissimilarity between Zeuxis, Aglaophon and Apelles, while at the same time there is not one among them who can be thought to lack any factor in his art."[122] In the same way, among orators "the ones admittedly deserving of praise nevertheless achieve it in a variety of styles." The discovery of Cicero's

Brutus in 1421, in which the relative merits of two entirely different styles are debated, must have reinforced the implications of this passage.[123] Cicero's view was adopted in the fifteenth century, most closely by Benedetto Accolti in his *Dialogus de praestanti virorum sui aevi* of the 1460s:

> Among orators one can hardly enumerate the varieties of speech and arguments, any more than among painters, architects, and masters of the other arts.... We are forced to acknowledge by nature and the experience of things that, without a doubt, the same law or custom does not suit in all places, nor in all times, nor among all peoples.[124]

Pius II also insisted that quality was the sole criterion in the selection of models, and recognized it in modern as well as ancient authors. For instance, he recommended studying the writings of Bruni, Poggio, and Traversari along with those of the Ancients, while he warned against the reading of Hungarian chronicles because they are "destitute of attraction in form, in style, or in grave reflections. For boys must from the earliest be made familiar only with the best, if we look for them to develop a sound judgment in their later years."[125] Excellence and decorum (appropriateness, that is) are judged to be more important than the imitation of authoritative models. Salutati had argued this point against Poggio in 1405. Times have changed, he said, and the Moderns must create their own style to suit present needs rather than practice a strict adherence to the style and models of the Ancients.[126]

Alberti describes and endorses modern eclecticism in his account of the making of the pavement at Ephesus, in his description of Agnolo's design process (at the end of *Profugiorum*), and in his discussion of his own procedure in writing *De Re aedificatoria*. Clearly, he was less concerned with the restoration of a pure Classical style than with the development of an architecture suited to modern needs. The quality of the model and its effective communicative capacity, rather than historical style, were the measures of worth. This view, once more, has its roots in Cicero, who says that given the variety of oratorical styles, the superior is distinguished from the inferior "almost more by capacity than by style, and everything is applauded that is perfect in its own style."[127] At about the same time that Alberti was writing his treatise on architecture, Ghiberti adopted a similar approach to the evaluation of artistic quality. Krautheimer pointed out the novelty of his willingness to acknowledge the excellence of Duccio and Cavallini, masters who had worked in the rejected *maniera greca*.[128] Leonardo Bruni expressed this same attitude, applying it to language: "It doesn't matter one bit whether you write in Latin or the vernacular, since each language has its own perfection."[129] Benedetto Accolti made the same point: "I do not think it much matters whether one speaks in his maternal language or in Latin, as long as he speaks soberly, beautifully and copiously."[130]

I have suggested that Alberti shared this attitude, and that he saw his task as distinguishing the superior from the inferior in architecture (as also in language), and not the Classical from the medieval. In order to do this, he sought to discover those principles by which quality could be determined, and it is in these principles that the kernel of his architectural thought lies.[131] I have dealt with these only summarily here; their investigation is the subject of the next chapter.

Alberti's classicism was of a more subtle and more elusive type than Niccoli's emulation of authoritative Classical models, and Vasari's preoccupation with rule, order, and proportion. His thought reveals a far more profound absorption of Classical attitudes toward the creative process and toward the nature of cultural development. The Ancients themselves accepted stylistic diversity, welcomed the development of new forms for new needs, recommended an eclectic approach to source material, and refused to acknowledge authoritative models. Their thought was relativistic and historical rather than idealizing. As the early Humanists struggled to define their relation to tradition and to evaluate the culture of their own times, some of them, including Alberti, found in Cicero, Horace, and Aristotle arguments that clarified their thoughts and justified their arguments. They also helped define the larger issues that surrounded the specific problems they faced. The problem of the vernacular, for example, is but a specific instance of the larger issues of whether a culture evolves, and whether cultural evolution should be accepted or resisted. Different positions taken on these issues led some Humanists to encourage the progress of modern culture, and others to demand that it be purged. The positions taken on the larger issues could be, and were, applied to a variety of specific manifestations. Alberti's thought on linguistic usage parallels his view of architecture; thus his acceptance of the vernacular, of neologisms, and of the eclectic use of source material for his writings is perfectly consonant with, and indeed paralleled by, his praise of the Gothic Florence Cathedral, his incorporation of medieval elements in his works, and the eclectic nature of his designs. The problem of the relation between tradition and innovation is resolved in the same fashion in each of these manifestations: modern innovations are acceptable, if they can be proved worthy by the sober yardstick of excellence.

FIVE

Alberti's Description of Florence Cathedral as Architectural Criticism

Certainly this temple has in itself grace and majesty; and, as I have often thought, I delight to see joined together here a charming slenderness with a robust and full solidity so that, on the one hand, each of its parts seems designed for pleasure, while on the other, one understands that it has all been built for perpetuity. I would add that here is the constant home of temperateness, as of springtime: outside, wind, ice and frost; here inside one is protected from the wind, here mild air and quiet. Outside, the heat of summer and autumn; inside, coolness. And if, as they say, delight is felt when our senses perceive what, and how much, they require by nature, who could hesitate to call this temple the nest of delights? Here, wherever you look, you see the expression of happiness and gaiety; here it is always fragrant; and, that which I prize above all, here you listen to the voices during mass, during that which the ancients called the mysteries, with their marvelous beauty.[1]

Alberti's description of Florence Cathedral in *Profugiorum ab aerumna* is the unique example of extended architectural description in his work. It provides the opportunity to investigate some aspects of his thought and expression for which we have little other evidence. Although the example of a single text is too insubstantial a base on which to build the image of Alberti as architectural critic, still it offers the occasion for an examination of how, at least in one case, he understood the precepts of good architectural style to be manifest in a specific building. Not only its length renders this description precious, but also its character, markedly different from the bits and pieces of description elsewhere in Alberti's work. It is in striking contrast with Alberti's use of one or more nouns to signify his approval of buildings in *De Re aedificatoria*; these are epithets, not descriptions.² In the treatise, his recommendations regarding good design are either practical, and limited to single aspects of a building, or abstract and theoretical. The treatise, concerned with truth and precept rather than opinion, abstains from subjective expression.³ While it might be supposed that Alberti's *Descriptio urbis Romae* could be a useful source from the point of view of architectural description, in fact, its short text is merely a guide to the graphic representation that it accompanied.⁴

Neither in *De Re aedificatoria* nor elsewhere in Alberti's writings do we find works characterized by the rhetorical device of sequential antitheses, as in *Profugiorum*. In his business correspondence, Alberti's language is simple and plain: describing his design for Sant'Andrea in Mantua in a letter of about 1470 he claimed it to be superior to that of Manetti because "questo sarà più capace, più eterno, più degno, più lieto; costerà molto meno" ("this design will hold more people, will last better, be more worthy, more pleasant, and it will cost much less").⁵ In *De Re aedificatoria*, Alberti explained that in order to make his meaning clear to the reader he preferred to write in a plain expository style rather than try to appear eloquent (VI.i, p. 155). *Profugiorum*, by contrast, is neither a business letter nor a technical treatise but a moral essay, whose purpose is to persuade the reader through argument. For it, Alberti uses a rhetorical style of writing, including the device of antithesis. This technique, we will see, is closely linked to his aim of persuading the reader that the style of Florence Cathedral is good, even ideal (although not in the Platonic sense). Alberti's use of diverse styles underscores the truth in Landino's comment that Alberti, "come nuovo camaleonta sempre quello colore piglia il quale è nella cosa della quale scrive."⁶

Not only is the description in *Profugiorum* unique in Alberti's work, but it has no precedent in earlier Western medieval writing on architecture and little in Classical ekphrasis. For this reason, it raises the problem of the origins of modern architectural criticism, especially in its descriptive vocabulary and the criteria of excellence that it proposes. If we understand stylistic criticism as the capacity to date and interpret monuments by means of visual analysis and comparison, and to recognize in their forms the spiritual content, then it must be admitted that Alberti's text is an early contribution to the formation of this genre.⁷ Although hardly a mature piece of criticism, the description addresses the problems of the verbal expression of architectural qualities and determination of their significance. Alberti's description first defines the stylistic qualities of the cathedral and, in the course of his essay, relates them to the values of the beholder.⁸ His approach, being critical and judgmental, is evaluative; it is not, as I showed in chapter 4, concerned with historical periodization.

Alberti's description departs from the medieval tradition of architectural description, based on measurement and enumeration, in that it aims at the evaluation and interpretation of visual experience rather than at definition.[9] Although, like a medieval writer, he presents his subject primarily as the stimulus for spiritual elucidation, this significance is derived directly from the visual and emotional perception of the observer, rather than from an abstract definition of the building in terms of measure and number. Alberti is not interested in giving an accurate account of the building's morphology, nor does he attempt to define its beauty with the abstract terms he uses in De Re aedificatoria (*firmitas, venustas, utilitas*). Instead, he seeks to formulate in words the spectator's perception of Florence Cathedral; thus he is concerned not only with intellectual but also with sensual, emotional, and moral judgment. Although, as we will see, the spiritual message is believed to be objectively present in the building, it is perceived by the senses rather than the intellect. Alberti's paired adjectives evoke the stylistic qualities of the whole building, rather than designate the forms to which these qualities adhere. Someone familiar with the cathedral will understand that "slender and graceful" may refer to the elongated piers or to the vertical proportions of the interior space, and that "robust" evokes the weight-bearing capacity of the piers and the thickness of the mural boundary. But these connections are not made within the text, which is concerned with perception rather than its objects.

The Art of Rhetoric and the Art of Architecture

Much recent scholarship has focused on the dependence of Alberti's artistic theory on rhetoric, most often in relation to his treatise on painting, *Della Pittura*.[10] Such work has seemed especially appropriate in light of the numerous connections drawn between painting (and also sculpture) and rhetoric by the ancient authors. Some work has been done on the relation between ancient rhetorical theory and terminology and Alberti's treatise on architecture, but this has proved more difficult due to the paucity of architectural analogies drawn in Classical sources and the few specific references to rhetoric in *De Re aedificatoria*.[11] Yet there is a fundamental connection between the two arts, since the purpose of architecture, as Alberti saw it, was like that of rhetoric.

The aim of rhetoric is to persuade the mind of the hearer and move his emotions. Precisely this aspect of the art made it vulnerable to attack by Classical theorists. Plato disapproved of such oratory, denying its status as an art and defining it as a habitude that produces gratification and pleasure (*Gorgias* 462C); Cicero, instead, spoke approvingly of "the supreme orator ... whose speech instructs, delights and moves the minds of his audience." Cicero adds that "the orator's virtue is pre-eminently manifested either in rousing men's hearts to anger, hatred or indignation, or in recalling them from these same passions to mildness and mercy."[12] In *De Re aedificatoria*, Alberti assumed that churches would pursue the same aim of persuasion: "There's little doubt that for the religious cult it is important to have temples which marvellously delight the soul, and fill it with joy and admiration" (VII.iii, pp. 543–45). For this reason, Alberti recommended that the temple be so beautiful that nothing more elegant could possibly be imagined; each part should be so designed that those who enter are stupified with admiration for such wonderful things and cannot restrain themselves from shouting aloud that what they see is worthy to be a place for God (VII.iii, p. 545). Alberti's connection between the persuasive effect of oratory and of architecture reveals his belief that these arts provide analogous aesthetic experiences for their audiences.[13] In both cases, that is, persuasion is closely linked to feelings of pleasure, themselves the measure of sensory beauty. Alberti's conviction that sensory beauty is necessary for persuasion was supported by discussions, such as Cicero's in *De Oratore*, arguing that while the content of oratory is directed to the intellect, its sounds and rhythms are related to pleasure (III.xxv.98).

Alberti's discussion of beauty as a sensory and emotional (and, therefore, as a subjective) experience rejects the Platonic position on this problem, and accepts the authority of Aristotle, Cicero, and Augustine. Plato, who believed that the judgment of the senses was an improper criterion for beauty, reproached contemporary musicians for placing the judgment of the ear higher than that of the mind.[14] In the face of this attempt to separate aesthetic experience from the senses and emotion, Aristotle defined the aesthetic experience as one of intense delight, in which the spectator seems spellbound (*Eudemian Ethics* III.ii.7–8). His formulation is close to Alberti's in *De Re aedificatoria*. It is within the context of this controversy over the nature of aesthetic experience and the roles assigned to intellectual and sensory knowledge that Augustine's discussion in *Soliloquies* can be understood (I.ch.4–8). Reason asks Augustine what kind of knowledge he requires of the God he wishes to see. Does he require the clear and certain knowledge that he has of geometric figures? No, says the saint, because such knowledge does not fill him with delight, as would the vision of God. Augustine affirms the primacy of sensory experience and emotion in the perception of divinity.

The aesthetic experience that Alberti recommends for churches corresponds to, and was perhaps inspired by, Augustine's description. The beauty of Florence Cathedral, then, not only persuades the spectator or reader, but also reveals to him a spiritual truth. Drawn in by his emotions and imagination, the reader comes not only to understand but also to experience the tranquillity of the soul as embodied in a great work of religious architecture. Its beauty moves the heart, mind, and soul of the spectator to cognizance of divinity.

The first portion of Alberti's description of the cathedral proceeds by pairs of opposites. The first of these, "grazia" and "maiestà," is borrowed from definitions of the stylistic differences between rhetorical styles. Alberti's choice needs to be understood within the atmosphere of excited discovery of previously lost or only partly known rhetorical treatises in the first decades of the Quattrocento.[15] The rediscovery of these rhetorical works permitted—in fact, demanded—a reexamination of the Humanists' understanding of this art in the light of new evidence. In particular, the new works raised questions about how many *genera dicendi* there were and how they should be defined, since the styles of oratory were not defined in the same way by all writers on the subject, or even by the same writer in different works. Although most writers define three styles, two of them are always characterized by the opposites beauty and dignity. For some, the three modes represented two extremes and a mean, whereas this antithetical structure is absent from other accounts. As example of the first type (two extremes and a mean), we may consider Aristotle, who recognized two kinds of expression—the plain and the adorned—and recommended as best the mean between these two (*Rhetoric to Alexander* I.iii.3). Cicero seems to follow him in *De Oratore*, where he defines the three styles as (1) the full, yet rounded; (2) the plain, which is not devoid of vigor and force; and (3) that style which stresses the middle course by combining elements of the other two styles (III.iii, p. 19).

The essentially antithetical structure of all systems is especially clear in *Brutus*, discovered in 1421, where Cicero says that there are really only two kinds of oratory: simple and concise, and brilliant and impressive (LV.201–3). The Humanists' discovery of this text (of which Alberti owned a copy) clarified an essential principle of stylistic differentiation, while bringing into prominence precisely that aspect of rhetorical theory that, I will show, was most usefully applicable in other contexts. It is this basic antithesis that Dionysius of Halicarnassus had already compared with style in the fine arts. He likened "the oratory of Isocrates, in respect of its grandeur, its virtuosity and its dignity, with the art of Polyclitus and Phidias, and the style of Lysias, for its lightness and charm, with that of Clamis and Callimachus."[16] Alberti applies the same categories in his discussion of the Orders in *De Re aedificatoria*, where the Doric is described as "plenius ad laboremque perennitatemque aptius" and the Corinthian as "gracile lepidissimum" (IX.v, p. 817). Here, however, he uses the three-part scheme of the *genera dicendi*, defining Ionic as a mean between the two other styles. In *Profugiorum*, by contrast, he uses the two-part scheme, which, by excluding the mean, emphasizes the antithetical character of the cathedral's stylistic qualities. This deliberate choice, we will see, is the device that enables his stylistic discussion to be read within a wider intellectual context.

Alberti's immediate literary model for the application of criteria of rhetorical excellence to architecture was Manuel Chrysoloras's description of Hagia Sophia in *Comparison of Old and New Rome* (1411):

> What sublimity and grandeur of speech could adequately evoke its sublimity and grandeur? What beauty of words could equal its beauty? Could the dignity of my words do justice to its dignity? Could I speak with as much simplicity as it possesses? What power and variety of words will be equal to its variety; or what precision of speech to the precision and excellence of all its parts? What harmonious composition can compare to its harmony and congruence of parts?[17]

All these diverse qualities produce harmony, or equilibrium. Chrysoloras's terms— "sublimity," "grandeur," "beauty," "dignity," "simplicity," "power," "variety," and "precision"—are examined in chapter 8. In chapters 2 and 3, I showed that Chrysoloras's text was probably known to Alberti, and that he drew from it in other of his works. In Alberti's version only two qualities, with their extensions, are opposed—grace and majesty.

In his next pair of opposites, Alberti transforms these general stylistic principles into terms of architectural description: "grace" becomes "charming slenderness," and "majesty" becomes "robust and full solidity." In this way, categories of linguistic style evolve into terms of architectural description, evoking on the one hand the elongated piers and tall spatial proportions of the cathedral and, on the other, its powerful masonry structure.

His last pair of opposites, "designed for pleasure" and "built for perpetuity," at first seem to bring the descriptive criteria within Vitruvius's principles of firmness and delight (*firmitas* and *venustas*), but may depend directly on Vitruvius's source in Cicero's *De Oratore* (III.xlvi). This fundamental antithesis—beauty and necessity—underlies, for Alberti, all architectural design; both criteria need to be satisfied. This is particularly the case in religious architecture. "One thing above all which a temple should have, in my opinion, is that all its visible qualities should be of such kind that it is difficult to judge whether ... they contribute more to its grace and aptness or to its stability" (VII.iii, p. 545). The antithesis of beauty and necessity is central to Classical discussions of rhetorical style, where its analogues are style versus content, and copiousness versus brevity. The pleasure received from beauty of expression is opposed to the necessity of speaking to the point. Here I will examine the requirements of beauty and necessity as criteria of eloquence revived with Classical rhetoric in the early Quattrocento; then I will consider their antithetical character.

In Ciceronian rhetoric, both requirements have to be satisfied. Thus in *De Oratore*, Cicero requires "that the style must be in the highest possible degree pleasing and calculated to find its way to the attention of the audience, and that it must have the fullest supply of facts" (III.xxiv. 91, p. 73). Elsewhere in the same work, architecture serves as example of the necessity of combining beauty and utility: "In temples and colonnades the pillars are to support the structure, yet they are as dignified in appearance as they are useful" (III.xlvi, p. 142). Tacitus also used an architectural analogy for the union of beauty and utility, criticizing Cicero's style for its lack of ornament: "It is just as in a rough-and-ready construction work, where the walls are strong, in all conscience, and lasting, but lacking in polish and lustre."[18]

This union, advocated by Classical theorists, was rejected in medieval rhetoric. Although brevity was praised as a virtue by Isocrates, Cicero, Quintilian, and Horace, only in the Middle Ages did its opposite, copiousness, become generally regarded as a fault.[19] Early Christian writers, opposing content to form, emphasized the importance of plain speech, equated with truth, over ornamental speech, equated with empty word plays that may affect the spectator by their form but are not sincere.[20] Thus in his early-fifth-century Life of Ambrose, Paulinus promised to tell the life of the saint

> in unadorned language—briefly and summarily, so that even if the language offend the mind of the reader, yet its brevity may encourage him to read it; and I shall not obscure the truth with florid words, lest while the writer seek the pomp of elegant diction, the reader, whom it does not befit to look for the trappings and pomp of expression more than the virtue of deeds and the grace of the Holy Spirit, miss the knowledge of so great virtues.[21]

Isidore of Seville, Alcuin, Robert of Melun, and William of Auvergne all praised brevity and condemned ornamental speech.[22] Matthew of Vendome, in his *Ars versificatoria* of about 1175, maintained that brevity characterized the modern stylistic ideal, whereas copiousness ("dilatatio" or "amplificatio") characterized the ancient.[23]

This attitude, suspicious of stylistic beauty as sacrificing clarity of content, helps account for Alberti's lengthy justification of beauty for architectural style in his treatise. Beauty is an extremely important consideration in architectural design, so he maintains, enumerating his reasons at some length (VI.ii, pp. 444–49).[24] In *Profugiorum*, Alberti applies this pair to eloquence: Nicola applauds Agnolo's speech for its combination of copiousness and brevity: "I observed in all of your argumentation and in the development of your position an incredible brevity joined to a marvellous abundance and fullness of the most weighty and appropriate sayings and thoughts" (p. 160). In *Della Famiglia*, Battista praises Lionardo for the great abundance ("copia") of his evidence and adds, "I like your marvellous brevity; your style of speaking seems elegant to me on account of its great brevity."[25] As in *Profugiorum*, Alberti's discussion of the principles of architectural beauty in Book VI of *De Re aedificatoria* is based on the antithesis of beauty and necessity (VI.iii, p. 453). The Greeks, he said, thought deeply about the principles of architecture and tried to discover what factors distinguished good buildings from bad. Realizing that good design was based on the conjunction of opposites ("quasi ex maris foeminaeque connubio tertium quippiam oboriretur"), they experimented with many kinds of pairs: straight and curved lines, light and shadow, left and right, vertical and horizontal, near and far. From this experience, they concluded that different criteria of excellence were to be applied to buildings destined to endure and to those intended primarily to be beautiful.[26] In other words, one set of conjoined opposites produces beauty, and a different set results in stability; excellence itself is composed of the opposites beauty and necessity. Alberti's position on the equal importance of beauty and necessity as components of stylistic excellence is consistent in his literary and architectural criticism and represents his acceptance of the Ciceronian ideal of rhetoric. His three pairs of antitheses in the description of the cathedral demonstrate that beauty and utility are equal partners in good architectural style.

Within the five *studia* that characterize Quattrocento Humanism, the investigation of literary style was the province of rhetoric, and its precepts furnished criteria within which style in other arts were evaluated.[27] This was possible because the rules of rhetoric, unlike those of the other arts, can be applied as general principles governing diverse kinds of subjects. Rhetoric alone, said Aristotle, is concerned with discovering the means of persuasion in reference to any subject whatever; its rules are not applied to any particular definite class of things (*The Art of Rhetoric* I.ii.2).[28] Recognizing this, Lorenzo Valla assigned to the art a preeminent role in regard to other disciplines. Rhetoric: "regina rerum est, et perfecta sapientia."[29] Alberti's descriptive vocabulary is rhetorical not only because it consists of adjectives rather than nouns, or because his criteria of excellence derive from rhetorical categories, but because his aim was to persuade the reader.

Antithesis and the Coincidence of Opposites

The harmony of which Alberti speaks in *Profugiorum* is a general quality; it can characterize spiritual states, music, or literature, as well as architecture. In chapter 1, I observed that the harmony of the cathedral is one term in the antithesis between spiritual tranquillity and disturbance of the soul. Spiritual tranquillity is an ideal relation between opposites—equilibrium—whereas disturbance is the more common, if regrettable, state of disequilibrium.

Alberti's term for harmony, *concinnitas*, is often thought to be identical with *mediocritas*—a mean between extremes.[30] *Mediocritas*, in this sense, expresses Aristotle's understanding of virtue in the *Nicomachean Ethics*, where virtue is said to lie between two faults of excess (II.ii.6–7) and of the perfect work of art (II.vi.7–11); and it corresponds to his third style of oratory, which combines qualities of the other two. Although in *De Re aedificatoria* Alberti speaks of *concinnitas* as a mean between extremes (IX.vii, p. 835), it seems clear that *mediocritas* and *concinnitas* are not identical terms in *Profugiorum*. The style of Florence Cathedral is said to possess both grace and majesty, opposites that retain their separate identities but are in equilibrium; their opposition is not resolved into that third term, the mean, in which they would become fused as one. While the stylistic qualities of the cathedral are presented as opposites in equilibrium, the building as a whole preserves *mediocritas* against the extremes of hot and cold outside it. Thus in *Profugiorum*, Alberti distinguishes between *concinnitas* as (1) a compositional principle using antitheses, and (2) a quality of the completed work, giving an impression of having achieved the mean.

In *Profugiorum*, Alberti not only asserts that both terms of his antitheses are equally present, but sees this as producing harmony. This directly contradicts statements in *De Re aedificatoria* where *gravitas* and *venustas* are kept apart, as in his discussion of the Orders (IX.v, p. 817; IX.iv, p. 805; VII.iii, p. 549). Alberti's description of the cathedral violates Classical notions of decorum, observed, instead, in the treatise, which required a decision as to which category an object belonged and the rigorous exclusion of qualities from other categories. Onians suggested that Alberti's definition of decorum in architectural style, in which *gravitas* and *festivitas* or *iocunditas* are presented as alternatives to choose between, derives from a passage in Cicero.[31] In *De Officiis* he illustrated the concept of decorum by defining the two possible types of physical beauty: "venustas" is appropriate to women, and "dignitas" to men (I.xxxvi). Decorum is restrictive, aiming at logical consistency. Since within its context what Alberti describes could only be considered disorderly and improper, the context of his discussion of antitheses must be another, or others, in which opposite qualities coexist.

The harmony Alberti describes in *Profugiorum*, a concord of opposites, is the essential structural principle in three contexts important to early Humanist culture: (1) rhetorical practice, (2) musical performance, and (3) the definition of the divinity and, by extension, definition of the structure of the cosmos. All three are implicitly present in Alberti's description of the cathedral, which (1) serves a rhetorical purpose within his moral essay (as churches, in general, are said to do in his treatise on architecture), (2) includes references to music performed in the building as part of the aesthetic and religious experience he is evoking, and (3) serves as an allegory for the ideal state of the soul, participating in divine perfection and order.

Rhetorical Practice

Alberti tells us in *De Re aedificatoria* that the function of *concinnitas* is to order the parts, which otherwise would be entirely separate from one another, in such a way that they manifest a reciprocal relation (IX.v, p. 815). *Concinnitas*, then, is not a quality, but a relation, connecting opposites in such a way that perfection is the result. Used in this sense, *concinnitas* is related to Classical rhetorical theory. Indeed, one of the few appearances of the term in Classical literature occurs in Cicero's *De Partitione oratoria*, in his discussion of the appropriateness of antithesis as a device in epideictic oratory.[32] He recommends intertwining small matters with great ones, simple with complicated, obscure with clear, cheerful with gloomy, incredible with probable. These are to be harmonized and brought together (that is, set in relation with each other) through *concinnitas* (IV.12). The use of antithesis was also recommended in regard to style: treatises on rhetoric consistently recommend that the best sort of oratory is that in which opposite qualities are equally present. I have already given examples of this view in my discussion of beauty and utility; as we saw, Cicero and Tacitus illustrated the principle of their union with architectural images.

The reason for which these opposites must be employed in good rhetorical style has to do with the persuasive aim of oratory. Assent is won by appealing to both the intellect and the senses of the audience. Content alone was not held to be persuasive in most Classical rhetorical theory (although it was believed to be sufficient by Plato and by medieval theoreticians). Cicero's discussion of the difference between the orator and the poet shows that the principle of decorum is not applicable when the aim is persuasion (*De Optimo genere oratorum* I.1–4). Whereas the poet achieves perfection within the boundaries of a certain poetic genre (epic, comic, and so forth), may not mix the genres, and must observe the principle of decorum, the orator is judged solely by his effect on the listener. Therefore, there are as many styles of oratory as orators, and the speakers may mix them at will. Moreover, the most successful orator will combine the most beautiful style with the greatest amount of information (*De Oratore* III.xxiv.91). Indeed, "as in most things, so in language, Nature herself has wonderfully contrived that what carries in it the greatest utility, should have at the same time either the most dignity, or, as often happens, the most beauty" (III.XIV.178, p. 140). Not only does the best style unite the opposite requirements of copiousness and brevity (that is, style and content), but for Cicero, contrary to what was believed in the Middle Ages, beauty of style is naturally connected to excellence of meaning.

> Even if one [orator] in pursuit of weight and dignity avoids simplicity, and, on the other hand, another prefers to be plain and to the point rather than ornate, though he is tolerable as an orator, he is not the best if it is true that the best style is that which includes all virtues. (*De Optimo genere*, II.6, p. 358)

The correspondence of these precepts to our case is well illustrated by Cicero's description of Catullus's oratorical style: "While weighty, its unique dignity nevertheless includes complete urbanity and charm" (*De Oratore*, III.viii.29, p. 24). His words could scarcely be closer to those of Alberti's description of Florence Cathedral. Cicero follows this passage by defining beauty as something from which nothing could be added or subtracted without making it worse. For Aristotle, as we saw, the same definition of artistic beauty meant seeking a mean between extremes. Thus these words, identical to those with which Alberti defined beauty, could mean either the coincidence of opposites (Cicero) or the mean between extremes (Aristotle)—entirely different things.

Quintilian seems to assume that qualities of opposed styles should be copresent when he recommends using figures (word substitutions) in order to add "force and grace" to the matter (*Institutio oratoria* IX.i.2). And Aulus Gellius certainly assumed this when, discussing Plato's style, he said, "One must penetrate to the inmost depths of Plato's mind and feel the weight and dignity of his subject matter, not be diverted to the charm of his diction or the grace of his expression."[33]

The early Humanists adopted the principle in their literary criticism that good style unites opposite qualities. Poggio Bracciolini praised Cicero's early orations, saying that none were more eloquent in their adornment of words or their weight of meaning.[34] Coluccio Salutati defended Petrarch, saying that Seneca, equaled by the Florentine in his thoughts, was exceeded by him in beauty of expression, and Cicero, although richer in store of language and weightier in gravity than Petrarch, was without contention lesser in invention.[35] Leonardo Bruni wrote of Plato's style that it had "urbanity, the highest order of disputation, and a subtlety of the most fruitful and divine *sententiae* joined with a marvellous charm in the protagonists and an incredible copiousness of speech."[36] This might seem little more than an elaboration of Aulus Gellius, but Bruni adds that Plato's style has "much of the most admirable quality which the Greeks call "χάρις.""[37] This comment should alert us that Bruni, who studied under Chrysoloras, also knew Greek treatises on rhetoric; in particular, he would have known Hermogenes, who maintained that there was only one "ideal" style, which included all possible qualities.[38] Demetrius, in his treatise on rhetoric, addressed the logical problems this raised. Although some writers think that there are only two styles (plain and elevated) and that they are in irreconcilable opposition and contrast, this is nonsense, since elevation can perfectly well be combined with vigor and charm. There are, he said, four kinds of style—plain, elevated, elegant, and forcible—and they may be combined with one another (*On Style*, II.36). In his view, the best kind of discourse will be "elaborate and simple at the same time, and draw charm from both sources."[39] Philodemus, Cicero's Greek contemporary, had also defined four kinds of style: copious, plain, grand, and elegant.[40] These four-part systems are close to some Humanist discussions of style—closer, in fact, than Cicero's three-part division. Pius II, for example, praised Virgil for his unsurpassed eloquence, manifested in four stylistic modes: brevity, fullness, simplicity, and elegance.[41] Bruni praised four qualities in Dante's *Divine Comedy*: "e veramente egli è mirabil cosa la grandezza e la dolcezza del dire suo prudente sentenzioso e grave, con varietà e copia mirabile."[42]

Valla, too, seems to have been aware of Greek rhetorical works when he claimed that the greatest merit of eloquence is "copia", which the Greeks call "εὐπορίαν."[43] But, he said, the more copious the discourse, the more it requires order ("ordinis") so that nothing is said that is not useful to the point and appropriate to the location, the occasion, and the situation of the audience.[44] Copiousness, in short, needs to be combined with brevity as equal terms in a pair. Valla used Quintilian's description of Homer's language to refute Aristotle's definition of excellence as the mean between extremes. Quintilian, said Valla, "praised in one man many different virtues, not just copiousness and brevity but sublimity and propriety, delight and control, mirth and gravity and excellence of both poetic and oratorical kinds."[45]

This is exactly the issue on which Alberti's description of Florence Cathedral also takes a stand. Alberti's antithetical description of Florence Cathedral, which has numerous parallels in the literary criticism of his contemporaries, follows Classical precepts for and examples of good rhetorical style. Both in *Profugiorum* and in *De Re aedificatoria*, Alberti depicts stylistic excellence as the union of opposite qualities.

Musical Performance

In *On the Soul*, Aristotle says that harmony is either a proportion or a composition of the ingredients mixed (I.iv). Thus the senses receive pleasure both from a mean between extremes and from a relation between extremes, as in musical harmony. While Alberti used the term *concinnitas* to describe the virtue of both *mediocritas* and the equilibrium of opposites, in *Profugiorum* he intends the relation that Aristotle associated with music. From the Pythagorean thinkers to Augustine, Boethius, Dionysius the Areopagite, and the theologians of the School of Chartres, music was defined as "dissimilium inter se vocum in unum redacta concordia" (or "musica sive harmonia est plurium dissimilium in unum redactorum concordia").[46] Thus Theon of Smyrna wrote that the Pythagoreans ... call music the harmonization of opposites, the unification of disparate things and the conciliation of warring elements.... They say that the effects and application of [musical] knowledge reveal themselves in four human spheres: in the soul, in the body, in the home and in the state. For it is these things that require to be harmonized and unified.[47]

The Renaissance did nothing to alter this definition: the frontispiece of Gaffurio's 1508 treatise on music shows the author saying "harmonia est discordia concors."[48] But if the definition of what musical harmony is did not change, opinions differed on what kind of aesthetic experience it did, or should, provide. As we saw, Plato believed that beauty in music resulted from the mathematical proportions it rendered audible, which were to be judged by the intellect rather than the ear. Aristotle, instead, was concerned with the sensory pleasure afforded by the equilibrium of opposites. The issue in question, in its simplest formulation, is whether beauty is an attribute belonging to beautiful things or something attributed to a thing by its audience. Is it, in short, an objective or a subjective quality?

Alberti's position on this problem is more complex than has been recognized. In *De Re aedificatoria* he says that both musical and visual harmony are based on numbers, as the Pythagoreans first discovered (IX.v, pp. 823–25).[49] This would seem to favor a Platonic interpretation. But the relevant passage is subtler than at first appears: Alberti explains that those numbers that provide *concinnitas* to the ears are the same ones that give pleasure to the eyes and mind (IX.v, p. 823). Beauty, then, while it may inhere in number, is recognized only by the pleasure it affords the senses and emotions. He says this quite explicitly: "Whenever the soul is reached through visual or aural or any other kind of perception we immediately recognize harmony" (IX.v, p. 815). Thus Alberti combined the Aristotelian definition of beauty as "that which gives pleasure through hearing and sight" with the Platonic (and Neoplatonic) notion that beauty is a quality possessed by beautiful things. This compromise, unsatisfactory from a philosophical point of view, must nonetheless have described Alberti's understanding of aesthetic experience. His formulation, I suggest, was much influenced by theological considerations. Perfect beauty, one of the attributes of the Divinity, was surely inherent in the Divine Nature and not in the eye of the beholder. On the other hand, as we have seen in other chapters, Alberti accorded sensory experience a privileged role in espistemology. His view is consistent with Augustine's discussion of the knowledge of God examined earlier.

The complexity of Alberti's attitude has not received adequate attention from scholars, who have emphasized the Platonic aspect to the exclusion of the Aristotelian. Naredi-Rainer, who has most deeply studied Alberti's musical knowledge, related his discussion of musical intervals to Ramos de Pareja's

Musica practica of 1482 and to Plato's *Laws*.⁵⁰ Mühlmann saw Alberti's connection among the well-being of the soul, musical harmony, and proportion in architecture as rooted in Platonic and Neoplatonic sources.⁵¹ These scholars assumed that Alberti saw architecture as related to *musica mundana*, the harmonious numerical relationships of the planetary and stellar orbits or of the spheres. Since both *musica instrumentalis* (actual musical sound) and architecture were believed to be based on the proportions of the universe, it has usually been thought that Alberti conceived of the relation between music and architecture in terms of numerical proportions. But is it true that Alberti was interested in an objective beauty based on laws of universal harmony in *Profugiorum*?

The cathedral itself is an allegory for that musical harmony which mirrors the ideal state of the soul; music brings the soul into a state of equilibrium. Thus Alberti wrote in *Profugiorum*, "Certo in questo convengo io colla opinione de'pittagorici quali affermavano che'l nostro animo s'accoglieva e componeva a tranquillità e a quiete revocato e racconsolato dalle suavissime voci e modi di musica" ("Certainly, in this I agree with the Pythagoreans, who affirmed that our soul welcomed, and was restored to tranquillity and calm, by the consolation of the very sweet voices and modes of music") (p. 178). For Platonic thinkers, it was not the sound of music but the underlying proportions that brought tranquillity, yet Alberti explicitly attributes this function to music as aurally perceived.⁵² The context of Alberti's discussion is certainly Pythagorean, but within the interpretation of Aristotle, Cicero, and Macrobius rather than Plato. For it was Aristotle who, discussing the beliefs of the Pythagoreans, drew together the definition of music and that of the soul: "It is said that the soul is a harmony of some kind; for, they argue, harmony is a mixture or composition of opposites, and the body is composed of contraries."⁵³ Although Aristotle presented this view only to reject it, it was accepted by Macrobius, in his commentary on Cicero's *Somnium Scipionis*: "All wise men admit that the soul was also derived from musical concords."⁵⁴ At the end of his description of the cathedral, Alberti illustrated the harmony of the soul with *musica instrumentalis*, rather than with *musica mundana*, describing voices joined in harmony: "Here you listen to the voices during mass ... with their marvelous beauty." The harmony of that which is heard is analogous to that which is seen; nowhere does he mention the numerical proportions of the building.⁵⁵ His discussion in the moral essay, in which the union of opposites is something heard, has a parallel in *De Re aedificatoria*, where he tells us that the harmony of high and low voices creates an equilibrium between the tones that gives pleasure and conquers the soul; the same dynamic occurs, he says, in every work that aims to persuade (I.x, p. 69).

Vagnetti showed, convincingly I believe, the importance for Alberti of

Cicero's use of *concinnitas* as signifying both the aurally perceived relation between words and that between voices singing music.[56] Cicero emphasized the analogous sensory effect of speech and music, rather than their common intellectual source in number, and associated this with aesthetic pleasure and persuasion, as does Alberti. That *concinnitas* was understood as audible harmony in the Quattrocento is further suggested by Cortesi's assertion that good Latin style called for a periodic composition that creates a "concinnitas ad sonum."[57] Indeed, the Classical and medieval sources that Naredi-Rainer cited for Alberti's notion of harmony all speak of *heard* music, not music theory.[58] Once again, the harmony evoked in *Profugiorum* must be understood within the context of rhetoric rather than philosophy: it concerns the senses before the intellect and seeks to persuade rather than to prove.

Manetti's oration describing the consecration of the cathedral in 1436 also juxtaposed architectural style and the sound of music performed in the building.[59] Of course, what Manetti had in mind was Dufay's motet *Nuper rosarum flores*, composed for the consecration ceremony.[60] Warren's elaborate comparison between the structure of the motet and that of the cathedral has not been generally accepted by music historians.[61] As an architectural historian, I would add that his derivation of the measurements of the building by a system of quadratura, although it might reflect the process by which the building was designed, has no relation to how its measurements were perceived in the Quattrocento. Manetti, for example, gives the total length of the building (260 paces), its width in the nave (66 paces), the height to the base of the dome (70 paces), and so on.[62] Goro Dati and Albertini measured the cathedral in the same way.[63] Their measurements do not yield ideal proportional relationships and cannot be related to musical intervals or mensurations.

Dufay's motet, however, as it was actually heard, possessed the same characteristics that Alberti attributed to the architecture of the cathedral. The motet was composed as a polyphonic work in which the tenor, based on the *introitus* for the consecration of a church, proceeded by sustained notes and the superius had a richly ornamented melodic quality. Not only the characters of the two voices but also those of the sung texts expressed the equilibrium between *gravitas* and *festivitas*, or *brevitas* and *copia*. The biblical text in the tenor is "terribilis est locus iste," while that in the superius is a poem especially written for the consecration ceremony full of gaity and charm. Such contrasts in text and setting are precisely what characterize polyphony, as it was first worked out in twelfth-century Paris and as it continued to be performed in the Quattrocento.[64] Although the tenor and superius do, at certain points, harmonize in consonant intervals that can be represented as ideal numerical proportions, this vertical correspondence is not the dominant structural impression in polyphonic, as opposed to harmonic, music.

When Alberti wrote of the beauty of harmonized voices in *Profugiorum*, he was describing the polyphonic music that he heard performed. By "harmonic," then, what he meant is that concordance of opposites that had always defined what music was in general, and the equilibrium between the antithetical tenor and superius voices in polyphony, in particular. The analogy between the optical and aural experiences of opposites joined in a harmonious relation furnished a second basis, in addition to rhetorical theory, from which Alberti constructed his architectural criticism of the cathedral.

Alberti was not the only one in early Quattrocento Florence to perceive the relation between artistic style and the antitheses of polyphonic composition. Del Bravo, trying to account for the antithetical styles of the cathedral's cantorie, made by Luca della Robbia and Donatello in the 1430s, suggested that they illustrated harmony and polyphony, respectively.[65] In his view, the severe architectural setting of the former embodied the idea of musical proportion, whereas Donatello's frieze represented melody. But it might be more accurate to identify the architectural setting of Della Robbia's cantoria with the tenor voice of polyphony, which was compared by contemporaries to pillars, and Donatello's melodic frieze with the superius. Together, the voices represent utility (the tenor is the voice that holds, hence its name) and beauty, the very qualities that underly Alberti's description of the cathedral. If this interpretation is correct, then the styles of the two cantorie would have corresponded to the style of music performed in them and, taken together, would have embodied the antithesis of beauty and necessity.

The Divine Nature and the Structure of the Cosmos

In Western theology, a strong tradition in which God is defined as the "coincidence of opposites" can be related to the influence of Dionysius the Areopagite. Dionysius himself discussed the question in *The Divine Names*, arguing that big/small, same/different, similar/dissimilar, and unmoved/moving are all true of God and to an infinite degree (IX.909b–917a). This perception was further developed by Dionysius's ninth-century translator and commentator, Johannes Scotus Eriugena, who not only presented God as the coincidence of opposites and his creation as the harmonization of opposites, but went on to conclude that beauty itself, therefore, consists of this harmony: "No beauty comes about except from the combination of like and unlike, of contraries and opposites."[66]

An important figure for such discussion was Alberti's contemporary Nicholas of Cusa, who owned a copy of Eriugena's *Periphyseon* and, like him, defined God as the "coincidentia oppositorum."[67] "For God himself is the likeness of like and the unlikeness of the unlike, the opposition of opposites, the contrary of contraries. For He gathers and composes all these by beautiful and inexpressible harmony into one concord" (*De Divisione naturae* 517). In *De Beryllo*, a work of 1458, Cusanus used the beryl, whose form is both concave and convex, as a symbol for how truth may be perceived: "If the intellectual beryl, which possesses both the maximum and the minimum in the same way, is adapted to the intellectual eyes, the indivisible principle of all things is attained."[68] Although the infinite (God) and the finite are opposites without relation, as Cusanus thought, it is man's task to talk about what he cannot know through talking about what can be known—that is, the world of human experience.[69] Therefore, even though the infinite is both more than and different from the sum of all the antitheses in the finite, these are the bases from which we form our idea of the *coincidentia oppositorum*, or God.

Alberti depicted the beauty of Florence Cathedral as structured according to the single principle of *concinnitas*, which, being descriptive of God, governs all beauty, whether of architecture, music, or language. Although Alberti's understanding of this principle owed much to discussions in pagan Classical authors, his conviction of its preeminence was based on theological considerations.[70] The concord of opposites is recognized by the pleasure it affords hearing and sight, but this is not why it is beautiful. It is beautiful because it corresponds to the divine nature, to the structure of the created world, and to the ideal state of the soul.[71]

Conclusion

Augustine described the structure of the whole world as made up of opposites: "and so you are to regard all the works of the Most High: two by two, one the opposite of the other" (*City of God* XI.18).[72] This perception of reality is revealed in the Humanists' writings. The dialogue, in which opposing views on a subject are presented, was a preferred literary form in the fifteenth century; this is the form Alberti chose for *Profugiorum*. Unlike most Platonic dialogues or Scholastic dialectic, the Renaissance dialogue is not a form of logical argument; rarely are the opposed views resolved in a final synthesis.[73] Camporeale has shown that early Renaissance thinkers, rejecting the philosophical-logical method of Scholastic theologians, returned to the rhetorical method of the early Patristic writers.[74] The discourse of rhetoric is metaphorical and interpretative, and, as we saw, the very premises of rhetorical style urged the necessity of combining opposites.

Cusanus remarked that we know truth "only through metaphor and through symbols."[75] Alberti chose such a symbol in his description of Florence Cathedral. But the image of the church does not merely allude to higher truths, but actually embodies them in its specific style. The visual experience of the cathedral is also an actual experience of divine harmony.[76] The ideal relation between opposites, *concinnitas*, is not the mean between extremes in this case, but an equilibrium of contraries. Alberti is not convinced, in *Profugiorum*, that this condition can be sustained by men, whose souls are born to mobility and change. But that it is within human grasp is proved by the cathedral, the product of human creativity. Here man united the opposites of which nature is made and placed them in relation to one another. His act of ordering is proof that man is, indeed, made in God's image.

SIX

Varietas and the Design of Pienza

WHILE MUCH is known about Pienza, little has been written about the aesthetic principles of its design (Figure 2). Problematic is its status as the first ideal city plan of the Renaissance to be (at least partially) built.

Renaissance planners, it is thought, rejected the irregularity or disorder of Gothic space in favor of a regularity and order manifested in regularly shaped piazzas with strong axiality, preferred monumental entrances to these piazzas, and favored either the distribution of classicizing structures around the piazza's periphery or the placement of a centralized church in the middle of the piazza.[1] What the designers sought was, as Le Mollé expressed it, "l'image de la perfection, de l'éternité et de l'idée platonicienne."[2] Such formulations do not differentiate between the aesthetic principles of urban design theory in the fifteenth and sixteenth centuries; instead, they treat Quattrocento projects as less fully articulated manifestations of Cinquecento ideals. Thus the historical evolution of city planning from 1400 to 1600 has been seen as the progressive clarification and realization of these ideals in both theory and practice; only Tafuri has suggested that city planning in the earlier Renaissance (from Alberti to Leonardo) is essentially anti-utopian.[3] Heydenreich was the first to refer to Pienza as an ideal city; Kruft is the most recent scholar to have done so.[4] Others have been more cautious: Murray regarded it as one of the first examples of regular town planning since Roman days, without claiming that it was also ideal.[5] Mack has acknowledged that Pienza is not so regular or gridlike as the new towns of the thirteenth and fourteenth centuries, and that these, rather than Pienza, point toward the ideal cities of the later Quattrocento and Cinquecento.[6] In a recent article, Gnocchi intuited that the design of Pienza was

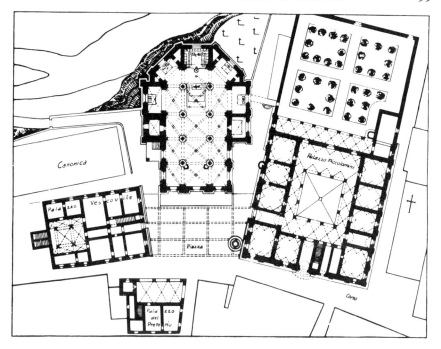

Figure 2. Plan, Pienza. (E. Carli, *Pienza* [Siena, 1966])

governed by aesthetic principles as different from those of medieval as of Cinquecento planning, although he did not treat the project in detail.[7]

This chapter seeks to clarify the aesthetic principles at work in Pienza. I argue that the urban complex is not a flawed or hesitant manifestation of Platonic and Pythagorean principles, but a fully realized expression of early, as opposed to high or late, Renaissance aesthetic values. The culture of the early Renaissance, as I have maintained in other chapters, was not a less fully developed or articulate version of that which came later; it was essentially different from, and opposed to, what followed. If the ideal city is the paradoxical attempt to realize a utopia, Pienza, instead, embodies what the early Humanists felt to be the underlying principles of reality. The design of Pienza is governed by the quality of *varietas*, understood as a virtue of all good style and, beyond that, as an essential principle of nature.[8] At Pienza, where the visual order is built up by contrast, the viewer is expected to recognize the differences between well-defined building types and to appreciate the ways in which the buildings have been harmonized through skillful manipulations of material and design.

The Building Project

The first extensive projects of Renaissance urban planning date from the middle decades of the Quattrocento.[9] The earliest of these, connected with the pontificate of Nicholas V (1447–1455), aimed at the renovation of the Capitoline, the Borgo, and the area leading up to Ponte Sant'Angelo in Rome, and the renewal of the main piazza at Fabriano.[10] These experiments in redesigning urban spaces provide the background for Pius II's rebuilding of Corsignano between 1459 and 1464,[11] and the planning of what has been called the first example of the Renaissance ideal city.[12] In recent years, Pienza (Pius's new name for Corsignano) has benefited from a good deal of scholarly attention.[13] The documents relevant to the project have been scrutinized, the urban layout has been accurately recorded, and the building histories of most of the component elements of the project have been clarified. The only important factual question that remains unclear is that of Alberti's participation in the project.[14] Documentation on this point has not, and probably never will, come to light since Alberti, as a Humanist at the papal court, would not have received payment for advising the pope. Whether or not Alberti designed Pienza is not central for my essay. We know that Alberti and Rossellino, the executing architect at Pienza, had been in contact for at least a decade before the Pienza project, and that Alberti and Pius were in rather close contact at the time the plan was conceived.[15] Thus both of the men known to have been directly responsible for the urban renewal knew Alberti's views on architectural planning and could have applied them at Pienza. While I believe this to have been the case, the argument of this chapter, concerning ideas widely current in the earlier Quattrocento and not unique to Alberti, does not rely on his authorship of the project.

The story of Pienza's rebuilding, told many times, need not be repeated in detail here. We know that a total of some forty buildings in this tiny town were either built or refurbished with the patronage or encouragement of Pius II. The main primary source for our knowledge of the project, Pius's *Commentarii*, leaves no doubt that Aeneas Silvius Piccolomini wished to honor his native town through extensive rebuilding, granting it episcopal status and renaming it after himself.[16] The extent of the original project, of 1460, is unclear: it surely comprised the Piccolomini Palace and the cathedral, probably the Canons's Palace, and perhaps also houses for members of the court. The most important of these structures were essentially complete in 1462, when the project entered a new phase of ampler scope, which may, or may not, have been foreseen from the beginning. That year saw the acquisition of land and existing buildings for the remaining three structures actually on the piazza: the Bishop's Palace, the Communal Palace, and a small house of uncertain ownership. The "curial row" east of the piazza on the Corso, five houses for courtiers, a hospital, an inn, and row houses in the northeastern corner of town belong to this second phase. Pius tells us that "many of the townspeople tore down old houses and built new ones, so that nowhere did the aspect of the town remain unchanged."[17] Whether this building by local citizens belonged to the first or the second phase of the project is not clear.

The urban complex, if seen as the first ideal city plan to be built in the Renaissance, does not display the kinds of features many scholars would like. The main piazza (Piazza Pio II) is not symmetrical and does not have an ideal form, such as the square. The buildings do not conform to a single ideal style, and *all'antica* elements appear only sporadically. There seems to have been little attempt at Pienza either to revive ancient Roman forms, whether of urban organization or architectural style, or to embody Pythagorean and Platonic concepts of ideal form and ideal order.[18] Since, for many scholars, it is precisely these two factors—the return to Classical models and the search for perfect beauty—that characterize the Renaissance, there has been little to say about the theoretical underpinnings of the Pienza project. Each of the buildings in the piazza conforms to a different building typology, each is eclectic in its sources, and each in some way combines traditional with innovative features.

The cathedral follows the model of Austrian or Bavarian Gothic *Hallenkirchen* and has the first *all'antica* façade of the Renaissance, even if its immediate sources are medieval rather than Classical (Figures 3 and 4). Scholars agree that Pius's models for the cathedral were works by Hans Stetheimer (ca. 1350–1432) in the Bavarian towns of Landshut, Straubing, and Wasserburg; the church of St. Otmar in Mödling; or the parish church of Gumpoldskirchen.[19] Rossellino was able to carry out the pope's wishes with the help of his familiarity with Italian hall churches at Todi and Perugia, his own design for the choir of old St. Peter's, the tribunes of Florence Cathedral, and the transept of Siena Cathedral. The façade, which Pius said was like that of an ancient temple, owes much to the twelfth-century San Rufino in Assisi.[20] Although Heydenreich attempted to excuse Pius's choice of the Gothic on the grounds that he had been away from Italy during the crucial years of Brunelleschi's classicizing revolution, the evidence presented in chapter 2, including Pius's own statements of admiration for the Gothic, suggests that he, like Alberti and other contemporaries, accepted the Gothic style.[21] Mack explained the pope's choice as arising from his "almost claustrophobic love of bright, open spaces," although he admits that Pius himself claimed to have asked for a German design because "it makes the church more graceful and lighter."[22] It should be emphasized that Pius's models were recently built structures, which he might well have considered to represent up-to-date design principles.

The Piccolomini Palace owes much to the Florentine tradition that favored

Figure 3. Interior, Pienza Cathedral.

Figure 4. Interior, St. Martin's at Landshut.

Figure 5. Piccolomini Palace, Pienza. (Photograph by Corey L. Lowenthal)

Figure 6. Rucellai Palace, Florence. (Alinari)

Figure 7. Canons' Palace, Pienza. (Photograph by Corey L. Lowenthal)

Figure 8. Vernacular housing, Florence. The drawing is attributed to Benedetto de Maiano. (Alinari)

the use of dressed masonry façades, and it takes from the recent Medici Palace in Florence its symmetrical design around a large cortile (Figures 5 and 6).[23] The palace also has Roman sources for its window type and the superimposed Orders. The superimposed loggie on the garden façade perhaps derive from the Vatican Palace in Rome or a recently completed villa in Tarquinia.[24] Whether it is the first Renaissance palace to use the Classical Orders on the façade depends on the still-unresolved question of its chronological relation to the Rucellai Palace in Florence; in any case, these are almost exactly contemporaneous works.[25]

The Canons' Palace follows a house type long established in late medieval Florence and elsewhere, characterized by an asymmetrically placed portal, round-headed windows, string courses between the stories, and an intonaco coating covering the whole façade (Figures 7 and 8).[26] Its sgraffito decoration, depicting masonry blocks, was already in use in the late fourteenth century, although it became popular only in the fifteenth. The Casa Davanzati (third quarter of the fourteenth century) is one of the earliest Florentine palaces to have a sgraffito façade; in the early fifteenth century, it was used for the Palazzo Ridolfi, Palazzo Morelli, Palazzo Bardi Busini, Palazzo Gerini, and Palazzo Lapi.[27] The Bishop's Palace draws much from fifteenth-century cardinals' pal-

Figure 9. Bishop's Palace, Pienza.

aces in Rome, especially for its rectangular cross-windows, its string courses, and the untextured plane of the façade (Figures 9 and 10).[28] The Communal Palace reiterates some of the most familiar motifs of medieval communal palaces, especially those in Tuscany; the ground-floor loggia to the left of the main block, double-lancet windows on the piano nobile, and an asymmetrically placed tower are all common in medieval examples of this genre (Figures 11 and 12).[29] Although most of the models are late thirteenth century, it is worth noting typological resemblances with Quattrocento examples such as the Palazzo Pretorio in Radda in Chianti, of around 1415, and the Palazzo Comunale of Viterbo, begun in 1448. Finally, the little brick house, sometimes said to have belonged to Tommaso Piccolomini, but more likely to have been the residence of the carpenter Magio d'Agnolo da Montefalonico, differs in material, scale, window type, and ornament from all the rest (Figure 13).[30] Its affinities with late medieval houses in Pienza itself suggest that it continues a local tradition of vernacular architecture (Figure 14). As such, it provides a fixed point against which the "foreign" architectural modes elsewhere in the piazza can be compared. It is one of the most important of all the structures, since it, together with the modest Canonicate, permits us to evaluate the scale, novelty, and diversity of the more monumental buildings. While most discussion

Figure 10. Palazzo Venezia, Rome. (Alinari)

of Pienza has ignored, or at least neglected, the two humblest structures on the square—the Canons' Palace and the house of Magio the carpenter—my interpretation of the aesthetic principle at work at Pienza does not privilege the larger and more expensive structures; the smaller works are essential for the full extension of *varietas* in the design.

Striking about these buildings is how little they draw on Classical models and, therefore, how poorly they correspond to notions of the ideal Renaissance city. Many of their models are either pure examples of Gothic style, such as the hall church and Communal Palace, or Gothic by reason of date, like the house types drawn on for the Canonicate. Virtually all the structures, even the most innovative, have strong links with medieval tradition. It should not be forgotten that on the roof of the classicizing Piccolomini Palace were originally twenty-three chimneys, described by Pius as like fortified towers, all gaily painted.[31] The *all'antica* façade of the cathedral is in fact based on a twelfth-century church façade in Assisi; the string courses and planarity of the Bishop's Palace reflect medieval Roman housing; and the large masonry blocks of the Piccolomini Palace recall the Palazzo Vecchio in Florence. Also striking are the strong relations with recently completed buildings in Gothic style, such as

Figure 11. Communal Palace, Pienza. (C. Mack, *Pienza: The Creation of a Renaissance City* [Ithaca, N.Y., 1987])

the Cathedral of Perugia (which Pius consecrated), the Communal Palace of Viterbo (begun in 1448), and Stetheimer's hall churches.[32] Two conclusions may be drawn from this: first, that there is certainly no "rejection of the immediate past" in Pienza, and no "condemnation of medieval forms as outworn or inappropriate for modern use."[33] These formulations, widely accepted as defining the relation of Renaissance architecture to the past, are inapplicable here. But this does not mean that the buildings at Pienza are uncritical reiterations of medieval models. Instead, and this is the second conclusion, the designs reveal careful study of medieval building types in regard to their layout, morphology, materials, finish, and trim. The architect proceeded, as does the modern architectural historian, by categorizing historical examples into functional groups, extrapolating common features, and employing these to evoke the group as a whole. His procedure recalls Alberti's recommendation that the architect proceed like the man of letters, whose knowledge is built up from the study of all those past authors whose work has won approval.[34] It would

Figure 12. Town hall, Radda in Chianti.

seem that at Pienza the architect intended the viewer to recognize the traditions he drew from and that he was guided neither by personal formal preferences nor by the goal of discovering ideal beauty. Moreover, he sought to define what made these groups distinct rather than what they shared; hence it was not his intention to obtain a uniform style in the piazza. The models were known to those who used the buildings; there are no revivals of monuments that did not form a part of the patrons' daily life experience. The use of sources at Pienza is highly intellectual and erudite, but it is not classicizing. My analysis is further supported by arguments about the nature of early Renaissance culture presented in earlier chapters: for the acceptance of the Middle Ages (chapter 4), the approving attitude toward eclecticism (chapter 5), and the postulation of contrast and stylistic diversity as positive aesthetic criteria (chapter 5).

Each of the buildings in the piazza at Pienza is easily recognized as belonging to a certain class of structure typical of a restricted geographical area and identified with a clearly defined class or function. Yet each is also in some way new: this is the first Communal Palace with a sgraffito-decorated façade; the Piccolomini Palace may be the first domestic structure to have pilasters on the

Figure 13. House of Magio d'Agnolo da Montefalonico (?), Pienza. (Photograph by Corey L. Lowenthal)

Figure 14. Medieval house, Pienza. (C. Mack, *Pienza: The Creation of a Renaissance City* [Ithaca, N.Y., 1987])

façade; unlike Roman palaces, the Bishop's Palace has a finely worked ashlar masonry façade. These new features acquire additional resonance precisely because the core of the design is familiar.

The governing principle of *varietas* at Pienza operates through the relationships established between the different buildings. They do not demonstrate mere diversity, since each repeats some aspect or aspects of others and shares with them materials and scale. For example, travertine columns with classicizing capitals appear on the cathedral façade, in the courtyard of the Piccolomini Palace, on the well in the piazza, and in the loggia of the Communal Palace. The windows of the Pope's Palace share their rectangular shapes and crossmullions with those of the Bishop's Palace, and the tracery lunette of their bifores with the Communal Palace and the nave windows of the cathedral. The channeled masonry of the palace is echoed in the fictive masonry of the Canonicate and Communal Palace, while the travertine ashlar of the cathedral façade is answered by the sandstone ashlar of the Bishop's Palace. To sum up, each building selects from an available vocabulary of features; like any vocabulary, the terms are limited in number, and it is their combination in different syntactical structures that lends them variety.

Aesthetic Principles

If the buildings are not ideal, at least the spectator's view of them has been said to be. Since Heydenreich, it has been assumed that the piazza was designed to be viewed by an immobile spectator from a single point, which Heydenreich located at the place where Via G. Marconi opens onto the piazza (Figure 15).[35] In fact, this dark and narrow alleyway with high buildings on either side prevents the viewer from seeing more than a glimpse of the right part of the cathedral façade until the last few meters of its length (Figure 16). Heydenreich, recognizing the difficulty that this lack of monumental access could cause to the notion of Pienza as an ideal urban complex, located the ideal viewpoint within the piazza, at the corner of a small brick house in the southeastern part of the square. It is true that this point of view offers the sight lines he indicated. Yet this ideal point does not correspond to the actual position of the viewer approaching, or even entering, the square. Thus I cannot agree with Mack, who suggested that "the central portal of the cathedral is on a direct axis with the spot at which this street enters the square."[36] This is true in plan, but not according to vision, since the street approaches the piazza on a diagonal. In reality, the approach to the piazza is made along the main street, Corso Il Rossellino; the first view of the cathedral is as a building set back from the road on the right. The spectator who turns toward the church at that moment is quite far from the ideal viewing point. Heydenreich, who certainly realized

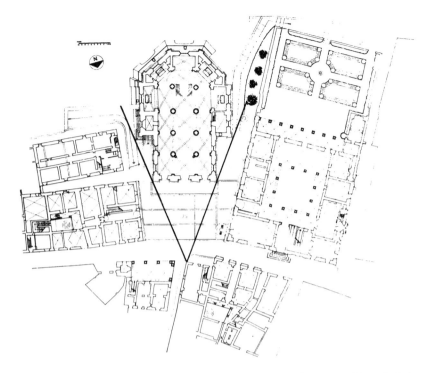

Figure 15. Plan, Pienza, with sightlines following Heydenreich. (G. Cataldi et al., *Rilievi de Pienza* [Florence, 1977], with sightlines added by author)

this, suggested that the piazza was designed to be on an axis with a small market square, the two squares being connected by the narrow alleyway already mentioned.[37] If Via Marconi were intended to become a main artery, perpendicular to and of equal importance with the Corso, it is strange that no palaces were built along it and that nothing was done to widen the street. Those structures that make it so narrow and dark as it approaches the piazza were only begun when the cathedral was complete in 1462; if Pius had wanted the spectator to see his piazza from the ideal point indicated by Heydenreich, he would not have permitted the small brick house to be built on it.[38] In short, since there was an opportunity to create a broader entrance to the piazza or one aligned with the center of the cathedral in 1462 and this chance was not utilized, there is no evidence that this approach was considered of special importance.

The fact that the piazza is trapezoidal has been explained either in terms of the disjunction between theory and practice or as a subtle manipulation of

Figure 16. Pienza Cathedral, as seen from Via Marconi.

the architect. The diverging sides of the piazza, constituted by the Piccolomini Palace to the west and the Bishop's and Canons' Palaces to the east, have been said to correct the optical illusion of convergence that would have resulted had they been strictly orthogonal.[39] Others have seen the divergence as a device to make the church façade seem nearer, as a "trick of perspective" intended to enlarge the impression of the square and façade beyond their actual dimension, or as a means to separate the structures visually and permit them to be viewed against the landscape beyond.[40] Mack explained the trapezoid as the by-product of topographical necessity, resulting from the need to align the façade of the Piccolomini Palace with the Corso.[41] Almost all critics have discussed the aesthetic effect of the square implicitly or explicitly as organized according to one-point perspective.[42] Thus, in general, the aesthetic effect of the main square of Pienza has been understood as the three-dimensional, if imperfectly realized, equivalent of those city views in the Berlin, Baltimore, and Urbino panels: mathematically ordered, static, timeless, not only Pythagorean but also Platonic (Figure 17).[43]

Just as the buildings at Pienza display more diversity than uniformity, so, I would argue, is the spectator's view of them governed by the principle of *varietas*. The actual experience of the urban fabric is optical rather than math-

Figure 17. Ideal city view. Palazzo Ducale, Urbino.

ematical, dynamic rather than static, and unfolds in a temporal sequence determined by the spectator's physical movement. The criterion of *varietas* at Pienza operates within four distinct aspects: decorum, natural law, theory of perception, and stylistic excellence.

Decorum

That the differences between the buildings on the piazza at Pienza are expressive of their diverse functions or the social status of their inhabitants has long been recognized and associated with Alberti's recommendations in *De Re aedificatoria*. For Alberti, "the use of edifices being various, it was necessary to enquire whether one and the same kind of design was fit for all sorts of buildings; upon which account we have distinguished the several kinds of buildings [and examined] what sort of beauty was proper to each edifice" (Prologue, p. 15). The close correspondence between this endeavor and the typological categorization undertaken by the designer at Pienza is obvious. Indeed, that the style of the buildings reflects social hierarchy in a manner corresponding to Alberti's recommendations in *De Re aedificatoria* is well known.[44] Yet *varietas* is not only a principle of social differentiation, but also a principle of style derived from rhetorical theory, where decorum required the accommodation of oratorical style to the demands of the occasion. Cicero wrote in *De Oratore* that "no single kind of oratory suits every cause or audience or speaker or occasion" (III.lv.210);[45] this was repeated word for word by Quintilian in the *Institutio oratoria* (XI.i.3–6).[46] Aristotle had urged the same principle: "One must not forget that different styles are suitable for different kinds of discourse."[47] Horace, too, had counseled the use of variety, as long as decorum is preserved: "Let each style keep the becoming place allotted to it" (*Ars Poetica*, 92).[48] Alberti transferred this notion to the art of architecture, claiming in his treatise that "the greatest glory in architecture is in the evaluation of what is appropriate [*quod deceat*]" (IX.x, p. 855). Just as the orator makes choices in response to variable occasions, so, in an analogous way, the architect takes into account the function of the proposed structure, the status and financial resources of the patron, the nature of the site, available materials, and the labor force. The related concepts of appropriateness, decorum, and τὸ πρέπον all presuppose the notion that value is relative and reject idealism.

Pienza demonstrates how tradition and innovation operate within the constraints of decorum. In some cases, such as the Communal Palace, it was decided that the traditional characteristics of this class of building were functionally and associationally apt, and few departures were made from tradition. Other examples, like the Piccolomini Palace, reveal a dissatisfaction with earlier solutions. The large size of the palace, its ample interior courtyard, the garden façade, and the *giardino pensile* have precedents in slightly earlier palaces and villas but break with the medieval tradition. The new typology is in large part a response to new trends in private living habits, including the housing of numerous people under one roof; the desire for different apartments for summer and winter; the enjoyment of larger domestic spaces together with an increase in number, kind, and cost of home furnishings; and the inclusion of amenities such as gardens as part of the "at home" experience.[49] The different relations of these buildings to their respective traditions suggest that whereas the range and type of functions proper to a communal palace had remained constant, private life had changed dramatically. The architect was obliged to be a sharp observer of social realities, aware of the available options in the storehouse of precedent and able to judge when continuity with them would be acceptable and when innovation was necessary. Once again, the architect was expected to follow the method of the orator, whom Cicero advised to deduce from experience a number of general categories that would define the type of case presented to him and the best method of dealing with it.[50] This procedure assumes the importance of memory for creative invention, a notion emphasized by Quintilian in *Institutio oratoria* (XI.ii.1–2). The architect also needed to possess a precise knowledge of the appropriate relation to establish between variables of size, style, and ornament, on the one hand, and social status and financial position, on the other. Who, for example, should have an entablature rather than a string course on his palace façade in the mid-Quattrocento? Today, quite modest homes may boast a columned entrance or a pedimented doorway; the Renaissance sense of decorum would have found this inappropriate and disorderly. This suggests that what is appropriate is culturally relative and changes over time. Evidently, in so sensitively attuned a social ambient, one in which the information conveyed by a structure's size, style, material, and ornament was essential to the smooth conduct of human affairs, utopian visions of ideal form in the sense of a single "superior" architectural style could have no place.

Natural Law

Alberti believed that the nature of reality is change, always moving "di varietà in nuove varietà"; people are either growing or declining, and all things on earth obey the "fatale e ascritto ordine della natura che sempre stiano in moto."[51] This belief in change and variety as part of natural law caused Alberti to recommend winding streets in *De Re aedificatoria*: "This winding of the streets will make the spectator discover a new structure at every step, and the facade and entrance of each house will face the middle of the street directly." (IV.v, p. 307). This passage very aptly describes what is, in fact, the viewer's experience at Pienza and what I propose was the primary viewing pattern intended for the complex. The logic of the design is better grasped by a moving spectator than from an ideal viewing point. The scenographic or perspectival view proposed in earlier interpretations constitutes the antithesis of the sequential experience and is secondary. Unlike it, the primary viewing pattern permits the viewer to control what he sees and create his own order of vision by moving from place to place. The emphasis is not on the object, with which the spectator must align himself, but on the continually changing relation between object and viewer.[52] Not Figure 15, but Figures 18 through 21 record the aesthetic order of the site.

The reading I am proposing produces a new understanding about how the buildings can be seen. For example, the Bishop's Palace, rather than the cathedral, is the first element of the piazza group to appear coming from the western city gate; and it is seen frontally, revealing itself bay by bay as the viewer moves up the Corso (Figures 18 and 19). The Piccolomini Palace is first seen at a right angle to this façade, and the two contrast in their materials, scale, and texture. But coming down the Corso from the east, the Piccolomini Palace also first reveals itself frontally: Figure 21 shows the play between the frontal view of one of the Piccolomini façades and the receding orthogonal of another. As the spectator moves forward, buildings come into focus in privileged ways and drop away as some new configuration forms. One of these configurations is, of course, that which is seen on entering the piazza when the space dilates back from the viewer. This experience incorporates an element of surprise, since up until the last few meters the Corso, both east and west of the piazza, seems to be enclosed by continuous walls. While this lack of axial approach to the piazza recalls the urban arrangements of the Campo at Siena and Piazza Signoria in Florence, unlike these the buildings accompany us along most of the length of the Corso, revealing more or less of themselves

Figure 18. View looking east along Corso il Rossellino. (Photograph by Corey L. Lowenthal)

as we progress. Not only are they visible, but pleasure derives from the orderly way in which they are perceived. For example, the Bishop's Palace and the Piccolomini Palace come into view bay unit by bay unit, hence in regular and measurable segments (Figures 18–21).

My reading finds support above all from the site itself: as I showed, the frontal view of the whole piazza is not easily obtained, whereas anyone who visits Pienza is positively obliged to experience the complex in the successive views I have described. Why would the designer plan his main effect to be seen from a dark alleyway rather than from the main thoroughfare of the city? Why would the principal doorway of the Piccolomini Palace be out of view from this ideal point? Common sense and the topographical reality of the town indicate that the designer accommodated his plan to the demands of the site and that the intended axis of his piazza is along that of the main thoroughfare, rather than perpendicular to it. Further, although the Corso existed before the urban intervention, it seems not to have been paved and in any case could have been moved. The site of the Piccolomini Palace was occupied before

Figure 19. View looking east along Corso il Rossellino. (Photograph by Corey L. Lowenthal)

1460, but by a number of structures and by gardens. Thus the decision to place the main façade of the palace along the Corso and the whole building at an angle to the square was not the only choice available. The only building on the piazza to have a fixed relationship to the road before the renewal was the Bishop's Palace, which is a renovation of an earlier building on the same site. The architect, then, was relatively free to have moved the Corso, especially in the stretch between the Piccolomini Palace and the Ammannati Palace, where the earlier structures were destroyed in 1460. Instead, it is precisely here that the curvature of the street was accentuated by the frontage of the new buildings. In this, as in the whole design, the principle of *varietas* applies. The spectator's experience in Pienza actualizes Cusanus's definition of motion as composed of nothing but rest: since to move is to go from one state to another, and motion is departure from one state, to move "is to pass from rest to rest and motion is nothing else than ordered rest, that is, resting places successively ordered."[53]

Figure 20. View looking west along Corso il Rossellino. (Photograph by Corey L. Lowenthal)

Figure 21. View looking west along Corso il Rossellino. (Photograph by Corey L. Lowenthal)

Figure 22. View into the Val d'Orcia. (Photograph by Corey L. Lowenthal)

Theory of Perception

Having arrived in the piazza and turning toward the church, the viewer is greeted with a new contrast. In the heart of the city, the viewer looks out, to left and right of the cathedral, over the Val d'Orcia (Figure 22). The prospect is not merely a scenographic backdrop for the church, or a means of visually separating it from other structures on the piazza.[54] It reflects the change from a communal to a territorial concept of the city. Leonardo Bruni's *Laudatio florentinae urbis* (1403–1404) is one of the earliest documents of a new sense of urban identity in which the city is defined as the walled town together with its *contado*.[55] Alberti's statement that Brunelleschi's dome was large enough to cover all the Tuscan people with its shadow reflects the same change, as does the later Chain Map of Florence (ca. 1480), showing the city and the surrounding countryside.[56]

Yet the primary function of the landscape view, connected with *varietas* as a principle of perception, is its establishment of the antithesis of city and country. I have discussed the importance of antithesis for Quattrocento aesthetic theory in chapter 5; it has also been the subject of a lengthy study by

David Summers, whose evidence need not be repeated here.[57] Antithesis, said to be a structural principle of nature itself (as we saw in chapter 5), is essential for human cognitive processes. As Aristotle expressed it: "We have no sensation of what is as hot, cold, hard or soft as we are, but only of what is more so, sensation being then a sort of mean between the opposites in things felt."[58] Thus at Pienza, the experience of being in the urban fabric is defined and clarified by the viewer's simultaneous perception of its opposite, the countryside.

Stylistic Excellence

If antithesis is a natural component of perception, it can also be artificially created as an artistic quality. This aspect of antithesis was explored in chapter 5 as it was applied by the early Humanists to literary style, to the style of Florence Cathedral, to music, and to the beauty of the created world. I extend the concept here to the early Renaissance perception of the built environment. Pienza presents antitheses to the viewer, as we have seen. I would emphasize that the pope's earliest project for Pienza, as recorded in his *Commentarii*, consisted of only the church and the palace. Thus the fundamental antithesis at Pienza is that between the travertine cathedral façade and the tufa Piccolomini Palace, the Gothic hall church and the *all'antica* private dwelling. But since the design consists of multiple, interrelated structures, it is not only antithesis (which operates in pairs) but *varietas*, extending over an unspecified number of elements, that controls the design.

In ancient rhetorical theory, the lack of *varietas* is consistently condemned and is equated with monotony. "This is one of the surest signs of lack of art," Quintilian tells us, "and produces a uniquely unpleasing effect, not merely on the mind, but on the ear, on account of the sameness of thought, the uniformity of its figures and the monotony of its structure."[59] Plutarch, whose rediscovered works were being avidly read at this time, commented that style became "dull and unemotional and wearisome from lack of variety," and added that "monotony is in everything tiresome and repellent, but variety is agreeable."[60]

There is no doubt that variety was a highly valued artistic quality in early Renaissance aesthetics. I do not agree with Klotz's objection to Gosebruch's argument about *varietas* that "es herrscht die Freude an der Vielfalt, nicht die bewusste Kultur der Mannigfaltigkeit."[61] There is, indeed, a culture in which *varietas* is an essential component, and this culture is opposed to another, concerned with the establishment of rules and uniform practice, as I showed in chapter 4. Leonardo Bruni, writing about Ghiberti's Gates of Paradise, explained that by splendor he meant "that they should offer a feast to the eye through the variety of design."[62] Alberti asserted in *Della Pittura* that what first gives pleasure in an *istoria* is "copiousness and variety of things."[63] In *De Re aedificatoria* he welcomed variety, providing it does not detract from the harmonious relations between the parts (I.x, p. 69). George of Trebizond seems almost to have foretold the principle of Pienza's design when, in order to emphasize his argument for the importance of variety in rhetorical discourse, he gave this example: "An architect builds his most beautiful buildings by using now arches, now a plain wall, and now bricks, now dressed stone—these being applied, of course, with art."[64]

Essential to this emphasis on *varietas* was the assumption that art should give pleasure to the spectator, an assumption that, as we saw, Alberti endorsed. "Without question," he says in the architectural treatise, "there is a certain excellence and natural beauty in the figures and forms of buildings, which immediately strike the mind with pleasure and admiration" (IX.viii, p. 843). The early Renaissance stylistic principles of *varietas* and *copia* derive, as Gosebruch rightly observed, from rhetorical theory and especially from Cicero.[65] At Pienza, the principle of *varietas* should be understood, at least in its rhetorical aspect, as relating to epideictic rather than forensic or deliberative expression, since it serves to provide pleasure rather than to persuade or convince.[66]

Not only was *varietas* associated with beauty and pleasure in the earlier Quattrocento, but some thinkers raised it to a metaphysical principle. Filarete believed that *varietas* was an essential principle of divine creation: "You know perfectly well that God could make all men look alike, and yet he doesn't do so."[67] And Nicholas Cusanus saw *varietas* as one of the ways in which man can begin to understand the notion of God's infinity: "Infinity, which cannot be multiplied, is best unfolded by reception in a variety of things, for greater diversity best expresses nonmultiplicity."[68]

The End of Early Humanism and the Rise of the Ideal City

I have been exploring the values of a culture that believed that art should give pleasure to its audience and that the absence of *varietas* was a fault. It was essentially a rhetorical culture. But it was to give way to a very different culture in which beauty would be measured not by the pleasure it gives, the form it has, or the task it fulfills, but by consonance with an ideal. This shift in the focus of Florentine Humanism from rhetoric to metaphysical philosophy, especially Platonism, is the subject of a recent work by Arthur Field.[69] One of the most dramatic episodes, which exemplifies this change of mood, is the controversy over the future of Carlo Marsuppini's chair at the Studium after his death in 1453.[70] His had been the "universal" chair, in the sense that he lectured on Greek and Latin literature, rhetoric, poetry, moral philosophy, and, perhaps, speculative philosophy. The controversy, centering on what subjects would be taught by his successor, was resolved when Argyropoulus was offered the post in 1456: he gave no lectures on rhetoric and perhaps none on Latin literature; philosophy, especially speculative philosophy, was what he made available to a new generation of students. Field has shown that this shift, although exemplified in Argyropoulus's teaching, had roots elsewhere in Florentine culture.[71]

The shift toward Platonism was accompanied by renewed interest in Scholastic philosophy. At the Studium, the *quaestio* was revived in its traditional four categories (*an sit, quid sit, quia sit,* and *propter quid*), and the search for definition was resumed where the Scholastics had left off. Only a year after Marsuppini's death, Ficino urged his correspondent Antonio Serafico to exchange letters with him "in the manner of philosophers, despising everywhere words and bringing forward weighty utterances."[72] Scholastic philosophy was defended not only by Ficino, but also by Benedetto Accolti and Alamanno Rinuccini in the third quarter of the fifteenth century.[73] The confrontation between Humanist eloquence and Scholasticism at midcentury is dramatically represented by Valla's eulogy of Thomas Aquinas in 1457, in which he condemned Aquinas's philosophy and exalted Patristic authors who, avoiding logic and metaphysics, had modeled their style on that of the Ancients.[74] Scholasticism triumphed: Aquinas's *Summa theologica* was adopted as the basic text for the teaching of theology in the second half of the century, and Vincent of Beauvais's *Speculum doctrinale* became a popular work, with printings in 1472 and 1494.[75]

Not since the thirteenth century were so many dictionaries, encyclopedias, compendiums, and systematic surveys written as in the second half of the Quattrocento. They included not only the treatises on architecture by Alberti (1452), Filarete (ca. 1460), and Francesco di Giorgio (1470s and later) but also works in which architecture occupies an important place, such as Giovanni Tortelli's *De Ortografia* (1453), Francesco Patrizi's *De Istituzione reipublicae* (begun in 1461), and Francesco Maria Grapaldi's *De Partibus aedium* (1494). The same preference for systematic treatment is reflected in the publication of Biondo Flavio's *Roma instaurata* (ca. 1470), Vitruvius's treatise (1486, 1495, and 1496), Frontinus's *De Aquis urbis Romae* (1486), and Alberti's treatise (1486). It is also evident in works not concerned with architecture, such as Luca Pacioli's *Summa arithmetica* (1494) and Gaffurio's survey of music in his *Theoricum opus* (1480) and *Practica musica* (1496). While these works may reveal a tendency to synthesize and organize the new material recovered in the first half of the Quattrocento, they also suggest that early Humanist culture, based on opinion and eloquence, was giving way to a new culture, one strand of which privileged objective fact.

This culture, philosophically rather than rhetorically oriented, considered beauty to be an objective property inherent in beautiful things.[76] Beauty, in other words, cannot be identified by an individual's reaction to it or defined by its capacity to give pleasure. The essence of beauty, in Platonic terms, lies in order, measure, proportion, consonance, and harmony; it is recognized by measuring the beauty of things against the Idea of Beauty that we carry in our minds. Against the pluralist view of art, fostered by Aristotle and later Cicero, the idealist view, in both its Pythagorean and Platonic formulations, argues that a single principle is common to all art and beauty. The task of the artist is to gaze at archetypal beauty and create material imitations of that perfect, invisible beauty.

This Platonic ideal had profound importance for late-fifteenth- and sixteenth-century architectural theory and practice. As Rackusin has shown, Luca Pacioli's recommendation of the seventy-two-faced polyhedron for architecture had little relation to real buildings or building practice; it was an embodiment of divine proportion and mathematical ideas.[77] Scamozzi recommended the use of the Orders on buildings of almost every type, whether sacred or secular, public or private, since "what is said well, it doesn't hurt to say twice."[78] He continued that the main task of man, as a reasoning animal, is to discover truth; once he finds it, he should follow it exclusively, disregarding tradition and authority.[79] The search for timeless rules began to produce treatises entitled *De Divina proportione*, *Dell'Idea dell'architettura universale*, and *La Città ideale*.[80] This Neoplatonic approach is in strong contrast with the purely phenomenal principles that govern the design of Pienza.[81]

Plato directly opposed the aesthetic principle of *varietas*: "Can we rightly speak of a beauty which is always passing away, and is first this and then that? ... Nor can we reasonably say that there is knowledge at all, if everything is in a state of transition and there is nothing abiding."[82]

This was accepted by sixteenth-century theorists like Francesco Giorgi: "Since we want to build a church, we must consider it necessary and right to follow this order [of numerical proportions], having as master and author the sublime architect God."[83] Of course, the notion that architecture should follow the divine proportions established by God was not a Renaissance invention. Von Simson, who showed its application to Gothic design, noted that this idea continued to be expressed in the sixteenth century.[84] It might be suggested, therefore, that the culture which is the subject of this book, that of the first three quarters of the fifteenth century, represents a temporary (and partial) rebellion against a Platonizing theory of architectural design that characterized the periods that preceded and followed it. Just as Scholasticism and Humanism were not sequential movements in fourteenth- and fifteenth-century Italian intellectual life but, as Kristeller has argued, rival alternatives, so the culture of philosophy constantly challenged the culture of rhetoric.[85]

Le Mollé rightly connected the ideal city views of the Berlin, Baltimore, and Urbino panels to the idealizing, Platonizing currents of the later Quattrocento.[86] It may be that the beginnings of this trend, in terms of urban design, can be traced to Chrysoloras's introduction of Plato's *Republic* to the early Humanists. Here Plato defines the ideal city:

> "I understand," he said, "you mean the city whose establishment we have described, the city whose home is in the ideal; for I think that it can be found nowhere on earth." "Well," said I, "perhaps there is a pattern of it laid up in heaven for him who wishes to contemplate it."[87]

The availability of this text has led many scholars to interpret Alberti's architectural thought in Platonic terms.[88] I do not deny that there is some influence from Plato, especially in *De Re aedificatoria*, which suggests both the persistence of earlier, medieval absorption of some Platonic ideas and Alberti's encounter with newly discovered works. However, my exploration of Alberti's thought, and that of the other early Humanists, in the first two parts of this book convinces me that theirs was a culture that looked to Aristotle, Cicero, and Augustine, more than to Plato, as authoritative figures. That early Humanist culture was one of discovery and change, in which quite opposed viewpoints coexisted, has been the thesis of the preceding chapters. Nonetheless, in the case of Pienza, the governing aesthetic principle not only is not Platonic, but is opposed to Plato's understanding of beauty.

Architectural theorists from about 1480 to 1600 wrote much, their views were logical and consistent, and they expressed them clearly with both words and images. By contrast, Alberti and other earlier Humanist writers expressed views that vary from work to work (or even within the same work), their terminology is often unclear, and their thought is eclectic. Part of the task of these chapters has been to try to look behind the veil cast over early Humanist culture by the imposition of later theoretical attitudes and to re-create important aspects of the culture that differed sharply from that which later developed. In this endeavor, the example of Pienza is of special importance, for it falls within one of the central topoi of Renaissance architectural theory, the ideal city. It is neither adequate nor accurate to interpret Pienza as a fusion of medieval and Renaissance culture, since its governing principle, *varietas*, is quite different from the simple diversity thought to characterize medieval urban organization. Like high and late Renaissance planning, Pienza is ordered by a single principle that harmonizes all the component parts. But it is not an ideal city, at least not in the Platonic sense.

PART III

Eloquence

SEVEN

Byzantine Learning and Renaissance Eloquence

MOST SCHOLARS ACCEPT that the recovery of Greek letters in the Italian Renaissance began with Manuel Chrysoloras's instruction in Florence in the last years of the fourteenth century (Figure 23).[1] Unlike earlier sporadic and often sterile encounters with Greek language and literature in the Western Middle Ages, Chrysoloras's exchange with his students stimulated the restoration of Greek studies in schools of higher learning, thereby fostering the recuperation and translation of the Greek classics as part of the Classical heritage.[2] After Chrysoloras, this movement gathered impetus and was largely accomplished in the course of the fifteenth century.

That is how we see it, but that is not how Chrysoloras's contribution was evaluated and defined by his closest pupils. At the Byzantine scholar's death in 1415, Guarino Veronese quite astonishingly declared that Chrysoloras had so greatly restored the splendor and dignity of the Latin language that the study of the arts and sciences could be cultivated not only in Italy but in the whole world.[3] Significantly, Guarino's assessment of Chrysoloras's role was modeled on Cicero's evaluation of the Greek contribution to Latin eloquence in *De Oratore*, (I.4), a source he acknowledged in a letter of 1452.[4] We will see that the Byzantine scholar possessed the tools with which early Humanists could practice the kind of eloquence that Cicero held up as an ideal. Of course, Guarino, who had an especially close personal relationship with Chrysoloras, might have exaggerated his importance in this letter of consolation to the scholar's nephew. Yet he substantially repeated this claim almost fifty years later.[5]

The association of Chrysoloras with the revival not only of Latin studies

Figure 23. Manuel Chrysoloras. The drawing is by an anonymous Florentine. (Louvre, no. 9849 bis. Courtesy of Documentation photographique de la Réunion des musées nationaux de France)

but with all Renaissance learning and eloquence was made by others in less emotional circumstances. The beginnings of the entire Humanist movement are attributed to the teaching of John of Ravenna and Chrysoloras in Biondo Flavio's *Italia illustrata* (1448–1453).[6] Poggio Bracciolini and Leonardo Bruni associated the revival of Latin eloquence with Chrysoloras's teaching.[7] If it had not been for Chrysoloras, wrote Vespasiano da Bisticci, Bruni, Traversari, and Guarino would never have acquired Latin eloquence.[8] Pius II claimed, in a letter to Sigismond of Austria, that Chrysoloras's nephew John (who came with him to Italy and remained for some years after his uncle's death) taught Bruni, Guarino, Poggio, Aurispa, Antonio Loschi, and others to speak with Ciceronian eloquence.[9] At the end of the fifteenth century (1490–1491), Paolo Cortesi essentially reiterated Guarino's claim:

> Indeed, after for a long time the study of the greatest arts lay sick and abandoned in the dirt, it was certainly Chrysoloras of Byzantium who brought across the sea that discipline to Italy, under whose guidance for the first time our men, ignorant of the entire training and art of eloquence, once they knew Greek letters, dedicated themselves with passion to its study.[10]

None of these or other authors claim that it was Chrysoloras's specific knowledge of arts and sciences that brought about a revival of learning, although we have many testimonies to his immense erudition.[11] They suggest, rather, that the critical and methodological tools with which he provided the Italians enabled them to pursue Latin learning. In particular, these were linguistic instruments that Cortesi summarizes with the single word "eloquence." Cortesi uses this term in its broad Ciceronian definition of "a good man speaking well" and implies by it the recovery of the Classical ideal of education, or *paideia*, meaning something acquired that elevates its possessor morally, intellectually, and spiritually.[12] That Chrysoloras was perceived as embodying this ideal is suggested by his students' evaluations: equal praise was accorded to his learning, manners, appearance, character, and piety; for them, these were all part of what made Chrysoloras a superior being. Guarino, in a letter to Vergerio of 1415, refered to the recently deceased Chrysoloras thus: "How much he had of liberality, constancy, faith, integrity, religion, modesty, holiness, greatness of soul, and knowledge of all the arts and of the greatest matters!"[13] Pier Candido Decembrio remembered that Chrysoloras's virtue and wisdom were such that he seemed to be an angel or even a deity rather than a man.[14] Particularly beautiful is Guarino's evocation: "How much cause for praise Manuel Chrysoloras sowed for himself on account of the integrity of his life; the sanctity of his character; the brilliance of his learning; his zeal for divine matters and vast experience of human affairs; and his prudence!"[15] That Chrysoloras's main contribution was perceived as assisting the Italians in their search for "eloquence" is explicitly stated by Guarino, lamenting the lack of a proper eulogy for him: "I think it unfair and a mark of ingratitude that he whose industry helped us not merely to speak but to speak with eloquence, should be immersed in silence."[16]

This search for eloquence arose from the Humanists' belief, inspired by Classical models, that the goals of education were the acquisition of good morals and the development of the whole person. It was especially closely related to Cicero's precepts. In *De Oratore* he urged the study of oratory "so that it may be in your power to become a glory to yourselves, a source of service to your friends, and profitable members of the Republic."[17] Chrysoloras's teaching was only an instrument for the realization of this aim. His culture and education constructed a bridge that enabled the Italians to pass over to that other, Ciceronian shore, which they had seen and desired; the Byzantine civilization that supported their crossing was only a path for them and was never a goal in itself. Yet his contribution was as fundamental as it was extensive. Thomson has shown how Chrysoloras's influence on the revision of educational curricula is already evident in Pier Paolo Vergerio's *De Ingenuis moribus et liberalibus studiis adolescentiae* (1400–1402), Guarino's translation of Plutarch's *De Liberis educandis* (1411), and Bruni's *De Studiis et litteris* (ca. 1425).[18] His teaching enabled Bruni to furnish Salutati with Saint Basil's arguments for the benefits of secular learning in his dispute with Giovanni da San Miniato in 1400.[19].

As discussed in chapter 2, dissatisfaction in Quattrocento Florence with the existing division of knowledge and the intuition that it no longer corresponded to the needs of the present led to a reassessment of the place of architecture as one of the mechanical arts. The Greek-language tradition, in which intellectual and manual activity could be seen as complementary aspects of the same branch of knowledge, was helpful in this reevaluation. Gille has sketched a more general picture of this decompartmentalization of learning and doing in the fifteenth century, showing close ties between scientists and artists in matters such as anatomy and perspective, and between scholars and artists in the development of archaeology.[20] My concern in this chapter is with a more fundamental level of fusion between disciplines, serving to connect the theoretical and the practical, but operating on the level of logic and language by applying to one branch of knowledge, or one activity, concepts, categories, and terms originally developed for a different discipline. And I am connecting this with the Late Byzantine approach to knowledge, particularly as we know it from Chrysoloras, in order to suggest that the Renaissance concept of eloquence, of which architectural description is a part, owes much to the Byzantine scholars.

To those aspects of the revival of eloquence that concern architectural description, Chrysoloras's most important contributions are his method of applying master terms expressive of general principles developed within one branch of knowledge to other areas of thought and his transmission of the Late Antique rhetorical tradition. By the latter, I do not mean to underestimate the importance of his role in the transmission of Greek Classical authors such as Plato, Demosthenes, Xenophon, and Aristotle, which has been well studied. However, an examination of his contribution to the Humanists' ability to speak, eloquently, about the built environment (whether ancient or contemporary) must above all focus on the transmission of Second Sophistic authors, of Hermogenes' rhetorical theory, and of the Byzantine training in literary genres such as the encomion, the ekphrasis, and the comparison. Chrysoloras's own *Comparison of Old and New Rome*, of 1411, the only *laus urbis* written by a Byzantine scholar in Italy and for an Italian (as well as Byzantine) audience, is a model of these elements; its quotations from Aelius Aristides, Libanius, and John Chrysostom, rather than from Plato and Demosthenes, underline the importance of Late Antique models for this genre.[21] While the *Comparison* had a certain amount of influence on Quattrocento descriptions of architecture and urban complexes, Chrysoloras stimulated the first Humanist efforts in this genre during, the immediately following, the period of his teaching in Florence (1397–

1400). We will see, in chapter 9, that Vergerio's innovative description of Rome (ca. 1398) was very likely an exercise undertaken at Chrysoloras' suggestion; Bruni's *Laudatio florentinae urbis* (1403–1404) was inspired by his reading of Aelius Aristides with the Byzantine teacher.[22] Thus Chrysoloras's contribution to the Humanists' new capacity to evaluate the Roman past and celebrate the present civic reality through architectural description occurred in the context of his teaching and conversation before that of his example. I believe that other of the Byzantine émigré scholars made analogous contributions and that the extraordinary flourishing of the *laus urbis* (and other kinds of writing about architecture, such as archaeological studies and ekphrases) in the Quattrocento owes much to their influence. It is insufficiently appreciated that Byzantine intellectuals of the last period were especially prolific in these areas. Bessarion's as yet untranslated *Encomion of Trebizond* (1436) is the most impressive, but we also have formal descriptions of Trebizond and Corinth by John Eugenikos (after 1444), a description of Constantinople by Joseph Bryennios (perhaps 1400), and a description of Hiera Pytna by Michael Apostolis (1469), as well as a wealth of accounts of the capture of Constantinople in 1453.[23] I do not think it coincidental that the most innovative Western descriptions (the subject of chapter 9) were written by Italians who were close to Chrysoloras and knew Greek (Bruni, Vergerio, Poggio Bracciolini), or who knew Greek and were in contact with the Byzantine intellectuals at the Ferrara/Florence Council or the Curia in Rome (Alberti, Giannozzo Manetti). Thus my discussion of Chrysoloras's contribution to Renaissance eloquence and analysis of his *Comparison* (the subjects of this and the following chapters) offers interpretations which could be applied to other Byzantine intellectuals and their works, insofar as they proceed from the same education and training. My translation of the *Comparison* in the Appendix makes this text available in English for the first time in its entirety, and is part of a larger project for the recuperation of these neglected Byzantine texts.[24]

The other aspect of Chrysoloras's contribution to Renaissance eloquence, the transfer of master terms and concepts, resulted from that decompartmentalization of knowledge characteristic of Byzantine learning in the last century of empire. This method placed an abundance of new terms, categories, and concepts in the hands of Italian Humanists, which they were free to apply to the subjects that interested them. Chrysoloras's approach emphasized the relations, rather than the distinctions, between branches of human learning, fostering the formation of the cultivated generalist, or *uomo universale* as he came to be called, rather than the narrow specialist or professional. The implications of this intellectual formation for the development of architectural description in the Quattrocento are decisively important, suggesting that architecture might become a part of general culture rather than remain the province of technicians and specialists. By making it possible to discuss architecture within the contexts of ethical, historical, and literary discourse, the subject was brought into the mainstream of topics that concerned Humanist writers.

The need to evaluate Chrysoloras's (and the other Byzantine intellectuals') contribution in terms of its influence on the Humanists' method and approach rather than the study of the Greek classics has been recognized by others. Kristeller, noting that the interests of the Italian Humanists seemed similar to those of the Byzantines, wondered to what extent the latter had influenced Latin studies. He also suspected that the new ideas diffused by Quattrocento translations of the Greek Fathers played a role in the theological discussions of the Reformation.[25] He outlined, in a general fashion, some of the elements I discuss in concrete detail in this chapter: the importance of the Byzantines' choice of authors, of their method of interpreting the texts, and of their tendency to decompartmentalization between disciplines.[26] Geanakoplos also intuited that the Italian Renaissance was influenced not only by the contents of Palaiologan learning, but also by its methods of teaching, curricula of study, and attitudes toward the corpus of disciplines.[27] His study of Chalcondyles substantiated some of these insights, for he was able to show that the Byzantine scholar himself believed that the importance of Greek for Latin studies consisted in linguistic and methodological advances, such as a better understanding of the fine points of Latin grammar, the acquisition of a more powerful and elegant style, and the development of a more meaningful content and ideas.[28] This evaluation seems very close to Guarino's definition of Chrysoloras's contribution. Geanakoplos concluded that the Greek revival of learning, carried out primarily by the Byzantine émigrés, "probably did more than any other single factor not to begin, but once it had begun on the basis of *Latin* literature, to widen the intellectual perspective of the Italian Renaissance."[29] Fryde, searching for evidence of Byzantine influence within the literary products of the Humanists, concluded that the results of exposure to Greek historians such as Herodotus, Thucydides, and Polybius (whose works reached Italy in the first quarter of the Quattrocento), together with the teaching of Chrysoloras, were evident in the tendency to more definite and clearly delimited subject matter, a more coherent arrangement of materials, a sharper sense of what constitutes evidence, and a sharper sense of what constitutes proof.[30]

Chrysoloras's Education and Second Sophistic Rhetoric

Perhaps because so little is known about education in late-fourteenth- and early-fifteenth-century Constantinople, there has hardly been an attempt to reconstruct Chrysoloras's education and intellectual development.[31] Yet this subject is crucial to an understanding of his approach to learning, his acquired mental habits, and his special intellectual strengths. The following, then, seeks to illuminate this aspect of his figure, as far as scanty evidence permits.

In Chrysoloras's time, two educational systems were available in the metropolis: the university, refounded by Michael VIII in the buildings surrounding Hagia Sophia; and private schools such as that of Nicephoros Gregoras at the Chora.[32] Manuel II brought about a major reorganization of the university in 1391 (the first year of his reign), moving the university (Catholicon Mouseion) to the monastery of St. John in Petrion and placing the Patriarchal Academy at St. John in Studion. Chrysoloras, who was born around 1350, too early to benefit from this new arrangement, studied with Demetrius Cydones (ca. 1323–1397/8) at the time when Cydones had embraced Western Catholicism, after 1360.[33] Chrysoloras's education probably followed the usual steps of the *enkyklios paideia*. This program began with grammar and composition; progressed to the study of Greek orators and to the manuals of rhetoric of Hermogenes and Aphthonius; broadened to include metaphysical philosophy, mathematics, and pure science; and culminated in the study of theology.[34] Yet Chrysoloras's education was unusual in Byzantine terms, since his choice of a literally un-Orthodox teacher, who was an accomplished Latinist and translator of Augustine and Thomas Aquinas, also acquainted him with Latin literature and Western theology. This familiarity with Latin culture, and especially the ability to speak without an interpreter, helps to explain why he was later chosen for diplomatic missions in the West. Moreover, his own experience in learning a foreign language might have guided him in the composition of a Greek grammar, the *Erotemata*, for his Florentine students. Probably, his interest in the problem of Latin diphthongs, a subject with which he would astonish Latin scholars in Italy, occurred while he studied the language in Constantinople if, as Gombrich suggested, it arose out of difficulties in transliterating Greek names.[35] Finally, we may suppose that Western theology was presented to Chrysoloras in a favorable light by Cydones and that this may help to account for his own request to take holy orders in the Catholic Church in 1405.[36]

Although Byzantines continued to receive the Hellenistic education in regard to the subjects covered, they studied grammar and rhetoric with textbooks written between the second and fourth centuries A.D..[37] This is of special importance for architectural description, which is above all a product of Second Sophistic rhetoric.[38] Late Byzantine students studied Hermogenes on categories of literary style and principles of literary criticism; Aphthonius's *Progymnasmata* was their model for rhetorical exercises. Although Aphthonius's work follows Hermogenes' *Progymnasmata* closely, it differs in providing models for all fourteen literary types, one of which, the ekphrasis, is illustrated by an extended description of the Temple of Serapis and Acropolis at Alexandria.[39] Thus every

Byzantine schoolboy studied and practiced the art of architectural description. In the West, by contrast, although Hermogenes' *Progymnasmata* was known through Priscian's sixth-century translation, it did little to develop skills in description, encomion, and comparison—the three literary types that govern architectural description.[40] In part this was due to the lack of examples in Hermogenes' text; but it seems probable that without the additional help of Hermogenes' analyses of the seven categories of style (which, with their subdivisions, produce twenty stylistic colors), the student would have been at a loss to create his own literary compositions.[41] In other words, the West knew the definition of ekphrasis without seeing how to apply it and without knowing the general principles governing panegyrical style.

Byzantine intellectuals not only learned how to compose works in each of the literary types, but practiced them as independent genres.[42] Their education prepared them to write epigrams, encomia, character studies, and descriptions—all forms especially developed in the Second Sophistic—but not epics or dramas of the sort composed by Classical authors.[43] Chrysoloras's *Comparison of Old and New Rome*, a comparison, encomion, and ekphrasis in epistolary form, is precisely the kind of composition for which his education had trained him. Its literary sources are predominantly in Second Sophistic authors, and its structure follows Second Sophistic principles. It has not been sufficiently appreciated that this constituted an absolute novelty for the Italian reader, who thought of Byzantine culture as preserving the Greek classics. We know, for example, that Leonardo Bruni went to study with Chrysoloras in order to study Homer, Plato, and Demosthenes; yet by the time he composed his *Laudatio* (about 1403), he was using Aelius Aristides as a major source.[44] Coluccio Salutati, one of the men responsible for bringing Chrysoloras to Florence, urged Jacopo Angeli da Scarperia to bring back copies of Homer, Plato, and Plutarch from Constantinople.[45] Vergerio, who said he wanted to learn Greek in order to read works he knew only from translations and to discover why these writers were so praised by Classical Latin authors, clearly had in mind the Greek classics.[46] Filelfo was quite cross about the teaching at the university in Constantinople, "for the Aeolian tongue, which both Homer and Callimachus most surely followed in their works, is quite unknown there."[47] Having studied in the Metropolis, he was the only early Humanist who could distinguish between the "Ancient" and "Modern" Greek authors, deploring the Byzantines' preference for the latter.[48] This preference for Greek classics persisted at Vittorino's school in Mantua where the principal authors studied were Homer, Demosthenes, Xenophon, Arrian, Herodotus, Thucydides, Plutarch, Isocrates, Plato, Aristophanes, Euripides, Sophocles, Aeschylus, Hesiod, and Pindar. The only Late Antique author read seems to have been John Chrysostom.[49]

A disparity existed between the Italians' notion of the corpus of Greek literature and what the Byzantines actually read and taught; Chrysoloras's major literary contribution not only was of a type unfamiliar to Italian readers, but also privileged authors of whom they had perhaps never even heard. Although Late Antique and Byzantine authors seem to have predominated over Classical ones in Byzantine libraries of the ninth and tenth centuries, my impression of the libraries of the fifteenth-century Byzantine émigrés is that they contained at least as many works by Classical as by later authors, suggesting a certain "return to the Classics" in the last period.[50] Nonetheless, Chrysoloras used works by Lucian (an author much admired also by Photius for his decorum and expressivity of vocabulary) for his introductory course rather than a Classical author.[51] That Theodore Metochites considered Aelius Aristides to be the equal of Demosthenes in oratory also suggests the continuing appreciation for Second Sophistic authors in the Palaiologan period.[52] Above all, Chrysoloras's notions of literary categories and stylistic criteria were founded on ideas totally foreign to Western rhetoric. Not until the 1420s was Hermogenes' *On Types of Style* introduced into Italy through George of Trebizond's Latin synopsis, his slightly later treatise on rhetoric, and copies brought from Constantinople by Filelfo; Aphthonius was not translated until the second half of the century.[53]

Although there is no equivalent in Greek literature for Vitruvius's treatise on architecture, there is a rich and continuous tradition of architectural description.[54] This was primarily developed as a special type of ekphrasis by Greek rather than Latin authors and within the Second Sophistic.[55] Among the first extensive descriptions are those of Aelius Aristides and Lucian; their practice was codified in Menander's two treatises on rhetoric in the late third century. These are not technical works but are governed by the rules of ekphrasis, which, by definition, seeks to describe so vividly that the reader (or listener) seems to see the object described.[56] This is not to claim that Byzantine descriptions are accurate accounts of the subject, but only that they purport to share a visual experience with the addressee. They are (or claim to be) eyewitness accounts, conveying a sense of immediate and personal experience. This impression is fostered by the use of structural devices, such as the description of objects in the order in which they would be physically encountered by the spectator. And it is enhanced by the interjection of personal emotions and judgments, as though the writer were reliving his moment of aesthetic apprehension. It is quite true that these effects can be obtained through the use of topoi, and that the author's wonder at some spectacle may be artificial and unconvincing. Nonetheless, the criteria of good description underwent relatively little change in Byzantium, and the challenge for the writer of an ekphrasis was always to strike a balance between personal observation and the traditions of the genre. A good description had to be personal, precise, and specific to the subject while relating the subject and the experience of it to authoritative values and literary models. My analysis of Chrysoloras's *Comparison* will show how he addressed these potentially conflicting requirements.

Although Latin authors shared in this development in the first centuries of our era, when culture was bilingual, their participation waned as knowledge of Greek declined in the West after the fourth century. The architectural description, which had, in any case, never enjoyed great popularity with Latin authors, fell into disuse and Greek models became increasingly inaccessible to the West. In chapter 9, I will discuss the revival of the genre in Latin literature, facilitated by the mediation of Byzantine tradition. My purpose in this and the following chapter is to introduce the reader to the genre as it was practiced in late Byzantium by Manuel Chrysoloras.

Decompartmentalization

One of the most striking differences between education in East and West about 1400 is the conception of how knowledge, taken as the whole body of existing disciplines, should be approached.[57] The West had increasingly compartmentalized knowledge since the twelfth century, defining clear boundaries between the various arts and sciences. Individual scholars were expected to specialize in only one branch of learning; further division occurred within each art. In rhetoric, for example, the *artes poeticae*, *dictaminis*, and *praedicandi* each had a different public, and the man who practiced one was unlikely to practice the others.[58] By contrast, such evidence as we possess indicates a much greater union of the disciplines in Byzantium, where learning was thought of as essentially two kinds, Inner (spiritual) and Outer (secular), understood to be complementary.[59] It seems clear that Chrysoloras was equally well trained in secular and sacred subjects. If, at the time of his death, he was considered to be a possible candidate for the papacy, as Vergerio's epitaph for him claims, his contemporaries must have considered him to be at least adequately versed in theology.[60] His students testify to his accomplishments in the secular disciplines.[61] His mastery of both the sacred and secular disciplines has to be understood in the light of Gregory Palamas's challenge to the usefulness of astronomy, foreign languages, rhetoric, history, physics, logic, and mathematics and Cydones' opposition to this position.[62] The double training is evident in another of Cydones' pupils, Manuel II (1350–1425), who wrote a number of theological works as well as works on history, politics, dreams, and the art of rhetoric. While the separation of the university and the Patriarchal Academy might seem to indicate a separation between the two kinds of learning, other evidence does not confirm this. For example, the priest Joseph Bryennios (ca.

1350–ca. 1431) seems to have taught at both institutions; John Chortasmenus (ca. 1370–ca. 1436/7) taught at the university before becoming metropolitan of Selymbria; the patriarch of Constantinople, Gennadius II Scholarius (ca. 1405–1472), both was educated at the university and taught there, as was true for the theologian Michael Apostolis (ca. 1420–1480).[63] The number of religious who were educated or taught at the university suggests that the Outer Learning (the trivium and quadrivium) and the Inner Learning (theology) were perceived as complementary in Late Byzantine education despite the objections of the Hesychasts. Chrysoloras must have been keenly aware of the question of the relation between the two kinds of knowledge and especially sensitive, therefore, to the Italian Humanists' discussions about the active and contemplative lives.

Not only were the sacred and secular studied together in Byzantium, but there were no strict barriers between various disciplines of secular learning. Thus Joseph the Philosopher (1280–1330) prepared an encyclopedia intended to correlate the various branches of Classical learning, showing how their interconnection and understanding could help in the study of theology.[64] This lack of specialization is evident from the Byzantines' writings: Nicephoras Gregoras wrote on theology, philosophy, astronomy, history, rhetoric, and grammar; Plethon studied and wrote on all these subjects plus astrology, geography, and music.[65] Chrysoloras, who wrote little, nevertheless taught philosophy, history, geography, grammar, and rhetoric. Decompartmentalization is evident even in specialized courses such as those on Aristotle given by Argyropoulus in the Florentine Studio. Covering twelve books over a two-year period, the Byzantine scholar (who had taught at Constantinople) began with Aristotle's dialectics, then covered natural science, and concluded with metaphysics. As Field has shown, what was new in this approach was the consideration of Aristotelian philosophy as a unified whole—understanding how the several parts of the corpus fitted together.[66] The sequence illustrates the view of Byzantine method I have been sketching in that the student first learns master principles and terms, which he will apply later to various subject matters. The hierarchical sequence reflects not only a Platonic influence, as has been suggested, but above all the relation between Inner and Outer Learning as conceived in Late Byzantine education.[67] The habits of mind embedded in Argyropoulus's syllabus are equally apparent in Chrysoloras's *Comparison of Old and New Rome*, where the material remains of the two cities are analyzed first in terms of outer knowledge (history, politics, culture) and then in terms of their spiritual message.

One of the advantages of decompartmentalization was that it enabled an educated person to speak meaningfully about a subject for which he did not know the specialized terminology, or for which no technical terminology existed. This ability was crucially important for the development of architectural description, since a Humanist was unlikely to know the technical jargon used by masters and masons; even if he knew it, it could scarcely have been used in literary composition. The art of describing architecture was so little practiced in Latin (whether Classical or medieval) that the stock of available categories and terms was meager. But even had the means been ample, the Humanists had little interest in architecture for its own sake. Only if it could be discussed within topics of general concern, such as moral virtue (chapters 1, 3, and 5) or cultural progress (chapters 2, 3, and 4), would it receive attention, at least during the first three quarters of the Quattrocento. And it could be inserted into these contexts only if it literally shared a language with intellectual concepts of a general nature. Thus the decompartmentalization of knowledge was closely tied to the extension of possible subjects about which an educated person might express opinions.

Once again, the motivation for the Humanists' desire to address a wide range of topics came from Cicero, who gave as his very definition of an orator he "who, whatever the topic that crops up to be unfolded in discourse, will speak thereon with knowledge, method, charm and retentive memory."[68] But this ideal was difficult to reach, given Western compartmentalization and professionalism. Chrysoloras had been trained in Second Sophistic oratorical theory, which prized the speaker's ability to speak extempore on any given subject. In order to do this, one must know a little about many things and have a good stock of general principles, arguments, and terms with which to construct a speech. The *Comparison* offers several examples of Chrysoloras's method of transferring general terms to new contexts, the most important of which, for this study, is the application of rhetorical categories of stylistic excellence to the description of Hagia Sophia, already touched on in chapter 5. It is my thesis that the development of Quattrocento architectural description was stimulated by the Humanists' absorption of Byzantine assumptions about the relationships between branches of knowledge as well as of their rhetorical techniques. An analogous process of decompartmentalization, in regard to painting, is in Bruni's *De Interpretatione recta*, where he compares the artist making a copy of a painting with the translator of a text. Baxandall observed that this parallel is drawn by using words that are applicable to both arts, such as *figura*, *status*, *color*, and *forma*, although their meanings differ in painting and literature.[69] By applying general terms to diverse contexts, Bruni was able to perceive relationships between literary and artistic processes, and to discuss artistic composition with a wider range of criteria than was usually applied. This technique, I suggest, he learned from Chrysoloras, whose decompartmentalization of terminology may be exemplified by the fact that one of his letters provides the unique instance in the Quattrocento in which the Aristotelian term for perception, *fantasia*, is applied to art.[70]

Evidently, decompartmentalization was also closely tied to the development of intellectual conversation, as opposed to the more formal disputation in which only specialists can participate. Conversation, derived from Latin terms meaning "to associate with someone" and "to turn around," is the mark of man in society; it aims not at proving the truth, but at sharing and persuading. Cicero defined conversation as that which is distinctively human.[71] A good deal of evidence suggests that learned conversation was of increasing importance in late-fourteenth- and early-fifteenth-century Florence.[72] The circles around Luigi Marsigli at Santo Spirito, Cino Rinuccini at Santa Maria in Campo, and, above all, Coluccio Salutati ("doctorum virorum quasi communis parens," according to Poggio's epitaph for him) were main centers of cultural and political discourse. Although this subject is at once too vast and too little studied to be treated here, it is worth suggesting that the practice of meeting at a private home to exchange ideas on intellectual subjects not only became increasingly popular in the early Renaissance, but could be seen as a means of sharpening the intellect, forming judgments, and developing criteria of value. Chrysoloras encouraged informal discussion; indeed, he refused to come to Florence unless his contract stipulated that he had the right to teach students in his own home.[73] The importance of conversation as the very mark of civic life had been urged by one of Chrysoloras's favorite authors, Libanios: "It seems to me that one of the most pleasing things in cities, and I might add one of the most useful, is meeting and mixings with other people. That is indeed a city, where there is much of this."[74] In 1399, Chrysoloras escaped the plague in Florence by retiring to Palla Strozzi's villa in the Casentino, where they discussed philosophy, presumably with other guests as well.[75] Vacalopoulos has shown that Argyropoulus's teaching method in the mid-Quattrocento also combined classroom lecture and discussion with informal discussion at his home.[76] Cortesi reports that learned discussion took place all day in Bessarion's home and that, even when he was old, he argued on both sides of the question.[77] There is some evidence that conversation was a traditional educational method in Byzantium. I think especially of Photius's nostalgic recollection of the years before he became patriarch:

> When I was still at home I was immersed in the most delightful of all pleasures, namely the zeal of those who were learning, the eagerness of those who asked questions, and the enthusiasm of those who answered. That is how the faculty of judgment is formed and strengthened among those whose intelligence is sharpened by scholarly pursuits.[78]

Thus there would seem to be a tenuous thread of continuity from Middle Byzantine to Palaiologan educational method and notions about social intercourse. Be that as it may, it seems clear that Byzantine intellectuals contributed to the development of conversation and to techniques of argument in the Quattrocento.

The utility of an education that sharpened the faculty of judgment was evident to the early Humanists whose concerns were, above all, civic, political, and ethical. That the Signoria of Florence perceived this connection between rhetoric and the *res publica* is suggested by the circumstances of Chrysoloras's invitation to teach at the Studio. In the Trecento, the Studio had been especially strong in canon and civil law, although medicine and the arts slowly gained ground. By 1389, there were nine professors of law, seven of medicine, and eight to teach moral and natural philosophy, grammar, and astrology.[79] Between 1396 and 1398, the Signoria interfered with the Studio's autonomy by making four appointments: Chrysoloras was asked to teach Greek for five years; Domenico d'Arezzo and Giovanni Malpighini of Ravenna were appointed in rhetoric; and Giovanni da Faenza was appointed to the faculty of medicine.[80] As a result, the faculty of arts had the largest number of professors, and the art of rhetoric was especially emphasized. As early as 1386, Rinuccini had his students pronounce orations for and against rhetoric; they argued either that it was necessary for civic life or that it produced civil disorder.[81] The debate in the early Quattrocento about who was greater, Caesar or Alexander, and the later debate between Poggio and Guarino about the relative merit of Scipio and Caesar were not sterile exchanges of erudition, but exercises in comparison aimed at defining criteria of excellence.[82] One cannot help but wonder whether training in this kind of argumentation influenced the development of the Renaissance dialogue, one type of which has recently been defined as "the conversation."[83] Of course, the dialogue has roots in the medieval literature of debate, and immediate precedents in Petrach's writings; its popularization in the Renaissance may be related to renewed interest in the works of Plato and Cicero, which furnished literary models for the Humanists. But the Byzantine teachers might have helped them to develop the intellectual skills necessary for the composition of an original dialogue. We know that Chrysoloras stimulated a debate on the merits of Homer versus Virgil and contributed to another on New versus Old Rome in his *Comparison*.[84] That his subjects debated cultural rather than political merit was important for the expansion of this genre of discussion. He had been trained to this kind of exercise, since the ability to argue on both sides of the question (*anaskeue* and *kataskeue*) was so highly developed in Byzantine rhetorical training that it became a mental habit.[85]

Leonardo Bruni's *Ad Petrum Paulum Histrum* (a dialogue written by, and for, a student of Chrysoloras) provides precious evidence for the Byzantine scholar's contribution to early Humanist conversational technique. In it, Niccolò Niccoli (another student) is made to argue for and against the achievements of modern Florentine culture.[86] This structure, which has long puzzled scholars, may be at least partly explained as an exercise in argument on both sides of the question of the sort advocated by Chrysoloras and demonstrated in his *Comparison*.[87] The dialogue is composed of two conversations on successive days in 1401 at the homes of Coluccio Salutati and Roberto de'Rossi. In it, Coluccio Salutati criticizes his friends for not engaging enough in conversation or exchanging views on what they had thought about privately or had learned.[88] Niccolò Niccoli replies that Chrysoloras had always exhorted his students to converse, but in a natural and simple way and without spelling out its advantages.[89] Thus he reproaches Salutati for presenting conversation as though it were a formal exercise, rather than spontaneous. Niccoli continues: in any case, they cannot converse because, having no real learning because of the dearth of books and teachers, they are not prepared to speak on any topic.[90]

In chapter 2, I discussed Alberti's solution to the problem of how to cope with the lack of books and teachers; here I want to discuss another aspect of this quintessentially early Renaissance problem. Niccoli takes the position that only specialists should speak; those with partial or flawed knowledge should keep silent. This may be a criticism of his friends' approach to education. Poggio tells us that those in the circle of Salutati were all autodidacts, who learned by just reading and writing: "legendo non audiendo."[91] Their freedom from the constraints of education at the Studio corresponds to their enthusiasm for informal discussion at private homes. Niccoli's attitude aroused hostility among his fellow Humanists, who mocked him for his pedantry.[92] What they objected to was that his concentration on specialized questions blinded him to those larger issues that were troubling intellectuals. Guarino, ridiculing Niccoli's study of manuscripts, exclaimed, "What a vacuous way to spend so many years if the final fruit is a discussion of the shape of letters, the color of paper and the varieties of ink."[93] Cino Rinuccini denounced those who care for "how many diphthongs the Ancients had and why today only two are in use," while "the meaning, the distinction, the significance of words...they make no effort to learn."[94] The Humanists attacked Niccoli's specialization as concerned with form rather than content and, ultimately, as being intellectually and socially irresponsible. By contrast, Chrysoloras's emphasis on conversation encouraged a manner of expression in which the divisions between branches of learning are obscured and the ability to apply general principles, categories, and terms to a broad spectrum of subjects is challenged. The Byzantine scholar was not responsible for the Humanists' desire to converse or to be able to express themselves on a variety of topics; rather, he furnished them with the techniques that enabled them to do these things better.

Real eloquence, to return once more to that concept, entails the ability to speak clearly and persuasively about topics of importance to the community; the eloquent man is a useful member of society. Cicero had emphasized this in *De Oratore*.⁹⁵ But this was also an accepted truth to Second Sophistic orators such as Libanios, who wrote about Antioch that

> among us the power of eloquence preserves the freedom of the senate in clear-cut fashion, and compels the administrators to appear in their tasks as what they are supposed to be, inciting sensible men to the discovery of what is the best, and resisting the lawlessness of insolent men through the compelling power of wisdom.⁹⁶

The Renaissance recovery of "eloquence" was sought in response to a perceived need, in the present, to examine and understand, but also to reform and perfect human society. The importance of eloquence for civic life was acknowledged by Pius II, Valla, Matteo Palmieri, and Francesco Patrizi, among others.⁹⁷ Without the availability of terms and categories appropriate to the reality that the Humanists experienced, it would not be possible for them to define themselves, their society, and their position in history. The connection between eloquence and civic affairs has up until now been discussed exclusively in terms of Latin models. While I agree with their primary importance, the contribution of the Greek-language tradition should also be considered.⁹⁸ For, as we saw, Chrysoloras's teaching was believed by his contemporaries to have assisted this effort.

It is often assumed that Byzantine literary compositions, especially ekphrases and encomia, are marked by a strange and elaborate formalism appreciated by only a small circle of refined intellectuals and devoid of significant content.⁹⁹ If this is so, then my argument that Byzantine educational method helped prepare men for civic life and for intelligent conversation about important subjects is incorrect. But I believe that the evidence, at least for the late fourteenth and fifteenth centuries, supports the view that the Byzantines believed that their learning should serve their country. Vacalopoulos suggested that the Byzantines in the fifteenth century had a greater sense of nationalism than Westerners and that they shared a feeling that they worked not for themselves, but for the sake of their country.¹⁰⁰ Thus, for example, Michael Sugliardus of Argos wrote that he did not transcribe for reward, but for the sake of his country; Constantine Lascaris exclaimed: "There is no higher or sweeter good than freedom. Rejoice therefore that you have a free country and live in the midst of your own people." In a recently discovered letter from Chrysoloras to Manuel II, he draws the connection between the revival of scholarship and the safety of the nation. He concludes with a plea to save the fatherland through the establishment of schools, support of scholars, and revival of the study of literary treasures, both Christian and pagan.¹⁰¹ After the fall of Constantinople, Bessarion took up the cause of national identity, writing to Apostolis that he wanted to collect manuscripts so that if the Greeks should see better times they would still have their literature, on the basis of which they could create new works; Greeks should not remain without a voice (again the concept of eloquence) like slaves and barbarians.[102] His translation of Demosthenes' *First Olynthic Oration* was intended to encourage a crusade against the Turks.[103]

Byzantine scholars on diplomatic missions to the West, like Cydones, understood the necessity of breaking down barriers of cultural difference that alienated the two capital cities in order to obtain military and financial aid, and they turned their eloquence to this task. Chrysoloras's teacher, Cydones, provides an example of how this could be done: in the following passage from his *Oration Requesting Aid from the Latins*, he used his rhetorical skills and knowledge of history to depict the Byzantines and Italians as fellow Romans: "What allies could be more appropriate for the Romans than the men of Rome? Or who more trustworthy than those who have the same native land? For their city was the mother-city of our city."[104] The opposite line—that both the Byzantines and the Italians were fellow Greeks—could also be taken. Johannes Canabutzes explained that Dionysius of Halicarnassus had written his *Roman Antiquities* in Greek so that the Greeks would know that the Romans were not barbarians but came of the same Hellenic culture.[105] Awareness of this diplomatic current sheds new light on passages, such as the following, in which the similarity—indeed, near identity—between Rome and Constantinople is charmingly expressed. The letter is from Chrysoloras to his nephew John:

> Roving every day through Rome and looking now at this part, now at that, I think that I am in my own native city and I sometimes forget that I am in fact so far from it, especially when I go up onto a height (you have heard of the seven hills of Rome from which by imitation we are accustomed to call our city the City of the Seven Hills), and I gaze out over the whole city, I think that I am not in Italy and Latium but in my own native city and I seek for our house in the part of the city where it is situated and its hanging garden and any other familiar sign of it and our cypresses.[106]

This diplomatic line was pursued as actively in works that seem to be mere ornamental displays of rhetoric as in negotiation, and they were very effective. Pier Paolo Vergerio's epitaph for Chrysoloras described him as an illustrious Roman knight, descended from those Romans who migrated to Constantinople: "eques Constantinopolitanus ex vetusto genere Romanorum qui cum Constantino imperatore migrarunt."[107] That the Byzantine scholar, whose ostensible reason for being in the West was to teach Greek language and letters, persuaded his students to view him as a Roman suggests that Byzantine eloquence was indeed a powerful instrument, serving the needs of an actual political and cultural situation. The early Humanists were not mistaken in their belief that Chrysoloras's methods could help them to apply theoretical knowledge to topics of general concern.

EIGHT

Manuel Chrysoloras's *Comparison of Old and New Rome*

CHRYSOLORAS WROTE the *Comparison of Old and New Rome* in Rome in 1411, after having embraced Catholicism and while he was part of the entourage of Pope John XXIII.[1] Although the work is addressed to the Byzantine emperor and therefore, ostensibly, intended for a Byzantine audience, a copy was sent to Guarino in Florence and, through him, disseminated among the Italian Humanists.[2] The choice of a comparison as a literary category was well suited to Chrysoloras's delicate position as emissary of the Byzantine government and loyal adherent of the papal court, since it permitted him to express admiration for the power of the Western Church in the same work in which he praised Constantinople as a seat of empire. It also enabled him to argue the common cultural heritage of the Latin and Byzantine peoples.

The *Comparison* takes the form of a letter and, therefore, purports to be a personal message rather than one intended for a wide audience. Its ostensible purpose is to inform the addressee about something he has not seen (Rome) and to share with him reflections about the relationship between Rome and Constantinople. But as Jenkins and others have noticed, Byzantine letters rarely have an exclusively personal message and it is common, in a long one, to insert an ekphrasis of a work of art, a building, or a piece of scenery.[3] Chrysoloras's letter is no exception. It was certainly intended for a wide audience of Italian and Byzantine intellectuals, and it has a diplomatic and apologetic intent, as Guarino recognized.[4] This kind of letter is rare in the West before the fifteenth century. As Vickers has shown, increased specialization in the *ars dictaminis* in the twelfth and thirteenth centuries produced utilitarian and formulaic letters that were not concerned with any wider cultural issues.[5] This practice began to change, especially with letters such as Petrarch's to Giovanni Colonna and Giovanni Dondi's on history; Salutati's letters, written when he was chancellor of Florence, were said to be worth more than a thousand soldiers.[6] Yet despite these precedents, the adoption of epistolary form for a formal rhetorical composition must have seemed novel to the Italians who were quick to appreciate its advantage in permitting a personal approach to the subject matter.

Chrysoloras was over seventy when he wrote the *Comparison*, having spent many years in the West. After the stay in Florence, he was at the court of Giangaleazzo Visconti until 1403. Following a brief return to Constantinople, we find him in Venice in 1404. Between 1407 and 1410, he traveled to Genoa, Paris, London, Salisbury, Barcelona, and Bologna. He was in Rome from 1411 to 1413, but visited Florence with John XXIII. In the last two years of his life, after the Roman sojourn, he would again visit Florence and then Constance, where he died and was buried.[7] Thus he had stayed at many of the courts of Europe and was thoroughly familiar with Western culture. Undoubtedly, Chrysoloras knew Petrarch's writings on Rome and was influenced by Pier Paolo Vergerio's description of the city (discussed in chapter 9); he may also have known the *Mirabilia* literature. His former student Leonardo Bruni was at the papal court when Chrysoloras wrote; we may suppose that he showed his teacher his own effort at urban description, the *Laudatio florentinae urbis*, and received a copy of the *Comparison*.[8] Another devoted friend, Poggio Bracciolini, was also in Rome; we will see that his later description of the city was influenced by Chrysoloras's work. A secretary to the Curia, Francesco da Fiano, expert guide to the Roman antiquities, may have been Chrysoloras's *cicerone*; Cencio de'Rustici, who studied Greek in Rome with Chrysoloras and later urged Fiano to defend the Roman monuments, may have accompanied them on their visits.[9] The close relationship between Poggio, Cencio, and Chrysoloras in these last years is indicated by the fact that his two former students stood vigil over Chrysoloras's body in Constance, in 1415.[10] After Chrysoloras's death, Cencio advised Bartolomeo Aragazzi, to whom he was teaching Greek, to copy the *Comparison*. Del Bravo has suggested that Chrysoloras may have conversed with Donatello and Brunelleschi, who were in Rome at the time he composed the work.[11] Nonetheless, I will argue that despite these Western influences, the most salient features of the description, and those most important for Western readers, derive from Chrysoloras's Late Classical literary models and from the perpetuation of their assumptions about the nature and interpretation of evidence in Byzantine culture.

Comparison between Chrysoloras's letter and the exactly contemporary anonymous account of Rome known as the *Tractatus de Rebus Antiquis* (or the *Anonimo Magliabecchiano*) shows the former to be the more sophisticated in approaching the ruins of Antiquity.[12] The editors of the *Tractatus* commented that its structure remains characteristically medieval and firmly rooted in the *Miribilia* literature, with "neppure un indizio della nuova erudizione umanistica."[13] The text, concerned mainly with topography and etymology, lists the Roman monuments without being able to interpret their significance for either Antiquity or the present. But it is more complete than Chrysoloras's account and, despite its shortcomings, aims at scientific exactness. The Western desire for precision and completeness is also evident in the first real corpus of Roman inscriptions, gathered by Nicola Signorili in 1409.[14] These examples show not that the Italians were more enthusiastic about the Classical past than the Byzantines, but that they approached it in an entirely different manner. Topography and etymology are all but absent from Chrysoloras's account of Rome; he describes only those monuments that illustrate some general point; and he conflates and summarizes the inscriptions he refers to. By Italian standards his work falls short. But in terms of the ability to communicate a visual experience and interpret it meaningfully, the texts surveyed in the next chapter are evidence that Chrysoloras's work would have no equal in the West for several decades.

The *Comparison* is not only a description of two cities, but also an encomion. Hunger has pointed out that in the case of cities, these two genres are almost always merged.[15] But an encomion, which requires the writer to claim that his subject is superior to all others of its kind, is intrinsically comparative. Chrysoloras had learned from Hermogenes' *Progymnasmata* that "the greatest opportunity in encomia is through comparisons. Comparisons must be employed everywhere," he says, "to distinguish between two persons of equal stature, two persons of unequal stature but both praiseworthy, and to compare total opposites of which one is praiseworthy and the other blameworthy."[16] Chrysoloras constructed his encomiastic comparison by arguing first the superiority of Rome and then that of Constantinople. He adduces the relevant arguments for the status of each city, and although he says he prefers Constantinople, he leaves intact the evidence on which Rome might still be considered superior. The Byzantine scholar's approach may be related to his training in arguing on both sides of the question, examined in chapter 7. Yet his purpose was not merely to display his virtuosic rhetoric to the Italians. If, as I have suggested, Chrysoloras's aim was to improve relations between Byzantium and the West, then his method in the *Comparison* was probably inspired by Aelius Aristides' example in *Concerning Concord*. Bad speakers, says Aristides, praise a city to its audience and then run off to slander it in a rival city; but since he does not wish to set cities at odds,

I shall praise them [the cities] all to the extent that each of them should be praised.... For if you accept being praised in common and none of you regards the praise of the others as an act of dishonor toward himself, but each of you is delighted by the attributes of one another, as if they were your own, first of all right from the start you shall give a demonstration of concord, and next you will gradually become accustomed also to praise one another and to have thoughts which are expedient for all of you in common.[17]

Chrysoloras's task was not only to provide clear mental images of the cities, but also to compare them. Both of these aims differentiate Chrysoloras's text from Western medieval city descriptions, which enumerate those aspects of the city from which superiority might be deduced (such as the number of houses and churches) without providing any visual image of the place or stating the criteria according to which value is being measured.[18] The reader must deduce that quantity, as manifested in a large number of edifices, is the criterion of worth; the author does not discuss, and perhaps is not consciously aware of, the values he is applying. One of the first, perhaps the first, Western city description to be organized according to clearly stated general principles is that of Chrysoloras's student Leonardo Bruni, in his *Laudatio florentinae urbis*.[19]

Gilbert Dagron complained that because of its strongly traditional character, Chrysoloras's description of Constantinople is imprecise and banal: "Il suffisait de replacer sur une vieille trame hermogénienne quelques extraits de Thémistios, de Paul le Silentaire ou de Constantin le Rhodien."[20] He found the description of Rome to be both more precise and more original, perhaps because there were fewer models from which Chrysoloras could have drawn.[21] Dagron, who was the first to offer a literary analysis of this work, is certainly right in distinguishing between the styles of the two parts of the *Comparison*. I would suggest, however, that the relative imprecision of the description of Constantinople reflects the fact that it was written from memory and with the help of literary sources, whereas, being in Rome, Chrysoloras could actually see what he described. Although his account of Constantinople is punctuated with "if I well remember" and "it seems," it is still more closely tied to the topography of the city and its thoroughfares than is the description of Rome, even though he had not seen the Metropolis for six years and had not lived there for an extended period for fourteen years. Therefore, I would propose that the difference between the two descriptions is an argument for the importance of eyewitness experience in the composition of late Byzantine ekphrases.[22]

Chrysoloras's Attitude Toward the Classical Ruins

Much attention has been given to the development of interest in the Classical past in the West, and it is therefore natural to suppose that Chrysoloras was influenced by it. For Dagron, the experience of Rome, but not that of Constantinople, provided Chrysoloras with the joy of archaeological walks; in Rome, Chrysoloras underwent a cultural "conversion."[23] He suggested that the Western enthusiasm for Classical Antiquity, coupled with the abundance of sculpture to be seen in Rome, opened the eyes of the Byzantine scholar. Alsop, noting that Chrysoloras's ideas about Classical art were very sophisticated and well thought out, proposed that they must have been formed in Italy since they have no equivalent in Byzantine thought.[24] In fact, however, very little research has been done on Late Byzantine interest in archaeology, and a good deal of evidence of Late Palaiologan interest in Antiquity still awaits analysis. This disparity cautions against too hasty an acceptance of conclusions such as those of Dagron and Alsop.

There is evidence in the *Comparison* that Chrysoloras's interest in archaeology can be documented some years before his arrival in Rome. He recalls his visit to London in 1409: none of the citizens knew the history of Roman Britain or the location of the Roman fortresses there; Chrysoloras knew the sites exactly.[25] In this case, it is quite clear that his interest in archaeology was not due to contact with English culture, but to his own education and interest. A well-known letter by Chrysoloras's school fellow (and the addressee of the *Comparison*) Manuel II manifests exactly the same attitude: dismay at the ignorance of the inhabitants and attempts to identify ancient sites.

> The small plain in which we are now staying certainly had some name when it was fortunate enough to be inhabited and ruled by the Romans. But now when I ask what it was, I might as well ask about the proverbial wings of a wolf, since there is absolutely nobody to inform me.... Admittedly, I seem unable to inform you clearly in what part of the world we find ourselves. For how can anyone spell out places which no longer have a name? Therefore, we shall have to try the next best way, that is, to take you as far as I can on an imaginary tour through the other places whose names have not been destroyed and in which we are spending some time.[26]

An interest in Classical sites, distress over the partial (or total) loss of the Classical past as a physical reality, and the practice of describing the places one visits are concerns shared by Chrysoloras, Manuel, and their teacher Cydones. The earliest example of an archaeological tour in Byzantium that I know of is recorded in the *Patria*: learned scholars visit and discuss sites in Constantinople.[27] Trips to visit famous ruins can be documented at least since the early Palaiologan period. Theodore Lascaris has left us a description of the ruins of Pergamon.[28] Planudes, seeing the ruins of the temple of Apollo near Miletus, but unsure of their identification, speculated that they could be those of the Mausoleum of Halicarnassus; recounting his visit to Hadrian's temple at Cyzicus, he expressed his annoyance that he was not told about the underground portion of the temple, which would have interested him.[29] Theodorus Pediasimus described the ruined metropolis of Serrai in the early fourteenth century.[30] The evidence suggests, then, that there was considerable Byzantine precedent for Chrysoloras's interest in Classical art and architecture. His readers have been deceived by his exclamations of delighted discovery; such expressions belong to the literary genre. To take just one comparison, Theodore Lascaris, concluding his account of Pergamon in a letter to George Acropolites with a description of the acropolis, wrote:

> With the sight of the acropolis comes a fresh surprise. The lower part of the hill is more beautiful than its peak, the city of the dead is more beautiful than that of the living. At this sight we were half despondent, half joyful, transported into a state of happiness and pain, tears and laughter.[31]

Finally, there is a hint in the *Comparison* that Chrysoloras considered the Greeks (by which he meant also Byzantines), but perhaps not the Latins, to be trained connoisseurs in art and architecture. Speaking of a man who judged ancient Rome to be superior even to Athens, he says: "He was a Greek, and an educated man, not one of the many who have never seen or heard of these things, but a man educated to them."[32] The notion that an educated man should take pleasure in seeing, evaluating, and talking about works of art and architecture goes back at least to the Greek Second Sophistic writers Lucian and Filostratus. Lucian's *The Hall*, an extremely popular text in the Quattrocento, is structured as debate about whether the orator's task is helped or hindered by the sight of something extremely beautiful.[33] To be silent is typical of the vulgar and tasteless man.

> But when a man of culture beholds beautiful things, he will not be content, I am sure, to harvest their charm with his eyes alone, and will not endure to be a silent spectator of their beauty; he will do all he can to linger there and make some return for the spectacle in speech.[34]

This injunction from one of Chrysoloras's favorite authors and the continuing tradition of ekphrasis in Byzantium suggest that the Byzantine scholar had considerable incentive to have studied and thought about Classical remains before he arrived in Rome. Why, then, does he describe the reliefs and statues of Rome, but not those of Constantinople in the *Comparison*? Perhaps the answer lies in his comment that the historiated columns of his native city were "made in precise imitation of the ones here in Rome."[35] In this case, his very impressive passages about the Roman reliefs may simply repeat material he had previously worked up in Constantinople. Whereas Chrysoloras identifies each of the monuments of Constantinople, the discussion of Rome is general—one does not know whether he refers to the sculpture of the columns of Trajan and Marcus Aurelius alone, or also to that of the arches of Constantine, Titus, and so forth.[36] This lack of precision, coupled with the absence of topographical indications in the Roman description, might suggest that he applied descriptive material developed for the monuments of Constantinople, which he knew well, to the Roman sculpture of which he had formed only a general impression.

Several considerations could have prompted this approach. First, the Roman description comes first, and if Chrysoloras thought the sculpture of Rome and Constantinople so similar, it made sense to describe the Roman material and then refer back to it in the second part of the comparison, rather than to describe it twice. Second, each of the city descriptions contains a major ekphrasis: ancient sculpture in Rome, the architecture of Hagia Sophia in Constantinople. This structure, in which secular and sacred, ancient and modern, are counterpoised, governs each of the separate descriptions as well as the *Comparison* as a whole. Thus the choice of these ekphrastic subjects strengthens the literary and conceptual organization of the entire work. Third, Chrysoloras may have desired to astonish the Italians with his ability to evoke the beauty and significance of their own monuments with words and concepts that far surpassed their capacities. Even if it were true that the Westerners were more enthusiastic about Classical monuments than the Byzantines—a thesis about which I have serious reservations—it is still the case that they did not yet have the linguistic and conceptual tools with which to express their admiration. Schlosser was right when, comparing the ekphrases of Manuel II and John Eugenikos with Western descriptions, he wrote, "wo wäre im Abendlande etwas ähnliches zu finden; wo wäre das aesthetische Moment, die Freude an der Sache selbst?"[37] The circle of Humanists and artists around Chrysoloras in Rome must have been delighted at his ability to verbalize and interpret the city.

Analysis of the Text

The *Comparison* may be considered as divided into the five parts of an oration: proem, narrative, counterproposition, resolution, and epilogue. Since the full text is in the Appendix and my analysis follows its order, I have not included page references.

Proem

By announcing that the purpose of his letter is to give pleasure rather than to serve necessity, Chrysoloras sharply distinguishes his purpose from that of the Western letter.[38] His declared subject is the city of Rome; his intention, to bring out those ways in which the physical experience of the city surpasses written descriptions. Since Chrysoloras proposes to do this with, precisely, a written description, the stage is set for an ekphrasis so vivid that the addressee will "see" rather than "read about" Rome. But first, the author must set his topic in its proper literary context. Like a modern scholar, Chrysoloras reviews the earlier literature on his subject before proposing a new interpretation. Not only Latin authors but also "our own"—Greek, that is—have written about it. In particular, he says (identifying the cultural milieu of the Second Sophistic within which his real models and competitors are to be found), Libanius, "that wisest of men, so loved by, and so close to us." And he quotes from a letter of about 355 to Jovianus: "Remember us, when, having arrived in Rome... you gaze upon such wonders as you never would have believed could exist on earth and which seem, rather, a part of heaven."[39] The choice of quotation is very subtle, alluding to Chrysoloras's own act of remembering his friends at home through the composition of the *Comparison* and establishing the antithesis of heaven and earth, which will organize his own description of the city. A further implication of the quotation is brought out by Chrysoloras's reminder to the reader that Libanius wrote the *Antiochikos*. An educated Byzantine, realizing that this is the most elaborate panegyric to a city ever written, would understand the hint: even Libanius, who loved his fatherland so greatly and knew so many cities, considered Rome to be more marvelous than anything on earth. The relevance to Chrysoloras's own position is again implicit.

Having cited a pagan rhetor, the Byzantine scholar now refers to a great Christian orator, John Chrysostom (a student, by the way, of Libanius). Chrysoloras's references, as far as I can reconstruct them, are to Homily XXXII, *On the Epistle to the Romans*, and to scattered comments in the *Homily on the Epistle to the Corinthians* (XXVI) and *Homily to the People of Antioch* (III).[40] Chrysoloras identifies his source only as having been written to someone in Asia, an assertion that serves the point he wishes to make about the superiority of Rome to the Seven Wonders: "For in Asia were the temple of Artemis at Ephesus, the Colossus of Rhodes, and the Mausoleum of Halicarnassus." In order to extend his comparison to other of the Wonders, Chrysoloras returns to Libanius's letter to Jovianus, suggesting that the recipient may have come from Egypt, and therefore been familiar with Alexandria (where the Pharos lighthouse was), Thebes (with its famous walls and gates), and the pyramids. Chrysoloras compares Rome with six Wonders; the seventh, the walls of Babylon, is reserved for Constantinople (see subsequent discussion). Underlying this sequence of comparisons is Ptolemy's geographic division of the world: Rome, having been said to surpass the most famous monuments of Asia and Africa, must now be shown supreme in Europe. Accordingly, Chrysoloras supposes Jovianus to have been a Greek and, knowing Athens, yet to have considered Rome greater. Chrysoloras concludes that while he used to think that what Greek and Latin authors wrote about Rome was hyperbole, now "my eyes confirm all that they claimed." The proem ends with the reassertion of the incontravertibility of sensory evidence: it is the final proof of truth.

Narrative

True to his premise, Chrysoloras begins the description of the Rome he saw with an assessment of its physical condition: it is almost all in ruins. Yet these remains show "what great things once existed, and how enormous and beautiful were the original constructions." Chrysoloras's suggestion that one could imagine the original whole on the evidence of a part was an entirely new concept in early-fifteenth-century Italy. For the Italians, whose concept of beauty was formed by Aristotle and Aquinas, a ruin was something deformed and therefore sad and ugly. In Aquinas's view, all things have two perfections: in regard to the whole form, "which form results from the whole having its parts complete"; and in regard to function or purpose. Therefore, "a thing is called perfect to which nothing is wanting that it ought to possess."[41] Chrysoloras's quite different approach for the imaginative reconstruction of Classical monuments, without which the Renaissance recuperation of Classical culture in its integrity would have been impossible, was fundamental. But although it constituted a novelty in Italy, Chrysoloras's attitude had a certain amount of precedent in Byzantium. The Byzantine observer tended to discuss ruined or damaged monuments as though they were intact. One notes this, above all, in the *Patria*, where the reader would suppose everything described to be in pristine condition, but it is also evident in other descriptions.[42] This Byzantine habit of mind permitted the spectator to see beyond the physical reality to the Idea and to read the remnant as signifying an ideal whole. That the mechanism of this approach was based on the relationship of part to whole is clear from Chrysoloras's explanation.

> These works were beautiful not only in their original composition and organization; they seem beautiful even in their dismembered state. Just as in a body that is beautiful as a whole, so the hand or foot or head is also beautiful; or, in a body of oustanding size, each of the limbs is large.

The ability to deduce the total size of a monument on the basis of an extant part (for that is the implication of Chrysoloras's statement for archaeology) only begins to be evident in Italian descriptions of the ruins in the third quarter of the Quattrocento.

Chrysoloras's simile rests on the conventional criteria of beauty and size, but his interpretation of the ruins brings to bear a far greater range of critical concepts: "These remains of statues, columns, tombs, and buildings reveal not only the wealth, large labor force, and craftsmanship of the Romans as well as their grandeur and dignity, if you will, ... but also their piety, greatness of soul, love of honor, and intelligence." Chrysoloras uses the ruins as evidence for understanding the practical and economic aspects of Roman culture as well as instruments for the delineation of Roman character.[43] There is some precedent for the relation of monuments to the character of a people in Aelius Aristides' *Panatheniac Oration* and *Panegyric in Cyzicus*, although this relationship is not developed very strongly.[44] Another source is the Greek historians of culture (ethnography), such as Herodotus (whom Chrysoloras acknowledges as a source) and Diodorus Siculus, who used physical products as evidence for the character and achievements of a civilization.[45] But Chrysoloras inverts their process, deducing an unknown character from the visible evidence instead of demonstrating a known character in a concrete example. He does this by applying a general principle—in this case, that there is a relation between maker and product—in a new way and to a new problem. The inversion, arguing from product to maker, was well established in philosophy and especially theology where the cosmos, emanating from the divinity, could be used in discussion of the divine nature. The application of this principal to the rhetorical genres of ethopoia and ekphrasis was facilitated by Chrysoloras's skill in decompartmentalizing intellectual disciplines. It was also aided by a long Greek-language tradition of considering buildings as manifestations of intellectual ideas.[46] While such ideas were usually located in the mind of the architect, Chrysoloras here extends it to the mind of a people. Thus his criteria for evaluating the Roman ruins were drawn from the methods of various disciplines—rhetoric, philosophy, theology, and history. He was trained to bring to bear on an experience any principle that would reveal its significance.

The role of the ruins for the reconstruction of Roman history was of special interest to Chrysoloras:

> These [imperial monuments] record in sculpture the wars, the prisoners, spoils and sieges, as well as the festivals, altars, and votive offerings.... Thus one can see clearly what kinds of arms the Ancients had, what kind of clothes they wore, what the devices of their rulers were, how they formed lines of battle, fought, laid siege, and built encampments.... Herodotus, and some other historians, are thought to have made useful contributions to our knowledge of such things. But these reliefs show how things were in past times and what the differences were between the peoples. Thus they make our knowledge of history precise or, rather, they grant us eyewitness knowledge of everything that has happened just as if it were present.

This is one of the first statements of the superiority of visual to literary evidence for understanding the past. I will present my research on this aspect of Renaissance historiography in a sequel to this volume. Here it must suffice to demonstrate the novelty of Chrysoloras's description by comparing it with that of the column of Trajan in the contemporaneous *Tractatus*: "Where there is a column of amazing height and beauty with a narrative of the emperor Trajan's history, near the church of St. Nicholas, where today they call it *'la Militia,'* on one side there was a temple said to be Trajan's."[47]

Chrysoloras concludes his discussion of ancient Rome, which he has made visible through the images of its monuments, with a reference to Aelius Aristides' *On Rome*. "One can truthfully say, repeating the words of one of our ancient authors, that such things as these could only be done by Ῥώμη."[48] Chrysoloras repeats Aristides' pun on the Greek word for "power" and the name of the city. His choice of this reference confirms the standard interpretation of Rome within its encomiastic tradition.[49] Most Classical and Late Antique accounts of the city interpret it as representing the dominion of empire, of military might and wealth; there is little description of the physical city and even less of its topography. But if Chrysoloras's conceptual structure reaffirms the authority of literary models, his descriptive approach is innovative, highlighting the importance of eyewitness evidence to prove the truth of traditional claims to grandeur.

The second part of the narrative, on Christian Rome, is distinguished from the account of pagan Rome by frequent recourse to the "inexpressibility topos"—protestations of the author's inability to describe so lofty a subject.[50] This device, paired with paralipsis (pretending to pass over subjects while actually mentioning them), serves to convey the greater dignity and richness of Christian Rome.[51]

> But what about the more recent city, the Christian Rome of our times? Who could describe it? How can one describe all that was made from such abundance of great stones and other materials, with the wonderful craft that flourished then, inspired by the love of God and by piety?

Chrysoloras contrasts the two periods of Roman greatness through language: ancient Rome is approached through the intellect, can be understood through reason, and is historically interesting; Christian Rome is approached through emotion, surpasses reason, and is spiritually enlightening. I am tempted to connect this with the apophatic element in Byzantine theology. The Outer Learning can explain, but only within limits; spiritual experience lies outside these limits. The contrast in the *Comparison* was prepared in the proem, not only by pairing pagan and Christian praises of Rome in the figures of Libanius and Chrysostom, but in the quotation from Libanius that he repeats here: "Thus what that author whom I mentioned at the beginning of my letter once said about his own times can be claimed with greater justice now—Rome can be called a part of heaven on account of its inhabitants."

The account of Christian Rome centers on the cults of Peter and Paul and on the apostolicity of the Roman Church. As a recently declared Catholic writing from the court of a schismatic pope to an Orthodox Byzantine emperor, he is now on very treacherous ground. His solution is to rely for his encomion on an unimpeachable Greek source, John Chrysostom's homily on Paul's Epistle to the Romans. Chrysostom had referred to Peter and Paul as shining eyes, great lights illuminating the whole world like sun's rays, rendering Rome brighter than any other city.[52] Here is Chrysoloras's version: "How, on the other hand, can I do justice to those two stars, or suns, or whole heavens; those witnesses to the glory of God—Peter and Paul—whose relics are here in Rome, and whose tomb is near me as I write?" The saints are depicted as triumphant rulers, whose subjects come on pilgrimage from all over the world to beg for absolution and salvation. The fame and power of Peter and Paul have eclipsed that of the pagan emperors (another borrowing from Chrysostom)[53], and their subjects smash the palaces and monuments of those persecutors of the faithful, striking them with their fists and spitting on them. Their power, moreover, is attested to by the long line of apostolic succession. Whereas Peter was poor and healed a cripple, saying "I have neither silver nor gold, but I will give you what I have,"[54] now he is so rich that "if all the rulers we know gave all the taxes and income over which they have control, this would not equal this sum."[55] This delicate hint at the Church's obligation to distribute wealth is followed by a diplomatic affirmation of the right of the Roman Church to impose penance and to judge the whole world. The message is couched in a rather inexact reference to Plato's *Gorgias*, in which Plato argues that a man who does wrong is more wretched if not punished than if punished. This section of the *Comparison* closes with another quotation from Chrysostom's *Homily*:

> He says there, and, as if inspired, he says it over and over, "I want to see this city" (meaning Rome) "in which there is so much" (meaning the deeds of the apostles, the tombs that hold their bodies, and their presence). "I want to see their feet, I want to see their hands, I want to see their heads." And he did not cease—just like someone desperately in love—enumerating all the things that were dear to him, and saying it over and over.[56]

It is remarkable that for Chrysoloras, there was no distance between his own feelings and those of the fourth-century orator, whose words could serve to express his own personal emotional experience of Rome. While aware of differences between his own culture and that of pagan Antiquity, and of the difficulty of grasping its reality, he was so perfectly at ease with Late Antique and especially Christian authors that he identified their emotions and desires with his own. The fourth century was as fresh and clear to him as his own times.

Antithesis

The second half of the *Comparison*, the encomion of Constantinople, is introduced by a transitional paragraph suggesting that Chrysoloras's admiration for the Roman monuments is really due to their similarity to those of Constantinople: "For I believe that never did a daughter resemble her mother so precisely as Constantinople resembles Rome." This brings him to the traditional comparison of "Old and New Rome," developed by Byzantine writers in favor of the latter.[57] Chrysoloras, anxious not to offend his Latin readers while yet asserting the superiority of Constantinople, offered as justification the arguments of cultural and biological progress. Whereas Rome did not derive its form from a model, Constantinople, using Rome as an archetype, "brought many things to greater perfection and splendor," for "the works of men competing with others can progress toward greater beauty." Moreover, since "the parents' beauty contributes to a more perfected beauty in their children, and their great stature lends greater stature to them ... it is not surprising, then, that such a great and beautiful city produced another even greater and more lovely." By appealing to natural law and the undeniable benefit of progress, Chrysoloras is able to argue for Constantinople's superiority without denigrating Rome.

Arriving at the description of the city itself, Chrysoloras had an abundance of literary models on which to draw. However, he had to find a way around admitting the sad state of Constantinople in the early fifteenth century, with a population of only some 50,000 and most of its statues and relics having been lost to the West after 1204.[58] Thus descriptions like Nicetas Choniates' *De Signis* and the seven "spectacles" of Constantinople in the *Patria* were of no use to him.[59] Perhaps the decrepitude of the city, coupled with his realization of its original richness, were influential in Chrysoloras's decision to compare the two cities not as they are but as they once were.[60]

The description adheres to Menander's guidelines for the *laus urbis*: one begins with the geographical situation of the region and the city's water supply and continues with the topics of local topography, harbors, and citadel. Accomplishments and virtues of the citizens compose the rest of the account.[61]

Accordingly, Chrysoloras begins by telling us that the city is situated on two continents (Europe and Asia) and at the conjunction of the northern and southern seas (Figure 24). Centrality of position, one of the criteria on which Menander recommended arguing for superiority, could be claimed for Constantinople but not for Rome.[62] The centrality of the metropolis had been a

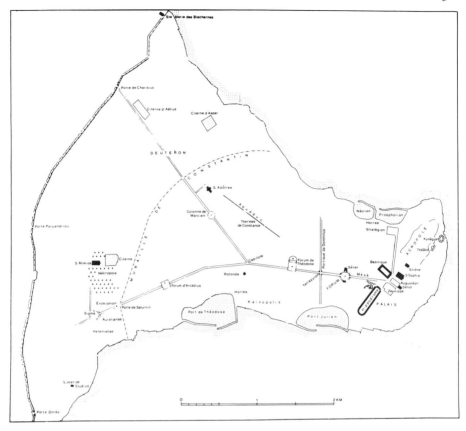

Figure 24. Plan, Constantinople, with sites mentioned by Chrysoloras. (Courtesy of C. Mango, from *Le Développement urbain de Constantinople* [Paris, 1985])

topos of description since the fourth century, when Gregory Nazianzen said that it linked the eastern and western shores and that the extremities of the world from every side meet there; Himerios noted that the city was in both Europe and Asia.[63] A further allusion may also be intended if, as would seem to be the case, one of the sources for this image is Metochites' *On the Hellenes*: the location described by Chrysoloras is the home of the Greeks. Thus Metochites wrote that "a cause of their nobility of character and natural dexterity may perhaps be found, among other factors, in the fact that the country they inhabit is in the middle of the world and actually midway both between East and West and between North and South."[64] This association is confirmed by Chrysoloras's elaborate discussion of the city's site as part of a continent but also in the sea, which derives from Aelius Aristides' *Panathenaic Oration*, a work he proceeds to quote: "Just as someone once said about Athens, you can sail past the city, around it, or come there on foot."[65]

Chrysoloras intends to argue, as Menander recommended, that his city is superior because it combines all good qualities: it is on two continents and two seas, on land and sea, and it was founded by "the union of the two most powerful and prudent peoples, the Romans and the Greeks." Patrinelis noted Chrysoloras's awareness of the Byzantines as descended from the Greeks and Romans in regard to his recently discovered letter to Manuel II, suggesting that this new conception of the Greco-Roman heritage, implying a keen sense of racial and historical continuity, fostered that nationalism later seen in Pletho and Bessarion.[66] Chrysoloras continues: the reason for this faultless origin is that the city was designed in order to be the metropolis of the whole world. Whereas other cities, founded for different reasons, later had to adjust their physical deficiencies to accord with exalted status (like Rome), Constantinople was planned with its supreme status in mind.[67] Chrysoloras criticizes cities that have grown up piecemeal, asserting that they can never be really beautiful because they always lack something useful or pleasurable. If I am not mistaken, this is the first discussion of rational—even ideal—city planning in the Renaissance. Garin has suggested that the Renaissance belief in the possibility of a rationally structured urban environment was indebted to Chrysoloras's translation of Plato's *Republic*; actually, Chrysoloras himself was the first to draw the implications from that text.[68]

The site having been chosen, the citizens began to build "houses that might have sufficed as cities on account of their size, and churches so wonderful that one cannot believe that they were made by human skill." Chrysoloras passes in brief review the notable structures of the city, as if casting his eye over the whole spectacle, before settling into an account of the water supply. His glowing praise of the system may be hyperbolic, since we know that the Turks found it completely broken down, and the city without water, in 1453.[69] The suburbs come next: they are so extensive as to form a kind of city on their own, including structures of such grandeur "that it would be difficult to find their like in any other city." Again, the source may be *The Panathenaic Oration* (351): "We can consider the country hamlets, some of which have been adorned more gloriously than cities elsewhere."[70] This topic is followed by a description of the Long Wall, admired for its size, excellent construction, and materials. Chrysoloras insists that the whole system of fortification was built for the sake of magnificence rather than for use because the status of a city is first assessed from its walls and the founders were building for future ages: "Thus the walls remain, not only as evidence of the power of those who built them, but also as testimony to their prudence and foresight for the future." Earlier in the text he compared them with the walls of Babylon, the only one of the architectural works among the Seven Wonders that he did not claim to have been surpassed by Rome. The suggestion that walls were built for future rather than present need may have been taken from Livy (I.viii.4) about early Rome: "They built their defences with an eye rather to the population which they hoped one day to have than to the numbers they had then."[71] The argument is intriguing, suggesting that building for mere necessity is not worthy of praise.[72]

The next topic, monuments to be seen within the city, seems to depend on Constantine of Rhodes's poem for its overall structure.[73] It begins with the tomb of Constantine at the Church of the Holy Apostles and other imperial tombs; the monument of Justinian is the next to be mentioned, together with other commemorative statues near it.[74] This brings Chrysoloras to the eastern section of the city, near Hagia Sophia. What follows is a description of columns that once supported such statues, proof that the city was once rich in this category of monument. Chrysoloras's discussion is not especially clear, and some of my identifications must be hypothetical. His list is important because he includes seven, rather than the standard five, columns mentioned in Constantine of Rhodes and represented in all the copies of Cristoforo Buondelmonti's view of the city of about 1423.[75] It has to be emphasized that Chrysoloras's subject is not the existing columns and bases, but the statues that they once supported, another demonstration of his intention to deduce the nature of the original whole from the remaining fragment. First are the statues of Theodosius the Great in the Forum Tauri and of Arcadius in his forum on the Xerolophos hill: "One can judge how grand and worthy of admiration and beautiful they must have been from the beauty, loftiness, splendor, and costliness of their bases."[76]

Evidently, Chrysoloras is walking through the city in his imagination, for he has taken us down the Mese through two of its forums. Now he says, "Thinking about them [the statues] brings to mind the former city gate, which is on the same road and in a direct line to the west." His description of the enormous gate topped with a shining portico must refer to the ancient Golden Gate (Porta Saturninus) in the walls of Constantine; it cannot refer to the Golden Gate in the Theodosian walls, since they are described later in the text.[77] The column before the gate is the Exokionion, mentioned in the *Patria* and perhaps of fourth-century date.[78] One "on the other hill, the one below which is the palace where you live now, called 'of the column'" might be the column of the Goths on the old Acropolis; if so, then Manuel must have been living in the Mangana Palace and not at the Blachernae.[79] The column at the Strategion could be the column of Phocas; the base "to the right of the Church of the Holy Apostles" might be the bronze group of the Archangel Michael and an emperor or the column of Marcian.[80] The account closes with the last and greatest of the columns, that of Constantine in the Forum of Constantine, wrongly identified by Chrysoloras as the courtyard of Constantine's Palace.[81] The itinerary of the columns has taken us on a complete circuit of the city, going from east to west along the Mese, turning north, and then returning to the area of Hagia Sophia by the same avenue (Figure 24).

Chrysoloras next addresses the topic of sculpture, mentioning the reliefs on the columns of Arcadius and Theodosius ("made in precise imitation of the ones here in Rome") porphyry statues of market overseers at the Philadelphion,[82] a reclining figure at the source of the river Lykos,[83] and the reliefs on the Golden Gate in the Theodosian walls.[84] Realizing that this is a meager list, he explains that religious sentiment in the fourth century discouraged the art of sculpture, and therefore the artists of Constantinople invented other forms of art, such as icons and mosaics. Now he is on stronger ground and remarks that mosaic art is the special achievement of Greece and Constantinople, whence come the techniques and materials for mosaics in Rome. Not only is this also true for sculpture, since the materials and techniques were passed on to Rome by ancient Greece, but "one might, perhaps, say this about a great many other things as well." Although the scope of this comment is left undefined, Chrysoloras surely intends to imply that Roman culture in general owes a fundamental debt to Greek (including Byzantine) know-how.

The Byzantine scholar has left his reader standing in the Forum of Constantine while he has summed up the main points of the tour. Now he turns (in his imagination) toward the east and exclaims; "Who, seeing that church, so well named after the wisdom of God (for, really, it is not the work of human wisdom), would be able to describe, or admire, or even remember anything else?" The writer himself is taken by surprise, transfixed by such a sight, aware of the impossibility of conveying it in words, and, in short, utterly overwhelmed. Within the literary genre, this is a clear preamble to the main ekphrastic effort of the piece, and the reader is eagerly poised to see with what words the image will be drawn. Chrysoloras does not disappoint. Many earlier descriptions of Hagia Sophia could have served as models, but Chrysoloras does something entirely new, using Hermogenes' categories of literary style as critical terms for the evaluation of the church. Perhaps his inspiration was Himerius's comment about the whole city of Constantinople: "How could the grandeur and beauty of words equal its grandeur and beauty?"[85] As in other cases, Chrysoloras takes an idea barely sketched by a fourth-century orator and develops it into a full literary structure. Of Hermogenes' seven categories, four are applied directly (grandeur, beauty, simplicity, and power, two indirectly (clarity and conciseness), and one is applicable not to the building but to the speaker's attitude toward it (frankness). Two further literary categories, sublimity and variety, are drawn from Longinus.[86] Here is the relevant passage:

What sublimity and grandeur of speech could adequately evoke its sublimity and grandeur? What beauty of words could equal its beauty? Could the dignity of my words do justice to its dignity? How could I speak with as much simplicity as it possesses? What power and variety of words will be equal to its variety; or what precision of speech to the precision and excellence of all its parts? What harmonious composition can compare with its harmony and congruence of parts?

The reader who knows Hermogenes' discussion of each of these qualities realizes that the description is structured in pairs of opposite qualities: grandeur/beauty, dignity/simplicity, abundance/conciseness.[87] These opposites are said to coexist within a unifying harmonious composition, as Hermogenes recommended. Of the ten terms that Chrysoloras applies to the church, many represent concepts new in architectural description not only in the West, but also in Byzantium. The description is a key example, of course, of decompartmentalization; by applying to architecture literary terms whose significance and connotations were highly developed, Chrysoloras immensely expanded its potential for expressive and aesthetic content. Architecture could now be thought of within conceptual categories familiar to the educated reader, and therefore it became interesting to experience and discuss; it became another kind of manifestation of the principles that governed cultural expression.

But Chrysoloras does not stop with this. Now he approaches the church, passing through the atrium and narthexes into the church and continuing to regret his inability to describe all the details he in fact notes. He stops beneath the dome to marvel at the overall design, technique of construction, and dimensions of the edifice—and then he looks up. He first speaks of the optical impression of the dome, of his incredulity at its miraculous floating quality and immensity. But he then reflects on the vision of the architect, who dared to imagine such a work, for which no model existed, and who felt confident in his ability to realize the dome as it was conceived in his mind. He must have been both an excellent craftsman and extremely skilled in geometry and mechanics, muses Chrysoloras. That is, the ability to create something so extraordinary requires knowledge of both theory and practice. But now he catches himself thinking in too restrictive a fashion about the cultural significance of the dome:

> And yet, what I am saying? For this work makes the spectator wonder at the ready intelligence and capacity to achieve great things not only of this architect and of the other builders but of the whole human race, as well as of this living person. The dome reveals inventiveness, elevated thought, dignity, and power such as no one before ever would have imagined.

This extraordinary articulation of the theme of the dignity of man, so important for the Quattrocento Humanists, has been discussed in earlier chapters. Chrysoloras looks behind the physical product to the mind of the maker and then realizes that he too, being human, shares this intelligence and capacity. Constantinople is superior to Rome because there is the visible proof, the demonstration, of human creativity at its highest level. The complete novelty of his analysis for the Western reader may be appreciated by comparing it with that of the Pantheon in the *Tractatus*:

> This same Pope Innocent caused the pinecone to be placed at St. Peter's in Almachia, having been brought from S. Stefano in Pinea, which took its name from the aforesaid pinecone. And it stood on top of a certain shrine of an idol which was in the oculus of the temple of Cybele, which is now called S. Maria Rotonda, and a windstorm lifted it from there and transported it to this piazza of S. Stefano a little after the death of emperor Phocas.[88]

Stangely enough, Chrysoloras's description of Hagia Sophia has its closest literary model in Aelius Aristides' account of Hadrian's temple at Cyzicus rather than in one of the earlier accounts of the metropolitan church. The image of the structure as completely dominating the spectator's attention and providing the key to the spirit of the inhabitants of the city is in Aristides' lines: "The Temple fills every vista, and at the same time reveals the city and the magnanimity of its inhabitants."[89] From the same oration comes the praise of the engineering achievement, praise due because of the magnitude of the task and the originality of the means: "If someone should forgo speaking about the temple itself, it is enough to express admiration for the engineering equipment and the transport, whose invention was prompted by the requirements of the temple, since they formerly did not exist among mankind."[90] Even the themes of absolute originality and superlative cultural achievement were suggested by this source: "I am close to declaring that you have shown all men who have attempted similar works to be like children, by having erected a work so great that it would have seemed to be an act of madness to have conceived it and beyond the power of man to accomplish it."[91]

Confrontation between the source and Chrysoloras's own lines shows how much further he developed the ideas implicit in Aristides' panegyric. Also interesting is the ease with which Chrysoloras applies a description of an ancient temple to a building of completely different structure, appearance, and function. This was possible only by ignoring the specifics of the description and focusing on its criteria of excellence.

Epilogue

The *Comparison* closes by returning to the writer's original statement—that the letter was intended to give pleasure rather than serve necessity. Although one may observe the ancient works in both Rome and Constantinople with pleasure, one may also draw a useful lesson from them: "For whoever, seeing these things, ponders on the empire of Rome, on the power and dignity of those men, their works and their efforts, will see what a death all these things have died." What follows is a meditation on the transitory nature of all human achievements, whether works or deeds, and on the uncertainty of fortune; only virtue endures. Interestingly, the only literary source that can be identified for this section is Aristotle's discussion of *tyche* in *Physics* (II.vi.197b), which Chrysoloras cites in reference to stones "which have lost their primal good fortune." Aristotle's point is that, properly speaking, only self-directed creatures can be said to have good and bad fortune. But the association of the theme of *fortuna* with descriptions of Rome was well established by Western authors, such as Petrarch and Vergerio. What is new about Chrysoloras's use of the topos is, first, that the moral lesson follows a description intended to give pleasure, rather than serving as pretended justification for it. The validity of literary description and of secular knowledge as ends in themselves is therefore asserted. Second, this order reflects the hierarchical nature of Chrysoloras's conception of knowledge: the same ruins that serve as evidence for the grandeur of Roman culture in the first part of the *Comparison* return as proof of the transience of human endeavor in spiritual terms. Moreover, it is precisely on the basis of secular knowledge of the ruins that Chrysoloras constructs his spiritual interpretation, unlike his Western models, in which the spiritual interpretation precedes and is imposed on the physical evidence. Finally, his meditation on the past has direct relevance to the present, for he concludes by wishing that the emperor may "save the best city under the sun and restore its good fortune." The Roman ruins become images of the destruction threatening Constantinople.

Chrysoloras invested sensory experience with many kinds of meaning: aesthetic pleasure, the use of visual fact as a tool for the clarification of literary sources and as an instrument for the recuperation of a lost civilization, stimulation to meditation on the nature of human life and the importance of spirituality, and the perception of the implications of the past for the present. In other words, he brought to bear on the experience his training in rhetoric, history, theology, physics, and politics. It is not his knowledge of these subjects that differentiated him from Western intellectuals of the time, but his ability to see how they could be applied to the experiences of life, rendering them far richer in significance. Although there are specific borrowings from his *Comparison* in Quattrocento descriptions, its real impact was on the Humanists' approach to their own experience. The Byzantine contribution to Quattrocento Humanism, in the more general aspect that is of interest in this section of the book, might be expressed in Cicero's words, when he wrote of Varro's importance for Roman culture: "We were like strangers in our own city, visitors who had lost their way. It was your books that, as it were, brought us back home, so that at last we could recognize who we were, and where we were."[92] The context of the comment underscores its relevance to Chrysoloras, for Varro has just told Cicero that "doctrines which nobody had been teaching up till now, and for which there was nobody available from whom those interested could learn them, I have done as much as lay in my power... to make known to our fellow-countrymen"(I.ii.8). If my analogy seems overdrawn, I would call attention to Guarino's almost identical assessment of Chrysoloras: "We should owe the greatest thanks, to you especially, who high above have enlightened the Italians, long sunk in darkness, as though the lamp of the sun had finally been brought near."[93]

NINE

The Recovery of Eloquence and a Problem in Historiography

THE RECOVERY of eloquence in Italian urban and architectural description began in what I will term the circle of Chrysoloras. This phrase serves to group men whose contact with Chrysoloras influenced their approach to writing about the built environment, without implying that their thought depended on him in other contexts. The works of these men are products of intense concern with the question of what lesson can be drawn from the built environment: Of what is it evidence? They reflect the conversations of a group of Humanists emulating the Ciceronian ideal of rhetoric in the service of the *patria* (as seen in the contributions of Salutati, Bruni, and Chrysoloras himself) but also interested in interpreting the significance of the ruins of Roman civilization for the present (Vergerio, Salutati, Poggio, and Chrysoloras). Thus their works favor the *laus urbis* and the development of the theme *de varietate fortunae*; both of these manifestations benefited from Greek ekphrastic models introduced by Chrysoloras. In particular, Second Sophistic orators such as Aelius Aristides and Libanius provided the Humanists with themes and literary images that they applied to their own subjects.

The political and historical engagement of their works was not lost on their contemporaries. Indeed, Leonardo Bruni's *Laudatio florentinae urbis* provoked a polemic over the question of whether panegyrical history was a permissible literary genre. The new eloquence was opposed by those who saw it as endangering truthfulness, distorting history, and fostering a style overly concerned with beauty. The conflict was resolved in favor of panegyric, but not within the prose forms that had originally been challenged. Instead, architectural description in the third quarter of the century is most often found in poems in praise of a patron. These texts, whose models are in Latin Classical and medieval poetry, describe buildings that exist only in the authors' minds. Whereas works produced in the circle of Chrysoloras used the built environment as evidence for the nature of its inhabitants, using the visible as evidence for the invisible, compositions after midcentury described nonexistent buildings in order to ensure the patron's fame for posterity.

Historical Background

In the West, architectural description had been removed from the sphere of rhetorical practice by the end of Antiquity. Some handbooks of rhetoric suggested that discussion of cities belonged to the realm of science rather than art because of its (assumedly) objective and factual nature.[1] Another trend interpreted the concept of city (*civitas*) as referring exclusively to the inhabitants; thus praise of the city need not include any physical description of the site.[2] A third direction, interpreting the physical city as signifying a higher, spiritual reality, made architectural description a subordinate part of theology.[3] These shifts mirror the development of medieval rhetoric itself. McKeon has shown how ancient rhetoric had been dismembered by the thirteenth century; some of rhetorical theory was made a part of logic, while other aspects became instruments for theology.[4] Commonplaces, definitions, and proofs were absorbed into dialectic, while moral and political questions became the province of philosophy. By the thirteenth century, the tradition of rhetoric was represented in three types of textbooks: the *ars dictaminis*, *ars praedicandi*, and *ars poeticae*.[5] This scheme left no scope for the practice of ekphrasis as a separate genre either in the school curriculum or in professional life. The description of buildings and cities, never very strongly rooted in Latin literature, almost ceased to exist in the Middle Ages.

A brief review of medieval descriptions will illustrate these generalizations and introduce the reader to some of the main Latin sources used in the Quattrocento. After the second century A.D., almost all "descriptions" of Rome (the most common ekphrastic subject) are in the form of a list or catalogue and are evidently intended to convey factual, usually topographical, information in as few words as possible. The *Notitia urbis Romae regionem XIIII* and the *Curiosum urbis Romae regionem XIIII*, both of the fourth century, are important documents in this development not only because of their novelty as complete lists of notable buildings in each of the city's *rioni*, but also because of their importance for Renaissance writers.[6] The numerous medieval sylloges, often in the form of itineraries of the holy places to be visited in Rome, are essentially lists with very little literary or descriptive qualities.[7] Altogether, before the eleventh or twelfth century, there are hardly any descriptions of the city or its buildings.[8] This situation changed with the compilation known as the *Mirabilia urbis Romae*, dated by its most recent editor to around the year 1000, but traditionally thought to date to the twelfth century.[9] This text, based in part on earlier sylloges and the *Curiosum*, is comprehensive in its approach to the pagan and Christian monuments and includes stories and legends as well as topographical indications.[10] Unlike the *Regionary Catalogues* and medieval itineraries, the *Mirabilia* is organized by subjects such as walls, gates, baths, palaces, and temples. This typological approach would dominate most early Renaissance topographical accounts of Rome; in chapter 7, I gave the example of the *Anonimo Magliabecchiano*, of 1411, and compared it with Chrysoloras's account of the city.

Undoubtedly, the *Mirabilia* was an important source for the *Graphia aureae urbis Romae* (1155), for Master Gregory's late-twelfth-century *Narratio urbis Romae*, and for the relevant portions of Canon Benedict's *Liber politicus* (1142–1143).[11] The twelfth century also saw the creation of Peter of Mallio's detailed description of St. Peter's (*Descriptio basilicae vaticanae*, 1145–1153) and Deacon John's description of the Lateran basilica (*Descriptio lateranensis ecclesiae*, 1159–1181).[12] These works, most of which have been associated with political propagandistic aims within the hierarchy of church or empire, evoked both past and present to legitimize claims or aggrandize stature. But they also compiled what was known and thought to be important about the sites of Rome. The texts reveal a certain amount of attention paid to inscriptions, inventories, and other documentary evidence together with the acceptance of legends about people and places.

Not until the thirteenth century was there any further development in the genre of architectural description in the West. One class of new texts, which revives the *laus urbis*, is closely related to the consolidation of the Italian city-states. Bonvesin della Ripa's *De Magnalibus Mediolani* of 1288, the anonymous *Liber de laudibus civitatis Ticinensis*, Benzo of Alessandria's *Chronicon*, and Villani's account of Florence, all three dated around 1330, belong to this genre.[13] Baron's assessment of the *De Magnalibus Mediolani* applies to all: "It makes hardly any effort to come to grips with individual traits; it is satisfied to attribute to the object of its praise the greatest quantity or largest number of what it considers good or honorable for a city state."[14]

Cantino Wataghin, following Roberto Weiss, suggested that a new attitude toward description may be detected in Giovanni Dondi's *Iter romanum* of 1375.[15] Although this account follows the usual structure of a pilgrim's itinerary, it is innovative in that Dondi cites the sources of his information and offers some description of the monuments. Instead of "marveling" in an uncritical fashion, Dondi observes a reality about which, by selecting some aspects over others, he forms a judgment. Yet what he selects shows that his categories of perception and of aesthetic evaluation are still those of the Middle Ages: enumeration and measurement are the means by which Dondi approaches the architectural experience.[16] He tells us, for example, that San Paolo fuori le mura has four rows of columns, as does St. Peter's, but only ninety of them; that two columns near the altar are as tall as those in the porch of the Pantheon; and that the church measures 123 by 83 paces.[17] Typical of both the innovative and conservative features of the text is his description of the Vatican obelisk: "The presbyter who lives near it said that he had measured its shadow with an instrument and that it was 45 braccie high."[18] While the account may be scientific in its factual precision and completeness, it creates no image of the structure in the reader's mind, nor does it relate the building to human concerns. The building remains a fact of small utility.

Humanist Descriptions in the Circle of Chrysoloras

Pier Paolo Vergerio

In contrast to these works, Pier Paolo Vergerio's description of Rome of around 1398 has not received the attention it merits.[19] Although it is pieced together out of medieval sources, its structural complexity and linguistic elaboration are those of a rhetorical ekphrasis. Vergerio begins with the familiar theme of *ubi sunt?*—where are the glories of the past? But he couches this theme in an altogether new form: although they say that the air of ruined cities is unhealthy, he begins, this may be true for the body but not necessarily true for the spirit (*animus*).[20] This introduction leads into a brief overview of the evidence of Christian piety in Rome, in which he mentions the major shrines, their locations, and chief attractions. His literary construction places the traditional devout itinerary in an ideated context: the ruin of ancient Rome, he implies, was necessary for the triumph of Christian Rome. However, he continues, while these things elevate the spirit, they depress the mind (*ingenium*), and therefore he has delayed in writing this letter to his friend.[21] Vergerio turns the medieval itinerary into a prologue for the main body of his letter and subordinates it to a more extended, rhetorically ordered whole. Moreover, he plays skillfully with the rhetorical device of antithesis. The body/soul opposition is decided in favor of the latter; the spirit/mind antithesis is left in parison; and the balanced pair, pagan Rome/Christian Rome, serves to structure the letter as a whole.

Having arrived at the main body of his text, Vergerio makes conventional rhetorical apologies for his inadequacy to the task he has undertaken. But he also blames the lack of informants and the distortion of the facts by legend.

THE RECOVERY OF ELOQUENCE   175

The section concludes with a reference to Petrarch: "Rome is nowhere less understood than in Rome."²² His solution is to take the monuments themselves as evidence: "Above all, for evidence of what Rome is and how great, there are the ruins and the *exempla* of things miraculously preserved."²³ After a brief survey of the sorts of ancient monuments still to be observed, he begins to survey the city region by region and, drawing on the *Mirabilia*, mentions the city gates and chief monuments in each topographical unit. The letter is incomplete and breaks off in the middle of this section.

Bringing together different kinds of medieval source material, Vergerio orders it in a new way, which enhances its significance. What appeared as factual lists in the original sources is made to serve as evidence for a point of view the author wishes to argue. Yet Vergerio acknowledges that despite his careful investigation of the subject he experienced great difficulty in composition and in finding adequate descriptive terminology.²⁴ His admission is not only a rhetorical topos, but also an honest statement of the fact that without rhetorical models, it is difficult to write an extended description.

The letter on Rome is not Vergerio's only attempt at architectural description: we know of a lost account of Florence, dated around 1400 to 1404; another of Capodistria; and an incomplete description of Venice, dated either from 1400 to 1403 or around 1412.²⁵ His interest in epideictic rhetoric is also proved by his discussion in *De Ingenuis moribus et liberalibus studiis adulescentiae* (1400–1402). Lamenting the absence of good rhetorical practice by his contemporaries, he says that epideictic works are not composed "according to the proper method. For in constructing speeches, just about everyone makes use of those arts that are diametrically opposed to the art of eloquence."²⁶ McManamon argued, and I think he is right, that Vergerio referred to Scholastic arts like dialectic.²⁷ But can his sudden outpouring of ekphrasis be accounted for by renewed interest in Cicero and Quintilian, and by his study of rhetoric under Giovanni di Conversino da Ravenna in the 1390s in Padua, as McManamon concluded?²⁸ Since all the descriptive works, except perhaps that on Rome, date immediately after his study with Chrysoloras in Florence, that experience would seem to be more immediately related to his innovation.²⁹ His trip to Rome occurred in the first months of 1398, and he began his study with Chrysoloras in the autumn of that year.³⁰ The precise date of the letter is unknown: it has been dated about 1398 only because that is when he visited Rome. Since in the letter itself he apologizes for his delay in writing, it certainly dates after the Roman visit and therefore from more or less the time he began his study of Greek. While there is no evidence that Greek sources were used for the Roman description, the form itself—an ekphrasis in epistolary form, so common in Byzantium and rare in the West—may have been suggested to him by Chrysoloras, who was, perhaps, the original addressee of the letter.

Robey and Law recognized the importance of Vergerio's encounter, as a student of Chrysoloras, with Plato's *Republic* for his *De Situ et conditione et republica grandis et aureae urbis Venetiarum*, the first Humanist encomion of Venice.³¹ And they noted that it may well date from the same time as Bruni's *Laudatio florentinae urbis*, a work that Bruni said had been inspired by his encounter with Greek models.³² However, they saw Vergerio's encomion as belonging to "the older and indigenous tradition of the medieval city *laudes*," especially in its treatment of the advantages of the Venetian site.³³ In this they were mistaken, since Vergerio's presentation of Venice as neither on the mainland nor really an island, and therefore avoiding the potential dangers of either situation, is drawn from Aelius Aristides' *Panegyric in Cyzicus* or *Panathenaic Oration*. Chrysoloras used the same passages for his description of Constantinople in the *Comparison*, as we saw in chapter 8.³⁴ Vergerio says that Petrarch described Venice as "totius orbis emporium," a reference that his editors were unable to identify.³⁵ Its actual source may be Libanius's *Praise of Antioch*.³⁶

Although their precise chronology cannot be established, it seems appropriate to view the descriptions of Florence by Vergerio, Bruni, and Salutati (all written sometime between 1400 and 1404), together with Vergerio's accounts of Rome, Capodistria, and Venice, as enthusiastic responses to the new techniques of eloquence that Chrysoloras taught. This is not to deny the importance of Cicero and Quintilian, both of whom Robey, Law, and McManamon have rightly identified as stimuli for Vergerio's revival; rather, as I argued in chapter 7, Chrysoloras offered literary models for, and verbal demonstrations of, the application of their precepts. A picture begins to emerge of a series of dynamic literary exchanges in the circle of Chrysoloras and his students.

Leonardo Bruni

Bruni claimed that his *Laudatio Florentinae Urbis* of 1403 or 1404 was based on Aelius Aristides' *Panathenaic Oration* (A.D. 155), and, indeed, some points of coincidence between these works exist.³⁷ Aristides' discussion of the superiority of the Attic dialect is reflected in Bruni's similar claims for Florentine speech. Aristides wrote that "it [Athens] also introduced, as a model for all Greek speech, a dialect which is clear, pure and pleasant" (15) and that

> emulation of your wisdom and way of life has spread over every land by some divine fortune, and all men have come to believe that this is the speech of the human race. And through you the whole of the inhabited world has come to speak the same tongue ... they have abandoned their native dialects and would be ashamed to speak in the old way even among themselves with witnesses present. (325–26)³⁸

This claim compares closely with Bruni's assertion that

all of Italy believes that this city alone possesses the clearest and purest speech. All who wish to speak well and correctly follow the example of the Florentine manner of speech, for this city possesses men who are so expert in their use of the common vernacular language that all others seem like children compared to them.[39]

The Greek orator's interpretation of the Athenians as saviors of Greece because of their defeats of Eurystheus and Xerxes may have inspired Bruni's treatment of the Florentines' victory over Giangaleazzo Visconti.[40] Other similarities, however, are not exclusive to these works. For example, although Aristides' image of the acropolis at Athens as the center of a series of concentric rings, like the boss of a shield, is close to Bruni's view of Florence as the central knob of a buckler, surrounded by rings of walls, suburbs, villas, and walled towns, surely both have as their source Plato's description of Atlantis in the *Critias*.[41] Aristides' version is:

> For as if on a shield, layers have been set on one another, in fifth place, the fairest among all fills the area up to the boss: if Greece is in the center of the whole earth, and Attica in the center of Greece, and the city in the center of its territory, and again its namesake in the center of the city.[42]

This is quite close to Bruni's:

> Just as on a round buckler, where one ring is laid around the other, the innermost ring loses itself in the central knob that is the middle of the entire buckler. So here we see the regions lying like rings surrounding and enclosing one another. Within them Florence is first, similar to the central knob, the center of the whole orbit.[43]

Again, whereas both Aristides and Bruni take up the topic of citizenship extended to exiles from other cities and other proofs of generosity, their common source is very likely Thucydides.[44] The theme of *romanitas*, so central to Bruni's *Laudatio*, has no connection with the *Panathenaic Oration* for obvious reasons. While some sources for this theme are Roman, certain parts of Bruni's discussion are based on Aelius Aristides' *Roman Oration*.[45] This oration is the direct model for Bruni's paraphrase of a Homeric passage; in both texts, Homer's image of snow blanketing a large territory (*Iliad* XII 278–86) serves as metaphor for the great extent of the state. Here is Aristides' version:

> Homer said of snow, that it poured over and covered "the high mountain peaks and the summits of the headlands and the fields of asphodel and the rich works of men; and it poured over," he says, "the harbors and the beaches of the grey sea," such as this city also does.[46]

Bruni's is:

And as Homer writes of the snow that it falls thickly on the mountains and hills and covers the ridges of the mountains and finally the fertile fields, in like fashion handsome buildings cover the entire region outside the city and all the mountains, the hills and the plains, so that they seem more to have fallen from heaven than to have been constructed by the hands of men.[47]

Bruni's discussion of Florentine foreign policy draws on the *Roman Oration*, and both works conclude with a prayer to the protector-gods of the state.[48] Bruni's extended discussion of why Florence's situation near, but not on, the sea is the best one is borrowed from yet another *laus urbis* of the Greek Second Sophistic, Libanius's *Praise of Antioch*.[49] It is interesting that while Bruni's own translations from the Greek are almost all from the "classics," such as Aristotle, Plato, Xenophon, and Demosthenes, his sources for the *Laudatio* are mostly Second Sophistic. In part this is because the *laus urbis* itself is really a development of the second century A.D. and later. But it also reflects Bruni's approach to the Greek language through Byzantine scholarship. Aelius Aristides was a great favorite in Byzantium; Photius considered him to be a model equal to Demosthenes. Theodore Metochites wrote a comparison of Demosthenes and Aristides claiming they had equal merit, although in different styles. He suggested that this difference stemmed from the fact that one was politically active, while the other was not, and that this, in turn, was the result of external circumstances. Bruni, who used the same argument in his comparison of Dante and Petrarch, may have known Metochites' work.[50]

What is most revelatory, however, is not the identification of Bruni's specific sources for his *Laudatio*, but their transformation. Whereas only one of the seventy-one pages of Aristides' *Panathenaic Oration* discusses the architecture of Athens, architecture plays a central role in Bruni's account. Indeed, whereas Aristides ends his oration with a discussion of the deeds and nature of the Athenian citizens, Bruni chooses to begin with this topic, claiming that the Florentines' qualities are analogous to those of the city fabric: "Just as these citizens surpass all other men by a great deal in their natural genius, prudence, elegance and magnificence, so the city of Florence has surpassed all other cities in its prudent site and its splendour, architecture and cleanliness."[51]

That this analogy underlies Bruni's entire description is confirmed by a second passage, in which he claims that although it is hard to believe that Florence could have repulsed the Milanese forces, the splendid architectural achievements of the city persuade all visitors of the high level of excellence and greatness within the Florentines' capability.[52] This assumption of a direct correspondence between the qualities of a place and those of its citizens might seem to derive from the *Panathenaic Oration*—"the nature of the country, as we see it, clearly will agree with the nature of its people"—or from the *Roman Oration*.⁵³ But Aristides did not develop this trope with anything like Bruni's consistency; the latter uses the physical appearance of the city as both analogy for the qualities of its citizens and emanation of their virtues. Moreover, Bruni used the city's monuments as evidence of its cultural achievement in his earlier work, *Ad Petrum Paulum Histrum*, which has no special relation to Second Sophistic sources. Bruni's development of this analogy would seem to be the fruit of a cross-fertilization between Quintilian's precepts for the description of a city and Chrysoloras's notion, developed in his *Comparison*, that buildings are visible evidence for invisible ideas. From Quintilian, Bruni took the idea that a city should be praised within the same categories as an individual: "Cities are praised after the same fashion as men. The founder takes the place of the parent, and antiquity carries greater authority.... The virtues and vices revealed by their deeds are the same as in private individuals."⁵⁴ That Bruni conceived of the *laus urbis* in Quintilian's terms seems clear from his statement that "as one usually does in discussing an individual, so we want to investigate the origins of the Florentine people and to consider from what ancestor the Florentines derived and what they have accomplished at home and abroad in every age."⁵⁵

Seen as an accomplishment of the Florentines, the city's appearance reveals their virtues and vices. For Bruni, the physical aspect of the city does not *symbolize* its qualities; it *embodies* them and is therefore evidence for them. The Greek and Byzantine sources of this new approach to the interpretation of moral and cultural status on the basis of sensory evidence have been encountered in my discussion of Alberti's letter to Brunelleschi and of Chrysoloras's *Comparison* (chapters 2, 3, and 8). Further, and most important, it is evidence of the Humanists' assumption that the external appearance and internal being of a person correspond; the one can be inferred from the other. Thus Alberti urges the sculptor to endow each statue of a god or hero with "a costume and gesture that express, as well as the artist can, his life and customs."⁵⁶ The expressivity of physical appearance was also discussed by Bartolomeo Facio in *De Pictoribus*.⁵⁷ The same approach could be used to assess living men: Valla had one of his speakers praise another for his consistency in speech, gesture, clothes, food, house, and gardens, all of which displayed the same qualities of ordered beauty.⁵⁸ Just as the physical appearance of an individual reveals his inner qualities, so the built environment manifests the nature of its inhabitants. Although this insight was not systematically explored in ancient authors, Latin or Greek, it was especially attractive to Leonardo Bruni and to many of his contemporaries, who trusted the evidence of the senses more than abstract reason. In particular, it is related to the Humanists' tendency to argue from example and to convey general ideas through the concrete and the visual, tendencies intimately related to their training in rhetoric.[59]

As far as I can determine, this use of the built environment as evidence of national character is a genuine innovation that first appears in Florentine circles in the first decade of the Quattrocento. This mode of argument was used in Bruni's *Ad Petrum Paulum Histrum* (as I showed in chapter 2) of 1401, where each of the conversations is introduced by a short description of Florentine buildings designed to illustrate the cultural achievement of the city. It also served Salutati for his *Invectiva in Antonium Luschum Vicentinum*, written sometime before September 11, 1403.[60]

Coluccio Salutati

Baron discussed the *Invectiva*, together with Bruni's *Laudatio*, as stimulated by the death of Giangaleazzo Visconti in 1402 and as a product of the consequent effulgence of civic pride in Florence.[61] In reality, the dates of both Antonio Loschi's invective and Salutati's reply are uncertain: Loschi's dates sometime between 1397 and 1401; Salutati's between 1397 and 1403.[62] Loschi studied with Chrysoloras in Florence before his appointment as secretary to Giangaleazzo in 1399. Therefore, this exchange of *vituperatio* and *laudatio* took place between men who either were in the same intellectual circle (if the works date 1397–1398) or had recently shared that experience. This is not to deny the political and propagandistic intent of the two invectives, but only to call greater attention to common elements in their literary and intellectual background

In Salutati's text, an ekphrastic encomion serves as an emotionally charged conclusion to his invective. Responding to Loschi's attack, which had claimed that Florence was "the dregs of Italy," Salutati wrote:

> I can't believe that my Antonio Loschi, who has seen Florence, or anyone else at all, who saw Florence, could deny that it is the flower of Italy and its choicest part, unless he were downright foolish. Which city, not only in Italy but in the whole world, is more safely guarded by its walls, more superb in palaces, more ornamented in respect to temples, more beautiful by virtue of its buildings, more illustrious in its porticoes, more splendid in its piazzas, cheerful in respect to the breadth of its streets; which is greater in population, more glorious in its citizens, more infinite in riches or more cultivated in its fields? Which is more pleasing in its site, healthier in its skies or cleaner; which has more wells, sweeter water, is more industrious in the arts and more admirable in everything? Which is more edified in villas, mightier in towns, more numerous in municipalities or more abundant in farmers?

(Non credam Antonium Luschum meum, qui Florentiam vidit, nec aliquem alium, quisquis fuerit, si florentinam viderit urbam istam, esse vere florem et electissimam Italiae portionem, nisi prorsus desipiat, negaturum. Quaenam urbs, non in Italia solum, sed in universo terrarum orbe, est moenibus tutior, superbior palatiis, ornatior templis, formosior aedificiis, quae porticu clarior, platea speciosior, viarum amplitudine laetior, quae populo maior, gloriosior civibus, inexhaustior divitiis, cultior agris; quae gratior situ, salubrior caelo, mundior caeno; quae puteis crebrior, aquis suavior, operosior artibus, admirabilior omnibus, quaenam aedificatior villis, potentior oppidis, municipibus numerosior, agricolis abundantior?)[63]

His description marshals an arsenal of comparative adjectives evocative of morally and aesthetically desirable qualities. Some of these terms have associations with Scholastic aesthetic theory (*formosior, clarior, speciosior*); others with Classical panegyric (*superbior, maior, gloriosior*) and literary criticism (*ornatior, gratior*). Of special interest is the characterization of the city in anthropomorphic terms, such as *laetior*, used in an ambiguous fashion that could refer to either a place or a person. For example, the adjectives *cultior* and *aedificatior* have literal sense in relation to fields and villas while implying figuratively the "cultivated" and "edified" condition of the citizens (or of the city seen as an individual). This elegant wordplay would seem to be an example of the transfer of master terms from one context to another that I associated with Chrysoloras's teaching in the preceding chapters; its direct inspiration is Quintilian's analogy between the praise of a city and that of an individual. Thus this short description reintegrates the built environment with human values, taste, and aspirations.

What has escaped notice is the literary connection between the two city descriptions by Salutati and Bruni. The chronological relation between Salutati's description and Bruni's *Laudatio* is problematic: if the latter dates to 1403 or 1404, then Salutati's work is earlier or exactly contemporary. There are specific points of coincidence in the descriptions: both introduce cleanliness as a new criterion of excellence. Undoubtedly, this reflects pride in the urban legislation of the city.[64] Yet the notion of raising cleanliness to a general virtue probably was inspired by a passage in Plato's *Laws* (778C–779D). Although Salutati was not one of Chrysoloras's students and read no Greek, he had been instrumental in inviting him to Florence and certainly participated in the learned discussions in his home. That the city of Florence, its character and achievements, was a favorite subject is suggested by reflections of these conversations in *Ad Petrum Paulum Histrum*, the *Invectiva*, and the *Laudatio*. Chrysoloras provided Bruni with literary models for his composition that were inaccessible to Salutati. But the principles and aims of their descriptions are almost identical:

both use physical structures to make general points; both treat urban description according to the rules for description of individuals.

To sum up: for both Bruni and Salutati, the authority on the *laus urbis* was an ancient Latin author. But Chrysoloras was instrumental in showing how Quintilian's precept that a city should be discussed according to the same categories as an individual could be applied, both by providing literary models and through his principle of decompartmentalization. These observations substantiate and confirm Baron's assessment of the importance of Greek models for Bruni's *Laudatio*: "He found in the Greek work conceptual patterns which he could use to impose a rational order upon his observations of the world in which he lived," and Greek abstraction helped him achieve a "concrete grasp of reality and a visual clarity."[65]

Guarino Veronese

Of all the early Humanists in the circle of Chrysoloras, Guarino Veronese was best equipped by education to write a *laus urbis*. The most devoted of his students, Guarino accompanied the master to Constantinople in 1403 and remained in the city for five years.[66] There he acquired the foundations of an impressive Greek library, including works by the two great exponents of the *laus urbis*, Libanius and Aelius Aristides, and the standard handbooks on rhetoric by Aphthonius and Hermogenes.[67] His studies enabled him to compose the first literary analysis of a *laus urbis* known to me, in response to Chrysoloras's *Comparison of Old and New Rome*, while he was in Florence in 1411. Guarino wrote how much he admired Chrysoloras's

> praise of both cities, of the first parent on the one hand and her daughter on the other, in which the golden oration mounts in a style of speaking so elegant, magnificent and noble that in it nothing which might pertain to the orator's task appears passed by; hence its delightful invention, its most suitable order of things, abounding in pungent thoughts and in the most elegant adornment of its words.[68]

Guarino's comments are clear evidence that he understood the *laus urbis* to be a rhetorical genre requiring evaluation under the categories of *inventio*, *dispositio*, and *elocutio*, the headings into which most Classical handbooks of rhetoric divide the art. Further, he knew that it is epideictic, aiming to give pleasure to the audience through elegant and ornate language. Thus his comments quite appropriately address the form more than the content of the piece. However, Guarino also knew that the aim of an ekphrasis is to enable the spectator to form a mental image of the thing described. Therefore, he assured Chrysoloras that "indeed, I attain no insignificant fruit in the midst of my reading, since I seem not only to be listening to you, but to the city of Byzantium herself, a sweet sight to me and a most kindly nurse during my five years under your tutelage."⁶⁹

Given Guarino's sophisticated analysis of Chysoloras's work in 1411, his advice to one of his students in 1427 comes as a shock. He recommends praising the country and condemning the city under the four topoi of utility, pleasure, virtue, and excellence, urging the student to remember this by memorizing the following couplet:

> Four things to praise all topics ample go
> Virtue and use, pleasure and goodness show.⁷⁰
>
> (Quattuor ista solent augere negotia cuncta:
> Utile, iocundum, laudes, iungetur honestas.)

The differences in language and in conceptual sophistication between these two characterizations of the art of description might suggest that Guarino's students were taught a simplified ekphrastic method, either because they were inadequately prepared to compose at a higher level or because Guarino felt this to be sufficient. Yet Baxandall has shown that the only descriptions of paintings in the first half of the Quattrocento are from Guarino's students or circle.⁷¹ It could by proposed, then, that Guarino's four topoi could be applied with varying degrees of complexity and sophistication. In creating general rules, Guarino left his students free to apply them as they wished; I do not agree with Grafton's conclusion that the teacher did not expect his students to develop original or independent ideas, to express their emotions, or to treat topics in a fresh manner.⁷² Instead, his general categories derive from Chrysoloras's procedure as examined in chapter 7.

Guarino himself did not compose architectural ekphrases as an independent genre (if we except his praise of his country villa), and his assessment of the role that should be accorded description in the writing of history was not very positive: the site of an action should be described, "non quidem ad satietem atque fastidium, sed quantum rei gestae usus postulat."⁷³ Given this attitude, it is somewhat ironic that he was the only one of Nicolas V's translators to be assigned a work that consists of place descriptions, Strabo's *Geography*. Guarino's warning may follow Lucian's *How to Write History*: "You need to use special discretion in descriptions of mountains, fortifications, and rivers, to avoid the appearance of a tasteless display of your word-power and of indulging your own interests at the expense of the history."⁷⁴

Avesani has linked Guarino's teaching with the numerous encomia of Verona produced in the Quattrocento.⁷⁵ From a letter of 1436, it is clear that Guarino conversed about the praises of Verona.⁷⁶ His influence, together with the local literary tradition of the *Versus di Verona*, may have contributed to the production of Giovan Maria Filelfo's *Oratio de iis quae requiruntur in civitate bene morata et de laudibus Veronae civitatis* of 1467, Francesco Corna da Soncino's *Fioretto* of 1477 (which records Guarino as poet and orator), Giovanni Antonio Panteo's *De laudibus Veronae* of 1484, and Bernardino Barduzzi's *Epistola in laudem civitatis Veronae* of 1489.[77] This role as advocate of architectural description can be substantiated with other evidence. On at least one occasion, Guarino asked a former student to send him a description of his new surroundings.[78] Guarino's influence can also be detected in Michele Savonarola's *Libellus de magnificis ornamentis regie civitatis Padue* (1446–1447).[79] The author, a professor of medicine at Padua, was invited in 1440 to teach at the Studio in Ferrara, where Guarino had been employed since 1429. Although the description follows medieval conventions and cites few Classical authors, the lengthy accounts of paintings and appreciation for their artists connect this work with those of Guarino's circle.

Guarino's contribution to the art of architectural description was important not only because he popularized the genre of the *laus urbis* but also because he made explicit those criteria on the basis of which ekphrases should be both composed and evaluated. That these criteria were applied in architectural descriptions is proved by at least one example. The so-called testament of Nicholas V, probably composed by Giannozzo Manetti after 1455, explains the pope's building projects, "worthy of memory and praise," within Guarino's four categories. The works were intended for the defense of the papacy (utility), for ornament (pleasure), for the healthiness of the air (virtue), and for the increase of devotion (excellence).[80]

Poggio Bracciolini

Poggio probably knew Chrysoloras when he was Niccolò Niccoli's protégé in Florence; he certainly was in contact with him at the Curia in Rome in 1411.[81] That they were close is shown by the fact that he delivered a eulogy for the Byzantine scholar in 1415. As late as 1452, Poggio collected all the works that had been written about the master and disseminated them under the title *Chrysolorina*.[82] In a letter to Guarino, he praised Chrysoloras for encouraging many to take up eloquence, as a result of which, he claimed, "the old highly wrought style was almost revived."[83]

It is usually thought that Poggio began his Greek studies only in 1443 or 1444, although Marsh has proposed the earlier date of 1428.[84] Marsh's argument centered on Poggio's translation of Lucian's *Iuppiter confutatus*, a work that, he observed, treated the same theme as Poggio's *De Varietate fortunae*, the text with which I will be concerned. Although there is no evidence that Poggio studied Greek with Chrysoloras, the Byzantine scholar's educational method is evident in his decision to make an *ad sententiam* translation of Lucian. We know that Chrysoloras gave beginning students texts by this author for translation, and that he preferred translation *ad sententiam* to the traditional translation *ad verbum*.[85] If Marsh's dating is correct, Poggio's effort may also reflect his response to Bruni's diffusion of Chrysoloras's principle in his treatise on the correct way to translate (1424–1426).

Sometime between 1424 and 1431, Poggio composed his *Ruinarum urbis Romae descriptio*, incorporated in 1448 into the *Historiae de varietate fortunae*.[86] Poggio brought a much broader literary experience to the task of description than Vergerio or Bruni due to his own manuscript discoveries, including a Vitruvius, an Ammianus Marcellinus, and the *Einsiedeln Itinerary*. He also drew on a wider range of Greek authors, including Dionysius of Halicarnassus, Herodotus, and Diodorus Siculus in addition to Libanius.[87] These experiences enabled him to compose a classicizing *laus urbis*, as his *In Laudem rei publicae venetarum* (1459) demonstrates.[88] His work on Rome, *De Varietate fortunae*, is not a *laus urbis*, but a meditation on Fortune in the tradition of Lucian, Petrarch, Vergerio, and Chrysoloras. Its thematic content, the importance placed on architectural description, and its rhetorical character all suggest that it is best understood as work of Chrysoloras's circle, despite its late date. Since Poggio borrowed a quotation from Libanius from Chrysoloras's *Comparison*, he surely used that work as a source.[89] Poggio's account is organized as a dialogue between himself and Antonio Loschi, taking place when the two friends, on holiday from the Curia, strolled up the Capitoline hill. Antonio recalls a line about the site from Virgil—"now golden, once bristly with thickets of thorn" (*Aeneid* VIII)—and suggests that now it would be more appropriately worded "once golden, now crammed with briars and brambles." There is no greater spectacle of the two kinds of Fortune (a reference to Petrarch) than to see this city. Once great, now "stripped of all decency it lies prostrate like a rotted giant corpse."[90]

The last section of Poggio's work is a long meditation on the nature of Fortune, developing the theme suggested by Lucian's work (which he was translating at that time), by Chrysoloras in his *Comparison of Old and New Rome*, by Petrarch in his *De Remediis utriusque fortunae*, and by others. Change, he concludes, is the essential law of human affairs. By comparing past and present, the past known primarily through literary evidence and the present through visible signs, Poggio draws ontological and ethical conclusions important for guiding right action. For Petrarch, the Roman ruins were examples of virtue and power that functioned as timeless yardsticks of excellence; Poggio's conclusions privilege, instead, the historical, the relative, and the contingent.[91] Above all, the sight of the ruins suggested to him the necessity of writing the history of his own time so that, just as his contemporaries knew the great deeds of the past through the ancient historians, future generations would remember his era. The deeds of the Ancients were no greater than ours, he claimed; they seem so only because they were exaggerated by the eloquence of the ancient historians.[92]

The Attack on Bruni's *Laudatio*

In 1437, Francesco Pizolpasso received a long letter from Leonardo Bruni in which the Florentine Humanist defended his *Laudatio florentinae urbis* against criticisms from Lorenzo Valla and Pier Candido Decembrio.[93] Both men had accused him of lying. Valla, calling Bruni an idiot, criticized the *Laudatio* on grounds that attacked the very nature of the work. Bruni had written, said Valla, as though no one would respond to him and even as though everyone would agree with his absurd claims.[94] This demonstrated his superficiality, vanity, and disregard for others' opinions. Valla's first criticism ignored the fact that Bruni was writing panegyric and condemned him for writing bad history. By refusing to recognize the genre, Valla dismissed it from the realm of permissibility. Evidently, he believed that one must write as though in a dialogue with the reader, anticipating objections and providing evidence to support one's statements. He refused to accept historical writing that aimed to give pleasure; all writing in this genre ought to be concerned with truth. His second criticism charged that Bruni had misrepresented Florentine history, pointing out (among other things) that it was ridiculous to say that the Florentines were heirs to the Romans, as though the Romans no longer existed.[95] Finally, he criticized Bruni's prose as slack, garrulous, and enervated, deficient in seriousness and inspiration, and full of grammatical errors.[96] Since spontaneity of expression was considered a virtue in panegyric, the direct, almost spoken quality of Bruni's prose may have been dictated by the genre. That Bruni was sensitive to the need to use different styles for different sorts of subject matter is clear from the preface to his translation of Aristotle's *Politics*, in which he explains that since it is a work of moral philosophy, its "civic subject matter calls for a terminology different from that used by the Scholastics in their disputations and a rather more elegant style."[97] But Valla's objections challenge the permissibility of reviving ancient epideictic rhetoric at all.

Decembrio's attack, in the opening section of his own *De Laudibus Mediolanensium urbis panegyricus*, focused the question precisely on this issue of eloquence versus truth: "Now he [Bruni] writes splendidly and elegantly, so that unless you consider very carefully you may easily believe that what he says is true. But nothing is more excellent than truth; the greatest eloquence must cede to and is conquered by that."[98] These attacks launched a polemic in which the newly recovered eloquence, with its tropes and topoi, was seen as antithetical to truth. The challenge to eloquence, which initially centered on Bruni's *Laudatio*, had strong political as well as literary motivations. Florence and Milan were again in conflict, and the propagandistic value of Bruni's work was understood in Milanese circles.[99] The polemic should be seen in the context of the pope's search for a new city to host the Council of Basel. Baron argued that Bruni sent his *Laudatio* to the fathers at Basel in the mid-1430s precisely as propaganda for the suitability of Florence and that Decembrio's panegyric of Milan aimed to further the candidacy of Milan.[100] Not only the *Laudatio* but other of Bruni's writings served Florentine interests. Lanza saw Bruni's *Vita di Dante e del Petrarca* of 1436, which claimed that eloquence declines under tyranny, as a challenge to the regime of Filippo Maria Visconti.[101] This context may help to account for the panegyric of Florence that Visconti sent to Poggio in 1438, a work actually written by Decembrio.[102] The letter, by showing how Florence should have been praised by Bruni—truthfully, that is—demonstrated the tyrant's own, acceptable, eloquence.

What is Bruni's defense? He begins his letter to Pizolpasso by explaining the circumstances in which he composed the *Laudatio*. He was a young man and had just begun his study of Greek letters; the composition was a child's game, an exercise in rhetoric.[103] He took as his model Aristides' beautiful *Panathenaic Oration*, which belongs to that type of rhetoric the Greeks call *panegyricum*.[104] This genre does not linger over subtleties, but seeks the applause of the crowd by taking every occasion to praise and by exalting the subject in words. Thus, continued Bruni, Aristides says many things that are not true of Athens but that gave pleasure to his audience, since, in praising a city, you are addressing the very people you praise.[105] After all, this genre is neither forensic nor deliberative, but seeks to give pleasure. Having neatly explained the place of panegyric within the divisions of rhetoric, Bruni rebuts the accusations that he has falsified history. History is one thing; praise is another: in history you must follow the truth, but panegyric praises far beyond the truth in order to give pleasure ("aliud est enim historia, aliud laudatio. Historia quidem veritatem sequi debet, laudatio vero multa supra veritatem extollit").[106]

Bruni's distinction between history and panegyric addressed a problem in Humanist historiography that has received little attention by modern scholars. The revival of eloquence—the art of persuasion—raised the problem of what role it might be permitted in historiography. As early as 1392, Salutati rejected the historiography of the Moderns, like Vincent of Beauvais, which made the reader meditate on moral precepts, in favor of ancient historians whose eloquence, by moving and delighting the spirit, stimulated morality by presenting examples to follow.[107] As we saw, the development of architectural ekphrasis in early Quattrocento Florence was closely linked with the desire to argue from example and on the basis of sensory evidence. Such description, rich in adjectives and literary figures, was evidently value-laden. Buildings and cities were described, not for their own sakes, but in order to argue a point. This was true not only of Bruni's *Laudatio* and Salutati's response to Loschi's invective, but also of Alberti's letter to Brunelleschi, of 1436. All these works served a useful historiographic purpose, attributing to Florence and its architecture a preeminent cultural and moral status, which has been convincing to modern historians. And all these works, of overtly rhetorical literary character, afforded pleasure to the reader. All are interpretations of a built environment actually experienced by the authors; all seek the readers' assent to the meanings proposed. Although Bruni claimed that his *Laudatio* was not history, it recorded the achievements of the city for posterity, and this, for many in the Quattrocento, was the very purpose of historical writing. Indeed, Bruni himself had the speakers of the second dialogue of *Ad Petrum Paulum Histrum* praise him for raising the Florentine historical consciousness with his *Laudatio*. But are subjective interpretation, persuasion, and a desire to delight the reader permissible within historical writing of any sort? In part, the argument centered on old polarities of utility versus ornament, necessity versus beauty, and content versus form; distrust of eloquence was by no means new, as we saw in chapter 5.[108] It also brought into focus the problem of defining the limits of various kinds of historical writing, some of which might be permitted an encomiastic character and others not. To some degree, this task was necessitated by the Humanists' discovery that the ancient historians were not always truthful.

Guarino wrote, on the first page of his copy of Pliny the Younger's panegyric to Trajan (discovered by Aurispa in 1434), that he wished Pliny had been more honest, so that a good orator would not be discovered to be a liar.[109] In a letter of 1446, he insisted that the first and only aim of history is utility—that is, to tell the truth.[110] Although praise was permitted, it must be moderate and appropriate. Guarino quoted from *De Oratore* (II.62): "The first law of history is this—not to dare utter any falsehood lest there be a hint of partiality in one's writing."[111] In his *De Varietate fortunae*, written, as we saw, in these years, Poggio distinguished between the ability of the Ancients to win credence for legendary events by means of eloquence and the failure of his contemporaries to do the same. If the Moderns tried to magnify recent events, which everybody knows, they would be rejected as liars; indeed, the Ancients had greatly exaggerated the deeds that they recount.[112] Therefore, he said, he preferred to use a plain and unadorned style.[113]

The immediate effect of the controversy on architectural description is striking. Aeneas Silvius Piccolomini wrote, in the letter accompanying his 1438 description of Basel, that everything he had written was strictly true, and that he avoided using fancy words and rhetorical flourishes to that end.[114] Widmer saw the description of Basel, a copy of which was sent to Pizolpasso, as a response to the attack on Bruni and his self-defense of the previous year as well as an attempt to keep the council at Basel.¹¹⁵ Michele Savonarola wrote, in his 1446 panegyric of Padua, that everything he had related should be believed, since there was nothing made up or lying in his work.¹¹⁶ Filippo de'Diversi de Quartigiani's description of Ragusa, of 1440, and Cyriacus of Ancona's *Anconitana Illiricaque laus* of the same year are truthful accounts of the history, site, and appearance of these cities.¹¹⁷

Despite the criticisms of Bruni's *Laudatio*, his work served as a literary model for the very men who reviled it. Goldbrunner has traced the impact of the *Laudatio* on later fifteenth-century examples of the *laus urbis*, such as Decembrio's on Milan, Lauro Quirini's on Venice, and Piccolomini's on Basel.¹¹⁸ In particular, these later works repeat—sometimes word for word— Bruni's architectural descriptions.¹¹⁹ This, I believe, suggests not only the authority of Bruni's work, but also the difficulty of the task and the paucity of ekphrastic models available to the fifteenth-century Humanists. Decembrio developed Bruni's assumption that description of a city and description of a person are essentially the same thing. He began his panegyric of Milan by drawing a parallel between the difficulty of deciding what aspect of Helen to describe first and that of deciding what aspect of Milan he should begin with; he resolved this problem by concluding that since his task was like that of a portrait painter, he would begin with the head and then go on to show the beautiful proportions between the parts of the body.¹²⁰ He also repeated Bruni's image, but with an inversion, still borrowing from Homer via Aelius Aristides, but denying that the city extended over a large territory like snow blanketing the hills.

> For not, as Homer writes of the snow, "fallen from heaven, it blankets the mountains and the hills and the darkness of the mountains and the rich farmlands," do these buildings blanket all the hills around and the plain; in fact, all the cities in Italy seemed as though by some fate to have congregated in one place.¹²¹

The polemic over the *laus urbis* took place at the moment when the genre was becoming popular and therefore controversial. For the moment, panegyric was accepted as a subgenre of history, with the proviso that nothing untrue was said. Decembrio himself titled his work on Milan a "panegyricus" and wrote approvingly to Pizolpasso about the latter's effort in the genre.¹²² Further, as we saw, the Milanese Humanist himself composed a panegyric on Florence. However, the prose *laus urbis* had little future in the second half of the Quattrocento. It was, instead, another aspect of Bruni's work that influenced the future direction of Quattrocento historiography: contemporaries realized that Bruni's eloquence was a powerful instrument for political propaganda. Architectural description became an instrument for the panegyric of individuals in the third quarter of the century.

History and Panegyric

Manetti's *De Vita ac moribus Nicolai Quinti*, of about 1455, had to take into account the controversy that began in the 1430s over the issue of truthfulness and the nature of historical writing.[123] His description of Nicholas's achievements is not truthful: the projects that he describes in such detail were not all completed and some were hardly begun.[124] He tells us that Nicholas built St. Peter's from the ground up ("a fundamentis magnifice renovare, ac totam usque ad tecta ipsa reaedificare constituerat"), which was hardly the case.[125] Unlike Bruni, Manetti did not claim the right of exaggeration on the grounds that he was writing panegyric and not history. The title of Manetti's work, and those of its individual parts, suggest that he intended to write history. The first book considers Nicholas's life and character; the second, his deeds; the third is the so-called testament of Nicholas. All three books constitute the manuscript in the Laurentian Library in Florence (MS Plut. LXVI, n. 23).[126] His comparison between St. Peter's and Philon's arsenal supports this interpretation of his intention. If the Greek historians considered the author of this achievement worthy of recording for posterity, he says, surely the same honor should be accorded to Nicholas.[127] The purpose of his panegyrical ekphrasis as a work of contemporary history was, therefore, made explicit. His comparison between Nicholas and the builder of Solomon's temple and palace serves the same point: Scripture authorizes the praise of famous men on account of their architectural works. Both Classical historians and Scriptural writers serve to justify Manetti's use of architectural patronage in the *Vita Nicolai*.

Manetti's sources for the creation of an encomiastic biography have been perceptively analyzed by Onofri, who showed his dependence on Plutarch, Cicero, and Macrobius, authors who distinguished biography from history as a genre that does not have to be 'true,' and that can create the desired image of the person through selection of facts and events.[128] I would add that there was also considerable precedent among later Classical historians to combine the two genres. For instance, Dionysius of Halicarnassus's conception of historiography as panegyric—that is, as the encomion of friends—in his *On Literary Composition* may have influenced Manetti's composition of the *Vita Nicolai*.[129] The genres are also confused in Tacitus's recently rediscovered *Agricola*, which is a biographical encomion.[130] Also important, I believe, was Cicero's lengthy letter to Lucceius (of which Manetti owned a copy), in which the orator begged the historian to make his name famous in both the present and the future and to elogize his achievements "with even more warmth than perhaps you feel, and in that respect to disregard the canons of history.'"[131] The *Epistulae familiares*, known since 1392 and used as texts in Guarino's school, revealed the hypocrisy in Cicero's statement in *De Oratore* (quoted by Guarino, as we saw) that history had to be truthful. Only now were the Humanists ready to confront the fact that Cicero himself had condoned the use of eloquence to distort history.

Cochrane saw Manetti's willingness to distort fact in order to praise the better as representing a decline in Humanist historiography.[132] Miglio, commenting on another panegyrist of Nicholas, Michele Canensi, deplored the fact that he praised Rome only as a reflection of the importance of the pope.[133] But Ianziti has recently proposed that such works are not empty exercises in rhetoric; rather, they should be seen as a new type of Humanist historical narrative in which "the relation of contemporary events [is] organized around a single personage who was to be glorified through the recitation of his deeds."[134] Ianziti saw, as the pivotal work in this development, Bartolomeo Facio's *De Rebus gestis Alphonsi I commentarii* of 1455, and its major source as Caesar's *Commentarius*.[135] Although neither of these works is important from the point of view of architectural description, it seems clear that Manetti's *Vita Nicholai*, quite different in its array of sources but almost exactly contemporary in date, belongs to this historiographic trend. Ianziti rightly observed that this laudatory literature was propagandistic in intent and that it corresponded to a need "for a type of political apology which could serve as a kind of legitimization of power newly acquired."[136] Most frequently, it took the form of an encomion of an individual's deeds.

This historiographic trend has roots in the earlier Quattrocento. Grendler associated the very revival of Classical rhetoric with the funeral oration for Francesco da Carrara il Vecchio delivered by Vergerio in Padua in 1393.[137] Its epideictic character and the use of eulogy to foster admiration for Carrara were developed for living subjects by others. For example, Bruni praised Carlo Malatesta in a letter of 1409, and Guarino wrote a panegyric on the count of Carmagnola in 1428.[138] Guarino's letter on history, of 1446, was addressed to a friend whose task was to celebrate Sigismondo Malatesta's deeds.[139] These encomia would seem to have been inspired by the desire to laud the great deeds of the present for posterity. Poggio, as we saw, considered this to be the most urgent task of contemporary historiography. Describing Sigismund's coronation in Rome in 1433, he wrote to Niccoli: "Although I know, dear Nicolaus, that you do not admire what goes on in our generation because, I think, you concentrate on deeds of earlier men, which were indeed magnificent and worthy of the highest praise, still I think that you should not despise some of the more recent happenings."[140] Alberti's praise of Brunelleschi is certainly to be understood within this trend. Nonetheless, the genre seems especially to have been developed by Humanists associated with the court of Milan, perhaps because this was the greatest tyranny in Italy. As early as 1429, the archbishop of Milan had warned Filippo Maria Visconti that his fame would not be handed down to posterity unless it were recorded by a poet or a historian: Achilles had Homer; Aeneas, Virgil; and the Florentines, Leonardo Bruni.[141] He urged Filippo Maria to appoint Panormita court poet. Francesco Pizolpasso's description of Branda Castiglione's architectural patronage, of 1431, has a strongly encomiastic character.[142] Maffeo Vegio praised Filippo Maria in his *Convivium deorum*.[143] Just after 1450, Filelfo wrote his laudatory epic poem *Sforziade*, at least the idea for which must have been inspired by the great examplar of panegyrical biography, the *Aeneid*.[144] Lodrisio Crivelli followed in 1461–1463 with *De Vita rebusque gestis Francisci Sfortiae*. In his review of the state of contemporary history, Crivelli acknowledged the importance of Bruni's *Laudatio* in creating a positive image of the Florentine state.[145] Decembrio closed his career with the *Annotatio rerum gestarum Francisci Sfortiae*.[146] These works are not biographies, but praises of selected actions, and they are overtly partisan. Such characteristics, however, ought not exclude them from consideration as historical writing insofar as their aim was to transmit a record of the present to the future. Their mendacious character derives, at least in part, from, first, the authors' suspicion that ancient historians had willfully distorted the truth in order to transmit an idealized version of their present to the future; second, their observation of the efficacy of Bruni's *Laudatio* in creating a positive, if exaggerated, image of Florence; and, third, their desire to transmit to posterity the same glorious record of achievement they had received from the past. In fact, this new encomiastic history reflects serious rethinking of the veracity of the ancient historians, the importance of eloquence for historical writing, and the necessity of projecting an idealized view of the Humanists' present into the future. These three aspects were inextricably intertwined in the Humanists' discussions.

After midcentury, political developments in Italy presented new opportunities for the praise of individual leaders. Ianziti noted the importance of Alfonso's court in Naples, but the newly strengthened papacy in Rome and the dominance of the Medici in Florence also provided fertile ground for the new historiography. The importance of eloquence in celebrating the deeds of modern men was urged by Landino in 1467.

> By means of eloquence the Ancients made known to us the wonderful things which were done at various times. With eloquence, those who live today will show to those born centuries from now all that has been done in our times which is worth remembering. And who is so stupid as not to understand that history is the true teacher of human life?[147]

Cortesi was convinced that without eloquence it was impossible to write history, not only because there was no sense in having the facts right if they were expressed obscurely, but also because history, composed of *res* and *verba*, served both *delectatio* and *utilitas*.[148] Benedetto Accolti, in his *Dialogus* of 1462–1463, saw history and rhetoric as inseparable, "because it was eloquence which enabled the historian to perform his essential function of rescuing the deeds of great men from oblivion."[149] Through panegyric, ancient historians made their subjects' achievements seem great:

> Their writers had such genius that they made mediocre and often even trifling deeds seem great through the force of their eloquence.... there was no limit on the falsehoods contained in history.... How great, how admirable, how similar to those of the Ancients would their [modern men's] deeds seem if worthy authors had only celebrated them!"[150]

This new confidence was nourished by the rejection of the notion that certain ancient authors were infallible within the particular disciplines assigned to them.[151] The growing realization that the authorities were historically conditioned, fallible human beings who might have exaggerated, or even lied about, what they recounted led to the notion that they might be corrected and even surpassed. Black has shown how, just after midcentury, Humanists began to defend the achievements of the Moderns over the Ancients across a wide range of disciplines.[152] Whereas Bruni's *Ad Petrum Paulum Histrum* had argued the question solely in respect to literature (and argued it on both sides of the question), Accolti's *Dialogus* praised modern warfare, morals, statesmanship, poetry, rhetoric, philosophy, law, religion—and architecture. Alberti argued that the Ancients had used buildings as a means for obtaining greater fame than they in fact deserved:

> Thucydides was right to approve the prudence of the Ancients who adorned their cities with every kind of building so as to seem much more powerful than they were. And who of the greatest and wisest princes did not consider architecture one of the most important means for gaining personal renown in posterity?"[153]

Filarete, at the court of Milan, urged the importance of architectural patronage as a means of obtaining fame:

> And just as great lords have fame, so do buildings. Each in their own way contributes to our recognition of their fame. We know from letters about the great fame deserved by many men on account of their great deeds, that is, on account of the great buildings made by these men.[154]

Such use of architectural evidence became extremely popular in the 1460s and 1470s. Indeed, one of the striking characteristics of architectural description in the second half of the Quattrocento is its 'privatization,' the tendency, that is, to take the form of encomia to patrons and to describe private palaces in addition to, and sometimes instead of, religious and public edifices. They also reflect changes in patronage in the later part of the century, since patronage of private works for private citizens is typical of the second half of the Quattrocento just as the patronage of public works of groups characterized the first half.[155] The new values are apparent in Benedetto Dei's *Descrizione di Firenze* of 1480, where he singles out thirty-three "muraglie grandissime e di gran fama e grandissima spesa"—a kind of "Wonders of Florence."[156] Of these, the first is Brunelleschi's dome, seven are churches, and nineteen are palaces of private citizens. There seem to be no equivalents for Bruni's *Laudatio* or Manetti's description of the consecration of Florence Cathedral in the second half of the century. One might make an exception of Cristoforo Landino's eulogy of Florence in the proem to his commentary on the *Divine Comedy*. But even here, the city's fame is largely (although not exclusively) seen as resting on achievements by private citizens.[157] The new importance accorded to architectural description within this historiographic trend corresponds to the shrinking scope of its subject matter. As Polybius observed:

> When dealing with a subject which is simple and uniform they wish to be thought historians.... they are compelled to magnify small matters, to touch up and elaborate brief statements of fact and to convert quite incidental occurences of no moment into momentous events and actions.... As for sieges, descriptions of places, and such matters, it would be hard to describe adequately how they work them up for lack of real matter.[158]

Whereas architecture played a small role in Humanists' discussions in the first part of the century, and was of interest only in connection with important moral, political, and cultural concerns, from Manetti onward the panegyrists mined the subject for their eulogies of individuals' achievements.

Pius II's extended account of his own palace at Pienza in his *Commentarii* and other accounts of the same project illustrate these trends. Of the six contemporary accounts appended to Mack's study of Pienza, many were expressly written as panegyrics of the project and its patron.[159] Four of them are poems; two are prose compositions. Many are not so much exaggerated descriptions as downright untruthful. Thus Porcellio Pandoni tells us that at Pienza "a hundred halls resting on pillars shine with gold," and that Cardinal Lolli's palace was built of "gold and Parian marble."

> Te quoque in urbe Pia fundasse palatia, Lolli,
> Aurea et Pario marmore fama fuit.

THE RECOVERY OF ELOQUENCE 195

> . . .
> In caelum summi surgunt fastigia tecti
> et centum frons est undique lata pedes,
> Irradiantque auro centum fundata columnis
> atria materiam nobile vincit opus.[160]

The literary models for these descriptions are of a totally different nature from those examined in the preceding two chapters. They do not come from public oratory or Second Sophistic authors, and they have no relation to civic concerns. Instead, they are inspired by imaginative descriptions of fabulous palaces, usually in poetic sources. Petrarch's description of a palace constructed of precious materials in *Africa* is important, as are Ovid's description of a similar structure in *Metamorphosis*, Prudentius's in the *Psychomachia*, and Martianus Capella's in *De Nuptiis Philologiae et Mercurii*.[161] Interestingly, two of the panegyrists—Porcellio Pandoni and Lodrisio Crivelli—had served at the court of Milan.

A similar group of encomiastic texts clusters around Cosimo de'Medici's building projects, as Hatfield has shown.[162] These include the so-called terze rime, letters from members of the Sforza court, and descriptions by Filarete and Alberto Avogadro da Vercelli. Again, many authors had connections with the court of Milan; many use poetic form. Avogadro, claiming that Cosimo had resolved to leave monuments through which his name would be remembered, felt no need to stick to the facts.[163] We are told that the façade of the Medici palace is of porphyry, serpentine, and emerald-colored stones, while in reality it is of a sober sandstone.

> Sed faciem non marmor habet, quod molle bitumen
>   Nectat, habet triplici colore petra.
> Conspicui loca summa tenet petra digna alabastri,
>   Dextraque porphyreus vult tenuisse lapis.
> At, quem serpentis vulgo dixere priores,
>   Laeva habet, O Clarri ianua digna modis,
> Et qua signa potes muris spectasse pilisque
>   Culta rubris, sex sunt nec numerata prior.[164]

To the texts discussed by Hatfield, I would add Niccolò Tignosi's encomion of Cosimo, of about 1459, which compares San Lorenzo with the Temple at Ephesus and Hagia Sophia, and Landino's poem *Ad Johannem Salvettum de laudibus magni Cosmi*, praising San Lorenzo, San Marco, and Santa Croce.[165]

While such works undoubtedly afforded pleasure to their recipients, they also served a useful purpose in legitimizing the subjects' roles as founders and builders, and therefore as owners. But this intention was not admitted by the authors. Michele Canensi, praising Nicholas V's building patronage, argued that the pope's pius example would persuade others to follow it.[166] Lauro Quirini defended the importance of history for the encouragement of right action, claiming that it not only was the custodian of the past and light of truth, but also had the task of praising virtue and blaming wrongdoing: "Indeed, history is the custodian of time, the light of truth, the life of memory, she who brings news of antiquity and praises virtue and damns vice and malice."[167] At the beginning of the century, Salutati had embraced the notion that history should guide moral action, but he had in mind the actions of citizens in the service of the *patria*. Now the recipients of praise (for there was no blame) were magnates and princes bestowing benevolence in return for adulation.

Praise and blame are, of course, the traditional province of epideictic rhetoric, and Quirini must have realized this when he urged the historian to acquire the linguistic skills for persuasion.[168] This idea depends above all on Cicero's discussion of the ethical nature of epideictic oratory as encouraging men to virtue or reclaiming them from vice through praise and blame (*De Oratore* II.ix.35). But even within Cicero's dialogue, other speakers associated the epideictic genre with pleasure, rather than persuasion or teaching. Aristotle's discussion of epideictic oratory is similarly undecided about whether the genre has an ethical function (*The Art of Rhetoric* I.ix.32–41).

The problems raised by these untruthful and encomiastic accounts were often resolved in the third quarter of the century by choosing to write poetry rather than prose. Encomiastic biography in poetic form had been sanctioned by Plato's approval: the poetry of praise, he said, encourages reverence, patriotism, and virtue while it confers fame on good citizens (*Laws* 80.1; *Republic* X.vii.607A; *Protagoras* 325–62). Quintilian had argued that the poet and the historian were closely related, but that the poet has greater liberty to embellish and ornament his material; moreover, the poet seeks to give pleasure by inventing what is not only untrue, but sometimes even incredible.[169] His argument was taken up by Poggio in *De Varietate* and Guarino in 1446, where the difference between poetry and history was said to reside precisely in the poet's freedom to describe what is unreal ("poetis enim fingere ornandi causa licet; redeo ad historiam").[170] Pontano, who continued this discussion in the second half of the century, saw the subject matters of history and poetry as close.[171] Both had the task of praise and blame, both aimed to persuade, and both could use the device of amplification; nonetheless, he says, history is more chaste. The association of poetry with moral persuasion seems to have been widespread. In *De Studiis et litteris* (1422–1429), Bruni defended poetry as an art that teaches moral philosophy; its powerful descriptions move men to do good deeds.[172] Maffeo Vegio went so far as to claim that this was the only duty of the poet ("quid . . . aliud poetarum est officium, quam hominum vitam instruere, a vitiis avocare, ad virtutem invitare quemcumque?").[173] Fifteenth-century panegyrists could escape the kind of criticism directed at Bruni if they wrote poetry rather than prose. But it is also true that the art of poetry acquired new stature in the course of the Quattrocento. Whereas in medieval schools it had been taught as a subsection of grammar and rhetoric, by midcentury it had become a separate discipline within the *studia humanitatis*. Cristoforo Landino taught poetry at the Florentine Studio from 1452 to 1497, and began to lecture on the *Aeneid* during the year 1462–1463.[174]

The panegyrical ekphrases of modern architectural works written in the third quarter of the Quattrocento are exercises in historiography, whose most common purpose was to confer on a patron the same enduring fame in future generations that contemporary Humanists attributed to the Ancients. To put this another way, the Humanists took the model of the relationship between past and present as their guide in shaping the relationship between present and future. Eloquence was an essential component of this endeavor, now that historiography itself had come to be understood as inextricably bound, rather than opposed, to panegyric. Although both Bruni's *Panegyric* and (to take one example) Avogadro's praise of Cosimo's buildings served as political propaganda, their differences in literary form, models, and purpose measure the difference between the place of architecture in the culture of early Humanism and its role in the new culture emerging after midcentury.

APPENDIX

Manuel Chrysoloras's *Comparison of Old and New Rome*

Most Excellent Emperor,

In other letters, which will probably arrive together with this one, I related at some length those matters I had to report to your majesty. Indeed, in earlier letters sent from here, and particularly in my last one, I have informed you about many things. Nonetheless, it seemed a good idea to set down on paper some thoughts for the sake of pleasure, rather than necessity, of which I harvest no small quantity from offering my words to your sacred person. I will choose as the subject of this letter the city of Rome, which I used to admire from what I had read in historical works. Now that it is before my eyes, I find it much more wonderful than anyone could have guessed just from reading about it. And this, even though the grandest things have been said about Rome, not only by those who have written in the native language, but also by almost all our own writers and by distinguished visitors, so that—as you know—entire orations and whole books have left behind its praise. Other authors captured the grandeur of Rome in a few words, like that wisest of men, so loved by, and so close to us, who expressed a very great idea in a few words right at the beginning of a letter to someone (if I well remember his words): "Remember us, when, having arrived in Rome... you gaze upon such wonders as you never would have believed could exist on earth and which seem, rather, a part of heaven. That we think of you is nothing to be wondered at, for we remain on familiar ground where no new experience might make us think less of our friends."

He wrote these words to a Greek or to a Syrian, a fellow citizen, whose name escapes me. We hear about the beauty of Antioch from many authors, including this one in his oration called *Antiochikos*, an encomium of his fatherland. And another of its citizens, now a citizen of heaven, whose mouth and tongue were called golden [John Chrysostom], said many things about the city in various of his works. He also wrote about it to someone then living in Asia, to which Rhodes and Smyrna then belonged, and where the glories of the whole inhabited world were found. For in Asia were the temple of Artemis at Ephesus, the Colossus of Rhodes, and the Mausoleum of Halicarnassus; remains of all—or, at least, some—of these were still be seen.

Or, if you will, let us suppose the addressee to have come from Egypt and Libya, where we hear were Alexander's city [Alexandria], and those wonders of the world, hundred-gated Thebes and the pyramids. We hear that the shadows cast by the pyramids between sunrise and sunset are as long as several days' journey. And they say that one of them—the smallest—has served Memphis and its surroundings as a quarry for all its building activity from antiquity to the present, without exhausting its own supply of stones. On the contrary, they say that even though whole cities and apartment blocks have been constructed from its material, hardly a small part of it has been used.

Probably the recipient of the letter (whoever he was) passed by the Islands and Greece sailing to Rome. Probably he had seen Greece before, and Athens, which was at that time Greece's eye. Or one must suppose that he was Greek, and an educated man, not one of the many who have never seen or heard of these things, but one educated to them—probably he wrote to a man of this sort, and one who takes pleasure in seeing such great wonders, and from such travels in foreign lands. What was it, then, that was about to amaze him so? So much so, that he would forget about Greece and its wonders once he saw what he had never seen before, and it, and his friends there, would be diminished in his eyes? What could convince him that the city was not of this earth, but a part of heaven? As I said before, Rome has received the greatest praise in both languages. I think that even the barbarians, the ones who are not totally unlettered, have written much about it in their histories. But I used to suppose they wrote what they did in order to exaggerate. Now, however, I know there was no hyberbole; my eyes confirm all that they claimed.

Yet almost nothing in Rome has come down to us intact: you will not find anything that is undamaged, either crumbling beneath the natural forces of time or due to the violence of human hands. Like our own city, Rome uses itself as a mine and quarry, and (as we say to be true of everything) it both nourishes and consumes itself. It is exhausted, so to say, in every way. Nonetheless, even these ruins and heaps of stones show what great things once existed, and how enormous and beautiful were the original constructions. For what in Rome was not beautiful? These works were beautiful not only in their original composition and organization; they seem beautiful even in their dismembered state. Just as in a body that is beautiful as a whole, so the hand or foot or head is also beautiful; or, in a body of outstanding size, each of the limbs is large. However, not a few of the monuments are still largely intact. Many things in Rome were made in Greece, as can be seen from their inscriptions: columns and remarkable stones; funerary monuments, reliefs and statues; and many Greek inscriptions of the most beautiful and ancient type. Many other works were made by Greek craftsmen in Rome, as their inscriptions demonstrate. It seems that, as the historians tell us, the Romans were extremely fond of Greek things, and many of our Greeks lived here.

These remains of statues, columns, tombs, and buildings reveal not only the wealth, large labor force, and craftsmanship of the Romans as well as their grandeur and dignity, if you will, and their ambition, love of beauty, luxury, and extravagance, but also their piety, greatness of soul, love of honor, and intelligence. They speak of Rome's victories, her general well-being, dominion, dignity, and deeds in war. Not only can one see aqueducts high in the air, bringing water from afar; the thickness of the city walls; and the abundance, grandeur, and beauty of porticoes, palaces, council halls, markets, baths, and theaters; but there are also many splendid and well-maintained churches, one after another and each with its own name; there are shrines, images, statues, and sacred precincts. There are also funerary monuments to those splendid men of antiquity, erected at public expense in gratitude for their service to the state; as well as monuments to valor and triumphal arches, preserving the memory of their triumphs and processions. These record in sculpture the wars, the prisoners, spoils and sieges, as well as the festivals, altars, and votive offerings. There are also images on them of battles on sea, on foot, and mounted, and of all classes and kinds (so to say) of battles and artillery. You see war machines and arms; you meet the subjugated lords of the Medes, Persians, Iberians, Celts, and Assyrians—each recognizable by his costume. There are the subjugated peoples and the generals who triumphed over them; their war chariots and quadrigas, charioteers, and spearbearers; followed by the captains and preceded by the spoils of war. All these things are represented as if alive by the images, and each is identified by an inscription. Thus one can see clearly what kinds of arms the Ancients had, what kind of clothes they wore, what the devices of their rulers were, how they formed lines of battle, fought, laid siege, and built encampments. One sees how they looked in different situations and surroundings—on expeditions, at home, in assembly, at the council chamber or the market; on land and on sea, marching or sailing, working, exercising, or thinking, at festivals or in the workshop—and one can distinguish the differences between the various peoples. Herodotus, and some other historians, are thought to have made useful contributions to our knowledge of such things. But these reliefs show how things were in past times and what the differences were between the peoples. Thus they make our knowledge of history precise or, rather, they grant us eyewitness knowledge of everything that has happened just as if it were present. Indeed, the skill of the artists rivals and contends with reality itself, so that one seems to see a real man or horse, a city or a whole army; a breastplate, sword, or a suit of armour; and people being captured or fleeing, laughing or crying, in a state of exultation or anger. On all these reliefs there are large letters saying: The Senate and the People of Rome to Julius Caesar (or "Titus" or "Vespasian" or whatever emperor it might be) on account of his virtue and bravery, having conquered amid terrible danger, or having protected his fatherland, or having defeated the barbarians, or some other such praiseworthy deed.

What do the myths and ancient Greek histories tell us about those ancient heroes, Meleager, Amphionas, and Triptolemos or, if you will, Pelops, Amphiaraos, Tantalus, and others like them? Here the streets are full of their statues; images of the ancient heroes cover commemorative monuments and sarcophagi as well as the walls of houses. All these works are of the best and most accomplished art; by some Phidias, Lysippos, Praxiteles, or other similarly skilled men. And so, walking through the city, one's eyes are drawn from one work to another, just as lovers never have their fill of wondering at the living beauties and gazing intently at them.

And the city's walls, and its site, and the flowing river! The charm of the fields, the suburbs, and the suburban villas! Nor is the city without harbors, dockyards, and docks. Some are inside its walls, others at the mouth of the river, and yet others are just outside the walls: one can still see their remains. I believe that it would be a difficult task for all the powerful lords of our time to destroy these—if something must be called difficult that one might say is impossible. Rome staged naval battles within its walls for the pleasure and gratification, but also for the education, of her citizens, not on the river, but in a sea or harbor created for this purpose in the city. Just as the city built gymnasiums, theaters for competitions, for gladiatorial combat, for contests between wild beasts, and for horse races, so also for battles between triremes. One can truthfully say, repeating the words of one of our ancient authors, that such things as these could only be done by mighty power [ʽΡώμη]. Such, then, was ancient Rome.

But what about the more recent city, the Christian Rome of our times? Who could describe it? How can one describe all that was made from such abundance of great stones and other materials, with the wonderful craft that flourished then, inspired by the love of God and by piety? How shall I recount the transformation of the ancient monuments, or tell how unholy and impure structures became holy shrines, and how temples of idols were made into holy churches? How can I enumerate the churches dedicated to the apostles, to martyrs, holy men, and saints, or to pious women and virgins? Who could count all the bodies of the saints, whose relics are preserved either whole or in part? Or those things that serve as remembrances of them, and of their witness to the Lord of all mankind? One might say that the whole city was nothing but a reliquary and a treasury of holy things.

Thus what that author whom I mentioned at the beginning of my letter once said about his own times can be claimed with greater justice now—Rome can be called a part of heaven on account of its inhabitants. But I will leave this point aside now. How, on the other hand, can I do justice to those two stars, or suns, or whole heavens; those witnesses to the glory of God—Peter and Paul—whose relics are here in Rome, and whose tomb is near me as I write? How can I describe the concourse of the whole world toward them, as toward satraps, dukes, or consuls? It is a marvel to see pilgrims from Spain and Iberia, from Galatea, the British Isles and from islands even farther to the west (if these exist) or farther to the north, from Germany, Sarmatia, Pannonia, and some from Greece and Asia. Who would be capable of enumerating all the places—the pilgrims come from everywhere! And not only do men of all ages and occupations come here, but women too, having traveled the long and difficult roads, exposed to terrible dangers and trials, enduring cold, mud, dust, stifling heat, and every hardship. They come in order to prostrate themselves before the coffin, the tomb, the reliquary shrine of the apostles even though at some distance and through curtains and latticed gates. Laying their faces and heads in the dust, on the pavement, against the stone enclosure and iron gates, they cry out to these saviors and lords—suns, as I called them—of the whole world in order to thank them, after all this time, for having preached the truth and for deliverance from error and darkness. And they do this here, in this city and this empire where, preaching these things, they died. How great and widespread was the superstition and error of those who opposed the true believers can be shown by the idols, statues, and temples in Rome. From Rome the edicts against the Christians were sent out to all corners of the earth, as were their executioners. Here one can see the ruined and razed mansions, and mutilated and half-broken memorials, of that accursed butcher Nero, of Diocletian, and of Maximianus. But the pilgrims pass by these physical remains, even though they once belonged to emperors and to those who persecuted and enslaved the Christians. But why do I say that they pass these by? Rather, they trample them underfoot, strike them with their fists, break off pieces, spit on them—as I said—and throwing them to the ground, smash them. With what ineffable pleasure and delight, with what reverence, the pilgrims beat their breasts and weep, bowing their heads to these paupers, foreigners, whose citizenship and race made them slaves and the servants of slaves, who were tentmakers and fishermen. For, as I said, these are the saviors and guardians; not only lords over the earth, but guardians of the gates of heaven. The pilgrims pray to them for salvation; for themselves, their friends and families, and for the whole world. These great saints are enthroned in images, looking down from the upper parts of churches and visible from afar, in glass or stone mosaic or in fresco. They are also depicted in pure silver or gold, in works wrought in the ancient style. They could be emperors, or princes, or the lords of all the world, holding the key that symbolizes the power to bind and to loose, but also the strength and dignity of those to whom it was given. The deeds of the Roman people, or of Julius Caesar, or Augustus, or anyone else, are unknown; even here in their own city they are unknown to most people. They withered away and faded from memory, while the deeds of Peter and Paul bloom and thrive all over the world.

I remember the procession and festival in honor of the apostles that I saw with my companions two years ago in London, which is beyond the sea and the ocean in Britain. The people there know nothing of the emperor who first conquered Britain, or of two forts he built there (one at the point of landing and one in London), which can still be seen. And I really believe that all these things, that is, the domination of Rome, the power of the emperors, and subjugation of peoples, took place in order to prepare the way for what followed. Why should we wonder at the dignity of the apostles, the reverent piety addressed to them, or the honor accorded to them by those who look at them, their bodies, or their tombs? For these can be seen and marveled at in the reign of the Church and the long line of apostolic succession. I believe that never during the rule of Rome—ancient Rome, I mean—were so many decrees of the emperors, senate, and people carried to so great a part of the world as now, under the Church. These decrees are always made through Peter and Paul, with the authority of their seals bearing the images of their heads. Thus the apostles can no longer say "silver and gold are not mine." Who could count all that they have given? Some will think that I am exaggerating, but I believe that if all the rulers we know gave all the taxes and income over which they have control, this would not equal this sum. But Peter and Paul give it, and must do so, for they distribute to others what they would not give to themselves. And the successors of Peter and Paul, and of their inheritance, must distribute the wealth. Those who receive feel that they are receiving from lords; those who give, give as princes. No one opposes this, except to dispute whether a certain gift should be made and to whom. And such matters are referred to Peter and Paul for judgment, since they are the arbiters and judges of those things done in their name. Frequently they are called on to judge the appeal of cases in which they had no part.

What shall I say about spiritual gifts, or about the judges of nations? In the basilica of the apostles one may see, on account of the diversity of languages, a priest hearing confession from pilgrims from the British Isles, who have come to Rome just for this. Another priest listens to those from Spain or France; yet another, those from Dalmatia or Dacia. On the roads, in the city, and in the churches, the sounds of various languages intermingle. And not only will the apostles sit on twelve thrones and judge the twelve tribes of Israel, as prophesized in the Gospel, but even now they are enthroned, on many thrones, judging the world. Not only in the early Church, but also now, they make themselves understood by each person in his own tongue. And, as I said, the laws and edicts of the apostles are proclaimed in every country. From every land pilgrims come here; no one forces them on this journey, which may last many days or even months—they are drawn by their own will. Plato, in his *Gorgias*, advises those wishing to become righteous to accept the judgments of the righteous and submit to the penalties they impose. I will not speak of those who hurry to Rome, or are led here from the farthest reaches of the world, in fulfillment of a penalty imposed on them. Who would not be amazed to see the pope listening in public council, administering justice, and uttering oracles? Who would not marvel, seeing such things? But why should I say any more about this subject? I was saying, that the importance of the apostles' relics to Rome is explained in John Chrysostom's exegesis, at the end of his commentary on Paul's epistle to the Romans (if I well remember). With the passage I mean, he brings to a close his whole discussion and the entire golden work (for not only the tongue, but also the soul of that man was golden). He says there, and, as if inspired, he says it over and over, "I want to see this city" (meaning Rome) "in which there is so much" (meaning the deeds of the apostles, the tombs that hold their bodies, and their presence). "I want to see their feet, I want to see their hands, I want to see their heads." And he did not cease— just like someone desperately in love—enumerating all the things that were dear to him, and saying it over and over.

But we must really leave these things aside, for it is not possible to treat them in a letter, and no matter how much I wrote, I would not be able to say anything about the apostles that is worthy of them. Besides, I began as if I were going to write only a short letter, and without realizing it myself, I greatly exceeded the measure. But in making such a long addition to my letter, I felt a certain gladness, for I wanted to write these things; the more so, since I find in your letters to me, among other things, this—that my letters are too short. I hasten to send this one, lest it reach Venice after the ships of that city have sailed. Really, it is difficult in all things to begin properly and finish as one ought. For the whole is defined by its extremities and gets its form from them. Many people are not able to judge from what point they should begin, or where they should end; therefore, they always begin by babbling. But I have refrained from writing in this manner to your Majesty for so long, that I find that having begun again I am scarcely able to stop. Just as you pardoned me for the brevity of my past letters, may you now forgive my wordiness. Besides, having told you about such wonderful things, I really must end my letter. But I do not want, on that account, to be unjust to our city, which might happen if I left off at this point. For seeing how impressive the buildings of Rome are, in the beauty of both their natural materials and their workmanship, I admire the magnificence and power that it once enjoyed. And most of all I rejoice because I see that they are very similar to our own monuments. As you know, whenever we see things that are very similar, the similarity itself attracts our attention. This is particularly true when we perceive the likeness of a son to a father, daughter to a mother, or brothers to each other, especially if they are somehow related to us. I think that in this city are many things, not just a few, that catch the attention of a man, particularly one who loves his fatherland and is far away from home. For I believe that never did a daughter resemble her mother so precisely as Constantinople resembles Rome. But also, seeing that this is the case (and I do not think that the mother is jealous of praise given to her daughter, just as it would not be fair, in return, for the daughter to envy praises of her mother), I think that our city is superior. I am comparing, as far as it is possible, the two cities as they were in Antiquity. What did Rome possess, either in natural advantage or through art, which we did not also have? Even the ruins of Rome are similar to ours. And if Rome seems to have the advantage in certain things, we can redress the balance with other things we have. For many things were made, and still exist, in Constantinople that Rome does not have. Moreover, we brought things to a fuller point of development. For Rome derived its whole proper form (so to say) from no model, and was satisfied by outdoing all other cities. But Constantinople, looking at this model as an archetype (for that is how Rome is correctly seen), brought many things to greater perfection and splendor. The works of men competing with others can progress toward greater beauty. The parents' beauty contributes to a more perfected beauty in their children, and their great stature lends greater stature to them. This is particularly true, so they say, in the case of mothers, whether they be human beings or other animals. It is not surprising, then, that such a great and beautiful city produced another even greater and more lovely. Having done this redounds to the praise of the mother. The beauty of the mother enhances that of the daughter, and rather than comparing one city with another, one compares the same city with itself—that is, the new with the old Rome. And so, there is no reason to be unfair judges in this case, especially not for us, who have inherited the good things of both as sons of the one and grandsons of the other. Rather, we should take pride in both of them, just like those who trace their noble ancestry can claim not only that their parents were great and admirable but that their forefathers were nearly as distinguished as well, and perhaps being able to recount that they were superior in these qualities. As someone said, it is the parents' prayer to be surpassed by their children, so that it is just for us to suspect that this is so in our own case.

Therefore, I say that the mother is extremely beautiful and in her prime, but that the daughter surpasses her in many points in regard to beauty. It is appropriate neither for the mother and metropolis to be placed before the daughter, nor for her colony and daughter to come before her, since, as I said, no mother and daughter were ever as similar as these cities. The beauty of Rome was transmitted to Constantinople as if kindled in her by a fire or the sun's rays.

Suppose we had nothing else to say about our city except that it is situated on both continents (Europe, that is, and Asia); or that it is located at the conjunction of the northern and southern seas so that it both joins and separates the parts of the world through the common bond of its continents or its seas— or, really, through both. Placed as a gateway to the whole world and those who people it, who would not consider Constantinople not only useful and beautiful, but also regal? The harbor, I think, surpasses all others anywhere in size and security; it is capable of receiving all the triremes and merchant ships that were ever built in the whole world. The city is entirely surrounded by both sea and land, so that it is both part of the continent yet also an island, on the mainland and at the same time in the sea. Truly, the crown and circuit of its walls yield nothing to the mighty walls of Babylon. The solidity of its towers, their grandeur and height, is such that even one alone presents a spectacle not less marvelous than that of an entire building. Indeed, even the stairs leading up to them elicit admiration for their size, solidity, and manner of construction. The gates and towers of the walls are indeed splendid. There is another wall in front of this one, which, to another city, might seem sufficient for its defense. Between these walls moats have been dug that are remarkable for their breadth and depth, as well as for the abundant water they contain. In fact, because of these moats, the city seems to be an island even on the one side that faces land, and this confirms the impression I expressed earlier that from the land it seems a city in the sea, whereas from the sea it seems to be on the mainland, because of its surrounding islands. Just as someone once said about Athens, you can sail past the city, around it, or come there on foot. There are two ways of arriving at Constantinople by land, and yet it is fully in the sea. To what city does Constantinople cede superiority? Rather, what city can even approach it, or think to equal it? All that is within the walls is really worthy of them, and of the paradigm on which the city was founded; worthy of the metropolis and mother, worthy of the dignity and power of the inhabitants.

Constantinople came into being through the union of the two most powerful and prudent peoples, the Romans and the Greeks. One of them ruled at that time, the other had ruled before, and both excelled in every art, in love of fame and in refinement. And the needs of the new city were served not only by the Greeks and Romans, but by all other peoples as well. First they chose an appropriate site for the future city, and this was done with careful thought with regard to its future hegemony, so that those who ruled there would be well placed for steering the whole world. They chose the site from which to govern the whole world out of the whole world, over which they were lords, keeping in mind that such had not been the case of Rome from the beginning, or of other cities that later became famous through good fortune. Cities had been founded for different reasons and because of different fortunes in many places and later expanded. They did not think it just to desert them in order to move to a better place. Instead, the citizens of these cities improved them as best they could, through skill, or loved them on account of their surrounding countryside or site. Some continued to put up with many inconveniences due to the nature of the place. As a result, these cities cannot be considered beautiful in all things, since they are deficient in some of the most important of them, just as is the case with houses that have grown up piecemeal.

But the very site of our city, as I said, was chosen from the beginning with careful judgment and deliberation so that it would be suitable for both the pleasures and necessities of life, and worthy of the future capital of the world. That the founders chose well is evident to us, now, on the basis of our experience. Next, the arts contributed from every side to its beautification, since the rich began building richly, the nobles nobly, and the emperors imperially. For how could the inhabitants, having come from such a great city as Rome, where they had beautiful homes, wish to live in a mean and shabby fashion? Those who arrived later continued to imitate the first patrons, as usually happens in the case of building. Nature had prepared for their use marble islands just outside the gates, as if she had known what a great city would be built and what noble inhabitants would settle there; she furnished ahead of time both the raw material and the accessory necessities for its beauty and dignity. But the white marble and many-colored marbles from Thessaly, Paros, and Peloponnesus did not satisfy the builders; they also sought their building materials in the quarries of Arabia, Egypt, Libya, Marmarika, and even from the torrid regions. Thus they immediately began to build houses that might have sufficed as cities on account of their size, and churches so wonderful that one cannot believe that they were made by human skill or created by men; indeed, it was with the help of their piety toward God that the architects succeeded. For if sacrilegious and unbelieving men have erected many and great temples, what kind and quantity of structures must those be that are made by men who fear the true God? And if great buildings were dedicated to the gods by those who serve them, what sort of thank-offerings should be given to the Savior, who delivered us from demons? But enough said about this.

I will not describe the covered and protected roads, once to be seen throughout the whole city, and which enabled one to go everywhere without the inconveniences of mud or the sun's rays. Nor will I discuss the monolithic columns, each of which could well be a tower or a fortress. I will also leave out the porticoes, the bases of monuments, and the obelisks; and I will pass over in silence the abundance, great size, and beauty of the palaces in all parts of the city, the theaters, barracks, sports fields, gymnasiums, and hippodromes. Nor will I describe the dockyards; the walls and moorings extending out to sea; nor how the stones of the foundations and walls project from the sea; nor the docks that receive ships from all parts of the world; nor the towers in the middle of the sea, rising up from the depths of the ocean amid the ebb and flow of the tide, and which contain sweet water. The aqueducts carry water in underground channels and lift it high in the air over the walls, so that one might call them rivers in the sky, arriving from great distances as far as many days' travel. Canals were cut through the mountains that barred their way, and underground cisterns prepared to receive the water; these once served the aqueduct system like seas or pools. The canals are roofed over and concealed, while the cisterns are open to the view of merchant ships sailing by. Some of the cisterns now nourish great trees, either below their vaults or above, on their roofs, and take the place of fields and orchards for their owners. They are also full of columns and arches. Other canals have been dug underground throughout the city, creating a hollow basement over which the city is suspended. There are, in addition, many underground public sewers, of which there are also many here in Rome. How could I find words to describe the baths, of which an unbelievable number is recorded? Or the wells, which are found in private homes as well as throughout the whole city?

With what words could I describe the suburbs, whose great size stretches out to a distance of not a few days' journey, and whose buildings compete in beauty, utility, and construction with those of the city itself? The suburbs encircle the city without interruption from Abydos and Sestos and the Hellespont up to the Thracian Bosphoros, from Hieron and from Kyaneas; from Asia on the one hand, and Europe on the other, they encircle the Propontis, and the islands around the city and the horn of the harbor, extending up to Pharos. If one wished to give a full account of the churches and secular structures in the suburbs, one would have to write a book, and no short one.

A good portion of the suburbs constitutes a kind of city (practically a province) protected by the so-called Long Wall, which extends from the lower to the upper sea, and to the Black Sea, and it includes structures of such grandeur that it would be difficult to find their like in any other city. The Long Wall might, with equal reason, be called the broad or lofty wall. It is so imposing that, even were it constructed simply of randomly chosen stones, of pebbles from riverbeds and mountain streams and of rough boulders, it would warrant admiration. In reality, it is so well constructed of such huge stone blocks that even if it extended only a few spans in length it would astonish all who saw it, seeming to be a mountain rather than a wall. And if it were suspended, supported only at its ends like a rope or a roof beam, the fact that its width is proportionate to its length would make one wonder why it did not break. This very wall, or rather the city's fortifications in general, as I said, at the time the empire was so large and at the height of its power, were built for the sake of magnificence, as were the other marvels throughout the city, rather than for necessity. To the degree to which smaller walls were needed, by that degree they made them larger and more extensive, both because the status of a city is first seen and assessed from its walls, and because they were not building only for the present but for future ages. Many things can happen over the course of time; since great men excite the envy of many, great changes may occur. Thus the walls remain, not only as evidence of the power of those who built them, but also as testimony to their prudence and foresight for the future.

Even if I wished to mention all the commemorative and funerary monuments, columns, and statues that now are, or once were, in the city, I might be at a loss for words. I should say that there are fewer of them than in Rome, but that some are much more beautiful and splendid than those here. For instance, the tomb of that emperor who was founder and is guardian of Constantinople, and the tombs of other emperors around it in the imperial burial area; just to see them is a wonder. Many imperial tombs are preserved around the Church of the Holy Apostles, many others in the same place are in a ruined state, and others are in the vestibules of churches elsewhere in the city. What about the commemorative monument of the law-giving emperor? And near it, to the east if I remember rightly, are other commemorative statues, many of which are supported by extraordinary columns. That there were once many other such statues in the city is shown by their bases, which are still to be seen, and by the inscriptions on them. Some are scattered throughout the city, but most were in the Hippodrome. And I myself once saw many others, which I know have since disappeared. There are those who say that the columns on the Xerolophos hill and on the Tauros hill, which faces it to the east, served as bases for statues cast in silver of Theodosius the Great and Theodosius the Lesser (as they are called). One can judge how grand and worthy of admiration and beautiful they must have been from the beauty, loftiness, splendor, and costliness of their bases. Thinking about them brings to mind the former city gate, which is on the same road and in a direct line to the west. Whole towers or forts could have passed through these gates, if such things could be moved, or a ship with all its masts and sails. And I think of the shining portico above it, visible from afar, and of the great size of the marbles in it. I remember, too, the column standing in front of the gate, which also once served as a base for a statue. Another column is on the other hill, the one below which is the palace where you live now, called "of the column." In addition, there are columns at the Strategion, to the right of the Church of the Holy Apostles, and many others, all of which served as bases worthy of the statues they supported. What can I say about the porphyry column, farther east on the same street, lifting up the cross as high as possible? It was set up and dedicated in the courtyard of the great Constantine's palace, and it surpasses all statues and all columns.

Besides these things we have many examples of glyptic art, such as columns carved with continuous friezes, made in precise imitation of the ones here in Rome. Yet they are surpassed by the porphyry statues between the columns; it is said that these figures, enthroned at a fork in the road, are called "the market overseers" on account of the office they held. A little farther along the same road, just above the source of the river that runs through the city, is a statue made of white stone or marble, which seems to rest on its elbow. There are many other such things that I have not seen myself, but that I heard are hidden in certain places. I will not attempt to describe the reliefs on the Golden Gate, for no one could adequately praise the gate itself with its marble towers, much less its marble reliefs of the labors of Hercules, the torture of Prometheus, and similar subjects, carried out with the best and most admirable skill.

The reason that the city does not have more of such things is that at the time of its foundation they were neglected—as also in Rome—for religious reasons; men fled, I think, from mimetic representation in both sculpture and painting. Who, after all, would want to create the kinds of images that had already been destroyed somewhat earlier in Rome? Therefore, they created and discovered other forms of art, such as panel paintings, icons, drawings, and mosaics, which they executed with the most splendid and enduring craft. Mosaic art is indeed rare, here in Rome, and can really be found only in Greece or Constantinople. Even if some examples of such work can be seen in Rome, or elsewhere, still the materials and the technique come from here. I think that the same is true of figural sculpture: it originated in Greece and then made marvelous advances in Rome. One might, perhaps, say this about a great many other things as well.

As I said, I do not intend to speak of these things, or of other features of the city that are unspeakably great and beyond compare. Yet who, seeing that church, so well named after the wisdom of God (for, really, it is not the work of human wisdom), would be able to describe, or admire, or even remember anything else? For I believe that nothing like this ever was, or ever will be, built by man. And indeed I too, remembering it, am constrained to omit all the rest of my description of the city's wonders, since neither is it possible to say anything worthy of the subject nor, with it in mind, can I speak about anything else. Indeed, when I think of it, it is impossible to turn my mind to anything else; and yet, as I said, I am entirely at a loss to describe it adequately. If someone began to describe it, he would have to spend longer on it than on all the rest of the city's monuments together, and yet, in the end, he still would not have said anything worthy of the subject. And I, as I said, am in a hurry to send this letter. Nor do I wish to describe the church at the end of my account, or as an afterthought, or in the spontaneous style I have used until now. On the other hand, even if having begun with this topic, choosing my words carefully, and intending to describe nothing else, I still failed to reveal anything of its qualities, I would still love it immensely. For what sublimity and grandeur of speech could adequately evoke its sublimity and grandeur? What beauty of words could equal its beauty? Could the dignity of my words do justice to its dignity? How could I speak with as much simplicity as it possesses? What power and variety of words will be equal to its variety; or what precision of speech to the precision and excellence of all its parts? What harmonious composition can compare with its harmony and congruence of parts? Not even if I limited myself to a single part, or its smallest member, could I succeed in describing it adequately. For it is not possible to give a precise and thorough description even of the gates alone, or of the paths, pavements, vestibules, or beautiful hangings in front of the doors; nor of the columns, intarsie, or marble wall revetement; nor the glass windows with their bronze grilles, or the lead, iron, lapis lazuli, and gold, or the glass mosaics— nor of anything at all. What could anyone say about the stones, or the design as a whole, or the manner of construction? Or of its breadth and height and vaulted covering? Not only am I unable to describe the dome with words, I cannot even grasp it with sight—or rather, beholding it, I cannot understand how it could have been built. For just as it amazes us that the heavenly sphere revolves by itself, so we are at a loss to understand how this inimitable and heavenly vault and ceiling was built and continues to stand. Even though the height diminishes the visual impression of great size in the building, or rather, the great size of the building reduces the impression of height, even so it appears to be inconceivably immense. Therefore, one should admire not only its great mass and skillful construction, but also those men who mentally conceived of such a great work without the help of any similar model and who, having formed a mental image of the work, believed themselves to be capable of actually realizing it and bringing it to completion. What I said about the work as a whole is particularly true of the dome, which the architect must have envisioned as we see it now from the moment the foundations were laid, or even before that; he must have felt complete confidence in his ability to contrive it, even though, looking at it now, it seems an impossible feat. Probably he relied heavily on craftsmanship and on the rules of right proportions; he must have been extremely skilled in geometry and the science of mechanics. And yet, what am I saying? For this work makes the spectator wonder at the ready intelligence and capacity to achieve great things not only of this architect and of the other builders but of the whole human race, as well as of this living person. The dome reveals inventiveness, elevated thought, dignity, and power such as no one before ever would have imagined. For I believe that none of the works of man have ever been like, or even came close to, this one. It has many features that are wonderful, either on account of their material nature or due to premeditated human wisdom. This superlative work is to be admired in silence; it renders those who see it, or even remember it, unable to think or speak—just like some people who, looking at the sun or some other shining thing, are blinded.

No one, I think, is completely ignorant of the bodies of saints and relics in Hagia Sophia and elsewhere in the city. Most of them have been here since the beginning, having been gathered from everywhere through the zeal of our famous emperors. And although many relics are no longer here, having been distributed throughout the world like water from a common well, many still remain in Constantinople.

In closing, I would like to say this about both cities: in Rome, just as in our city, one may observe with pleasure those things that the Ancients made, but one may also derive something useful from them. The works of man may be admired, but on the other hand they may be disdained and reckoned as nothing. For whoever, seeing these things, ponders on the empire of Rome, on the power and dignity of those men, their works and their efforts, will see what a death all these things have died. And not only the men themselves have died, but also their dominion, hegemony, and almost their city itself (for, as someone said somewhere, cities also die). How, afterward, would he reckon the works of man? How would he feel proud in good fortune or humbled by bad? In what would he pride himself? What human achievement would he think has absolute value?

Often, as I walk along streets on which triumphal processions passed, I think of the captive kings and rulers and other prisoners who were led along them. Many of them came from Armenia, Persia, and other extremely remote places. I think about the generals who defeated them, of their rejoicing and pleasure in their victory, and about the downcast spirits of their captives. I imagine the crowds of Romans surrounding them on all sides and the spectators watching from the windows of their houses; I hear the harmonies of musical instruments, the noise of the crowd, the shouts of praise and applause. I think about their pleasure at the victory, at victory in places so far away; and then I think of the sorrow of the prisoners and their families back home about the defeat and about a procession of this kind. How different the customs of those times were: to themselves and to others, the defeated seemed thrice miserable and unfortunate, the victors happy and blessed. But now all these things have been equalized; all lies in the dust, and no one knows the fate of Pompey or Lucullus any better than that of Mithras or Tigranes.

Even the houses, commemorative monuments, and great constructions lie in the dust, visible above the earth that holds them only to the extent that the ground, instead of being level, is raised in mounds. One can see many statues of those fortunate and blessed victors, along with their trophies and wreaths, wallowing in mud and mire, having fallen into ruin. Other statues have been mutilated and their members scattered about. Not a few have been made into lime and mortar, or have provided stone for construction. Those that have had a better fate, having lost what Aristotle considered to be the primal good fortune of stones, have become mounting blocks, building foundations, or mangers for asses and oxen. Most, however, have been split into pieces and broken up, as I said. Others are no longer visible, having been hidden by the thistles and thorns or bushes that have grown over and around them. Countless statues, once erected in splendid and elevated places where everyone could see them such as the Capitol, the acropolis, and the forum, are now beneath the ground.

Those fortunate enough to have had their deeds and rank commemorated in stone rejoiced at their destiny. We know such things from what Pindar said about the victors in the Isthmian, Nemean, and Olympian games in his odes. He makes them the equals of the gods, pronouncing them blessed, those who after great labor were crowned with celery leaves, oak, or wild olive. Yet those commemorated by the statues of Rome have not won enduring fame. Thus it is, that the deeds of man are indeed vain, whether performed by cities or individuals, princes or private citizens—or even by the emperors themselves—unless they possess the true and godly virtues. The deeds of men may give us pleasure or comfort us; not a few, however, give us pain.

I feel as if I were in my own country, and in order to strengthen this impression I try to contrive that which is most important and pleasant to me at home: through thinking about you, writing you letters, talking about you to others, and so on I imagine that I am seeing you face to face and talking with you. So you see that I have conversed frequently with Your Majesty through my other letters, both numerous and frequent, as through this one. I always think that those who serve their fatherland deserve an eternal reward not only for their actions, but also for what they suffer. Remember my reward, o emperor! And may you and yours save the best city under the sun and restore its good fortune. May we work together with you in the negotiations and labors on its behalf, for it is proper for well-wishing subjects to dedicate their strength to support philanthropic and virtuous rulers. This shall be my joy whether at home or abroad, whether seeing or hearing, in life or in death.

And now, may you forgive my wordiness, Your Majesty. Farewell.

Notes

ABBREVIATIONS

AB	Art Bulletin	PG	J. P. Migne, ed., *Patrologiae cursus completa. series graeca* (Paris, 1857–1868).
Bolgar	R. R. Bolgar, *The Classical Heritage and Its Beneficiaries* (Cambridge, 1954)		
Buck and Guthmuller	A. Buck and B. Guthmuller, eds., *La Città italiana del Rinascimento fra utopia e realtà* (Venice, 1984).	PL	J. P. Migne, ed., *Patrologiae cursus completa. series latina* (Paris, 1844–1864).
		Pros. Lat.	E. Garin, *Prosatori Latini del Quattrocento* (Milan, 1965).
Convegno	*Convegno internazionale indetto nel V centenario della morte di Leon Battista Alberti* (Rome, 1974).	QFIAB	*Quellen und Forschungen aus italienischen Archiven und Bibliotheken*
IMU	*Italia medioevalia et humanistica*	Rabb and Siegel	T. Rabb and J. Seigel, eds., *Action and Conviction in Early Modern Europe: Essays in Memory of E. H. Hardison* (Princeton, N.J., 1969).
JHI	*Journal of the History of Ideas*		
JSAH	*Journal of the Society of Architectural Historians*		
JWCI	*Journal of the Warburg and Courtauld Institutes*	RIS	*Rerum italicarum scriptores*
Memoria	S. Settis, ed., *Memoria dell'antico nell'arte italiana*, vol. 1: *L'Uso dei classici* (Turin, 1984); vol. 2: *I Generi e i temi ritrovati* (Turin, 1985); vol. 3: *Dalla Tradizione all'archeologia* (Turin, 1986).	Smyth	A. Morrogh, F. Superbi Gioffredi, P. Morselli, and E. Borsook, eds., *Renaissance Studies in Honor of Craig Hugh Smyth* (Florence, 1985).
		VZ	R. Valentini and G. Zucchetti, *Codice topografico della città di Roma*, vol. 4 (Rome, 1953).

NOTES
INTRODUCTION

1. Chapter 6, which offers a new reading of the aesthetic principles of the town plan at Pienza, is the only exception.
2. A recent discussion of this problem for historical method in general is in D. La Capra, *History and Criticism* (Ithaca, N.Y., 1985).
3. H. M. Goldbrunner, "Laudatio urbis: Zu neueren Untersuchungen über das humanistische Städtelob," *QFIAB* 63, (1983): 325.
4. E. J. Storman, "Bessarion Before the Council of Florence: A Survey of His Early Writings (1423–1437)," in *Byzantine Papers: Proceedings of the First Australian Byzantine Studies Conference, Canberra, 17–19 May 1978*, ed. E. Jeffreys, M. Jeffreys, and A. Moffatt (Canberra, 1978), p. 141.
5. C. Gilbert, "The Earliest Guide to Florentine Architecture, 1423," *Mitteilungen des Kunsthistorischen Institutes in Florenz* 40 (1969): 34.
6. J. Huizinga, *The Waning of the Middle Ages*. (1924; rpt., Garden City, N.Y., 1954), p. 225.
7. J. Onians, *Bearers of Meaning: The Classical Orders in Antiquity, the Middle Ages and the Renaissance* (Princeton, N.J., 1988), p. 5.
8. G. Holmes, *The Florentine Enlightenment, 1400–1450* (New York, 1969), p. 174.
9. R. Wittkower, *Architectural Principles in the Age of Humanism* (London, 1952) (orginally in *Studies of the Warburg Institute* 19 [1949]). The quotation is from Wittkower's introduction to the 1971 edition of the work. He is quoting from Kenneth Clark's review of the first edition, which he says "defines my intention in a nutshell."
10. E. R. de Zurko, "Alberti's Theory of Form and Function," *AB* 39 (1957): 145.
11. E. Gombrich, "From the Revival of Letters to the Reform of the Arts: Niccolò Niccoli and Filippo Brunelleschi," in *Essays in the History of Art Presented to Rudolf Wittkower*, ed. D. Fraser, H. Hibbard, and J. Lewine (London, 1967), pp. 71–82.
12. P. Murray, *The Architecture of the Italian Renaissance*, rev. ed. (1963; London, 1986), p. 28.
13. H. Burns, "Quattrocento Architecture and the Antique: Some Problems," in *Classical Influences on European Culturec, A.D. 500–1500*, ed. R. R. Bolgar, (Cambridge, 1971), p. 277

CHAPTER ONE

1. I cite the edition in C. Grayson, ed., *Leon Battista Alberti. Opere volgari* (Bari, 1966), 2:107–83. All translations from the Italian are mine; page references to the work will be given in the text. I have checked quoted passages against the new critical edition of the work, which appeared while I was preparing my text for publication (G. Ponte, ed., *Leon Battista Alberti. Profugiorum ab erumna libri* (Genoa, 1988). The work has received very little attention from scholars. See however, the sensitive analysis in G. Santinello, *Leon Battista Alberti. Una visione estetica del mondo e della vita* (Florence, 1962), pp. 132–49. Tateo's slightly earlier dating, of 1439, has not been generally accepted (F. Tateo, *Alberti, Leonardo e la crisi dell'Umanesimo* [Bari, 1971], p. 5).
2. For Agnolo Pandolfini (1363–1446), see his Life in Vespasiano da Bisticci and the commentary by A. Greco (Vespasiano da Bisticci, *Le Vite*, ed. A. Greco [Florence, 1970], esp. p. 280). Agnolo is a principal speaker also in Matteo Palmieri's *Della Vita civile*, ed. G. Belloni (Florence, 1982), with biographical information on p. 7, n. 2.
3. Nicola or Niccolò di Vieri de'Medici (1385–1454). See Poggio Bracciolini, *Lettere*, ed. H. Harth (Florence, 1984), 2:113–14; R. de Roover, *Il Banco Medici dalle origini al declino; 1397–*

1494 (1963; rpt., Florence, 1970), p. 54; and G. Mancini, *Vita di Leon Battista Alberti* (1882; rpt., Florence, 1911), p. 200.

4. Theogenius is published in Grayson (n. 1). The prefatory letter is on pp. 54–55.

5. Mancini (n. 3), pp. 181–88.

6. For the *certame coronario*, see G. Ponte, *Leon Battista Alberti. Umanista e scrittore* (Genoa, 1981), pp. 182–83; Santinello (n. 1), p. 132.

7. For these works, see C. Trinkaus, *"In Our Image and Likeness": Humanity and Divinity in Italian Humanist Thought* (London, 1970), esp. pp. 173ff., and Trinkaus, "Themes for a Renaissance Anthropology," in *The Scope of Renaissance Humanism* (Ann Arbor, Mich., 1983), pp. 364–421.

8. Plato, *Phaedrus*, 246. Aristotle, *Nicomachean Ethics*, I.xiii.9, also distinguishes between the rational and irrational parts of the soul. Although Grayson believed that Alberti's definition derived from Xenophon's *Cyropaedia*, VI.1.41, there it is presented in a shorter and more simplified form ([n. 1], p. 110).

9. Here his discussion parallels and expands Aristotle, *Nicomachean Ethics*, I.x, a text fundamental for Alberti's essay and said by Garin to be the seriously studied philosophical work in the first half of the fifteenth century (E. Garin, *Medioevo e Rinascimento* [1954; rpt., Bari, 1980], p. 202; Garin, "La Fortuna dell'*Etica* Aristotelica nel Quattrocento," in *La Cultura filosofica del rinascimento italiano* [Florence, 1961], pp. 60–71).

10. Proclus, *Commentary on Euclid's "Elements,"* X–XI, p. xxiii. Other uses of columns as metaphors for moral virtue in New Testament and Patristic authors are discussed in J. Onians, "Style and Decorum in Sixteenth Century Italian Architecture" (Ph.D. diss., University of London, 1968), p. 200.

11. Aristotle, *Nicomachean Ethics*, II.i.4. Interestingly, the philosopher's example of how virtue is acquired through its practice is architectural: "Men become builders by building houses."

12. See G. Boas, *Essays on Primitivism and Related Ideas in the Middle Ages* (1948; rpt., New York, 1978), and G. Costa, *La Leggenda dei secoli d'oro nella letteratura italiana* (Bari, 1972).

13. Horace, *Epistles*, epist. X, trans. H. R. Fairclough (London, 1926), p. 315. See the discussion of the topos of the *locus amoenus* in E. R. Curtius, *European Literature and the Latin Middle Ages*, trans. W. Trask, (Princeton, N.J., 1953), pp.195ff.

14. Leon Battista Alberti, *De Re aedificatoria*, ed. G. Orlandi (Milan, 1966), prologue, p. 9.

15. Vitruvius, *The Ten Books on Architecture*, II.i.1–3. Vitruvius draws on both Lucretius's *De Rerum natura* (civilization begins with the discoveries of fire and speech) and Cicero's *De Natura deorum* (man differs from other animals because of his upright stature, enabling him to look at the starry firmament and on account of his manual dexterity).

16. Cicero, *De Officiis*, II.iv, trans. W. Miller (London, 1938), p. 181.

17. Cicero, *De Natura deorum*, II.ix, trans. H. Rackham (London, 1933), p. 269.

18. Garin touches on the development of a moral dimension in Patristic writers, such as Lactantius and Nemesius of Emesa, whose works were of special interest for Renaissance concepts of the dignity of man (E. Garin, "La' Dignitas hominis' e la letteratura patristica," *La Rinascita* 1 [1938]: 102–46). More recent discussion is in Trinkaus, *"In Our Image"* (n. 7), pp. 182–83.

19. Boccaccio, *De Genealogia deorum*, IV.4. See E. Cassirer, *The Individual and the Cosmos in Renaissance Philosophy*, trans. M. Domandi, (1927; rpt., Philadelphia, 1983), p. 95.

20. Most recently discussed by L. Panizza, "Active and Contemplative in Lorenzo Valla: The Fusion of Opposites," in *Arbeit, Musse, Meditation. Betrachtungen zur 'vita activa' und 'vita contemplativa,'* ed. B. Vickers (Zurich, 1985), pp. 181–224.

21. For example, in *Idiota de mente* and *De Beryllo*. See P. M. Watts, *Nicolaus Cusanus: A Fifteenth-Century Vision of Man* (Leiden, 1982), pp. 135–36, 180.

22. Seneca, *Epistulae morales*, letter XC. The letter is discussed in relation to Valla by Panizza (n. 20), p. 195.

23. Innocent III, *De Miseria humanae vitae*, pt. 13, in *PL*, 217.

24. J. Bialostocki, "The Renaissance Concept of Nature and Antiquity," in *Acts of the XX International Congress of the History of Art*, vol. 2: *The Renaissance and Mannerism*, ed. M. Meiss (Princeton, N.J., 1963), pp. 19–30. Bialostocki notes that whereas Augustine considered nature, as a divine creature, superior to human art, Alberti's new idea was that art can surpass nature. However, when Alberti's approach is seen in the context I have suggested, there is nothing new about it. (C. Westfall, "Society, Beauty and the Humanist Architect in Alberti's *De Re aedificatoria*," *Studies in the Renaissance* 16 [1969]: 61–81). Westfall seeks to define a difference in attitude between Vitruvius, who thought that the artist gives form to matter, and Alberti, who, believing that matter already had form, sees the task of the artist as the re-formation of matter by means of intellect. This is a valid difference, but one that would characterize Boccaccio, Cusanus, Giannozzo Manetti, and other Renaissance thinkers besides Alberti who see man as second creator.

25. Aristotle, *Rhetoric*, I.xi.25–26.

26. Trinkaus's translation in *"In Our Image"* (n. 7), p. 247.

27. Baxandall, too, has noted this almost total lack of descriptive criticism of works of art in Alberti's writing, as opposed to what he calls "prescriptive criticisms," in which the rules for art are stated (M. Baxandall, "Alberti and Christoforo Landino: The Practical Criticism of Painting," in *Convegno*, pp. 143–54.

28. Lucian, *The Hall*, 2, trans. A. M. Harmon, 8 vols. (London, 1913), 1:179.

29. Aristotle, *De Anima*, III.viii, trans. W. S. Hett (London, 1935), p. 181.

30. Aristotle, *On Memory and Recollection*, trans. W. S. Hett (London, 1935), 1:287. On the definition of imagination as the image-forming capacity, see M. Kemp, "From *Mimesis* to *Fantasia*: The Quattrocento Vocabulary of Creation, Inspiration and Genius in the Visual Arts," *Viator* 8 (1977): 347–98.

31. Cicero, *Tusculanarum Disputationum*, V.xxxviii, trans. J. E. King (London, 1927), p. 536.

32. Recorded in the Life of St. Bernard by Guillelmo, I.iv.20, and published in V. Mortet, *Recueil de textes relatifs à l'histoire de l'architecture et à l'histoire des beaux-arts et des belles-lettres pendant le moyen-age* (Paris, 1911), 1:23.

33. C. Salutati, epist. II.292, is discussed in N. Struever, *The Language of History in the Renaissance: Rhetorical and Historical Consciousness in Florentine Humanism* (Princeton, N.J., 1970), p. 76.

34. Leonardo Bruni, *Ad Petrum Paulum Histrum*, in *Pros. Lat.*, pp. 44–49. For the most recent discussion of its date and Classical sources, see D. Quint, "Humanism and Modernity: A Reconsideration of Bruni's Dialogue," *Renaissance Quarterly* 38 (1985): 423–45.

35. A. de Laborde, *Les Manuscripts à peinture de la Cité de Dieu de Saint Augustin*, 3 vols. (Paris, 1909). These are also discussed in Westfall (n. 24), p. 86; J. Ruysschaert, "Miniaturistes 'Romains' sous Pie II," in *Enea Silvio Piccolomini. Papa Pio II. Atti del convegno per il quinto centenario della morte e altri scritti raccolti*, ed. D. Maffei (Siena, 1968), p. 268; and E. Battisti, *Rinascimento e Barocco* (Turin, 1960), p. 84.

36. A. Chastel, "Un Episode de la symbolique urbaine au XVe siècle. Florence et Rome Cités de Dieu," in *Urbanisme et architecture. etudes écrites et publiées en l'honneur de Pierre Lavedan* (Paris, 1954), p. 78. The manuscript, of about 1495 to 1500, was executed in Naples in the workshop of Reginald de Monopoli and is now Vienna Cod. phil. graec. 4.

37. The work is lost.

38. Discussion of this point is in Cassirer (n. 19), esp. p. 84.

39. The most recent treatments of the problem are in Vickers (n. 20), particularly P. O.

Kristeller, "The Active and the Contemplative Life in Renaissance Humanism," pp. 133-52; V. Kahn, "Coluccio Salutati on the Active and Contemplative Lives," pp. 153-80; and Panizza (n. 20). See also F. Schalk, "Aspetti della vita contemplativa nel rinascimento Italiano," in *Classical Influences on European Culture. A.D. 500-1500*, ed. R. R. Bolgar (Cambridge, 1971); J. Seigel, "Civic Humanism or Ciceronian Rhetoric? The Culture of Petrarch and Bruni," *Past and Present* 34 (1966): 3-48; A. Mazzocco, "Leonardo Bruni's Notion of *Virtus*," in *Studies in the Italian Renaissance: Essays in Memory of Arnolfo B. Ferruolo*, ed. G. P. Biasin, A. N. Mancini, and N. Perella (Naples, 1968), pp. 105-15.; and H. Baron, "The Memory of Cicero's Roman Civic Spirit in the Medieval Centuries and the Florentine Renaissance," and "The Florentine Revival of the Philosophy of the Active Political Life," both in *In Search of Florentine Civic Humanism*, (Princeton, N.J., 1988), 1:94-133 and 134-57. Discussion of the problem often centers on the following texts: Coluccio Salutati, *De Nobilitate legum et medicinae*; Leonardo Bruni, *Isagogicon moralis disciplina*; Lorenzo Valla, *De Vero bono*; and Cristoforo Landino, *Camaldulensian Disputations*.

40. Garin, *Medioevo* (n. 9), pp. 13-99; Trinkaus, *The Scope* (n. 7), p. 33.

41. M. Clagett, *Archimedes in the Middle Ages*, (Canton, Mass., 1984), 3:1329-41. Archimedes is discussed in greater detail in chapter 2.

42. As in the biographies by Vincent of Beauvais, John Waleys, Walter of Burly, Robert Holcot, Giovanni Colonna, and Francesco Petrarca. The relevant texts are in Clagett, n. 41. See the discussion in chapter 2.

43. Kristeller (n. 39), p. 135. Particularly important for my analysis are Panizza's arguments about Valla's thought (n. 20), suggesting a general tendency to reject dualistic patterns such as body versus soul, material versus spiritual, and so on.

44. Lorenzo Valla, De Vero bono, trans. Radetti, in Trinkaus, "*In Our Image*" (n. 7), p. 123.

45. Here he agrees with Seneca, *De Otio*, VI.1-3, that if the actualization of virtue is prevented for external reasons, the fault is not in the man, whose contemplation is still virtuous.

46. See A. Doren, "Fortuna im Mittelalter und in der Renaissance," *Vorträge der Bibliothek Warburg* 2 (1922-1923) and 3 (1924).

47. St. Augustine, *City of God*, trans. H. Betternson (London, 1972), p. 880.

CHAPTER TWO

1. I follow J. Spencer's English translation of the letter in *Leon Battista Alberti: On Painting* (New Haven, Conn., 1966). The Italian text is in Leon Battista Alberti, *On Painting and on Sculpture*, ed. C. Grayson (London, 1972). Grayson dates the vernacular version of *De Pictura* "in or before 1436" (p. 3). A manuscript of *Della pittura* in the Bib. naz. in Florence (cod. Naz. flor. II.iv.38) bears the date 17 July 1436. See G. Mancini, *Vita di Leon Battista Alberti* (1882; rpt., Florence, 1911), p. 128.

2. R. Krautheimer, *Lorenzo Ghiberti* (Princeton, N.J., 1956), 1:234. For Remigio dei Girolami's *Divisio scientie*, see E. Panella, "Un Introduzione alla filosofia in uno studium dei frati predicatori del XIII secolo. 'Divisio scientie' di Remigio dei Girolami," *Memorie Domenicane* 12 (1981): 27-127.

3. M. Salmi, "La Prima operosità architettonica di Leon Battista Alberti," in *Convegno*, p. 10. For the chronology of the dome, see H. Saalman, *Filippo Brunelleschi: The Cupola of Santa Maria del Fiore* (London, 1980).

4. G. C. Argan, "Il Trattato 'De re aedificatoria,' " in *Convegno*, pp. 48-49.

5. L. Benevolo, *The Architecture of the Renaissance*, 2 vols. (London, 1978), 1:44.

6. See H. Weisinger, "Ideas of History During the Renaissance," *JHI* 6 (1945): 415-35.

7. E. H. Gombrich, "A Classical Topos in the Introduction to Alberti's *Della Pittura*," *JWCI* 20 (1957): 173.

8. His opening words, "io solea maravigliarmi insieme e dolermi," reveal that the letter makes use of conventional literary phrases. In this case, the source is Cicero's *De Oratore*, as David Marsh has shown in *The Quattrocento Dialogue: Classical Tradition and Humanist Innovation* (Cambridge, Mass., 1980), p. 82. Alberti uses almost the same words in *Della Famiglia libri III* when, lamenting the decline of great families (in the prologue) he says: "solea spesso fra me maravigliarmi e dolermi se tanto valessi contro agli uomini la fortuna essere iniqua e maligna."

9. Gombrich (n. 7), p. 173.

10. VI.xxi.1–2: "Neque enim quasi lassa et effeta natura nihil iam laudabile parit. Atque adeo nuper audivi Vergilium Romanum paucis legentem comoediam ad exemplar veteris comoediae scriptam, tam bene ut esse quandoque possit exemplar" (*Pliny: Letters and Panegyricus*, trans. B. Radice [Cambridge, Mass., 1972], 1:446–48).

11. R. Sabbadini, *Le Scoperte dei codici latini e greci ne'secoli XIV e XV*, 2 vols. (1905; rpt., Florence, 1967), 1:96.

12. L. B. Alberti, *De Re aedificatoria*, VI.xxi.4, ed. G. Orlandi (Milan, 1966), pp. 448–49.

13. Ibid., pp. 446–47.

14. Lucretius, *De Rerum natura*, II.1150–53, trans. W. H. D. Rouse (London, 1924), p. 166.

15. E. Flores, *Le Scoperte di Poggio e il testo di Lucrezio* (Naples, 1980), p. 34.

16. On the theory of the degeneration of nature, see G. Boas, *Essays on Primitivism and Related Ideas in the Middle Ages* (1948; rpt., New York, 1978), p. 6. This view is not unique to Lucretius; the idea is found in Esdras 5:55 and 14:10–18, in Augustine's *City of God*, XV.9 (where Pliny and Homer are cited as well), and in medieval writers like Heinrich of Suso. For Heinrich of Suso, see N. Gilbert, "A Letter of Giovanni Dondi dall'Orologio to Fra Guglielmo Centueri: A Fourteenth-Century Episode in the Quarrel of the Ancients and Moderns," *Viator* 8 (1977): 319. Particularly telling, I think, is the contrast between Heinrich's complaint, in 1333–1341, that the love of God had grown cold in a senescent world (*Horologium sapientiae*, prologue) and Alberti's concern with declining human achievement, in 1436.

17. Lucretius (n. 14), II.1164–66, p. 166.

18. A useful discussion of early Humanist views on whether nature is in decline and of their Classical sources is in I. Kajanto, *Poggio Bracciolini and Classicism* (Helsinki, 1987), pp. 28–29.

19. Leonardo Bruni, *Ad Petrum Paulum Histrum*, in *Pros. Lat.*, pp. 44–99. The most recent discussion of its date is D. Quint, "Humanism and Modernity: A Reconsideration of Bruni's Dialogues," *Renaissance Quarterly* 38 (1985): 423–45.

20. *Pros. Lat.*, p. 44.

21. Ibid. p. 76.

22. Ibid., p. 60 (trans., E. H. Gombrich, "From the Revival of Letters to the Reform of the Arts. Niccolò Niccoli and Filippo Brunelleschi," in *Essays in the History of Art Presented to Rudolf Wittkower*, (ed. D. Fraser, H. Hibbard, and J. Lewine [London, 1967], p. 73).

23. C. Grayson, ed., Leon Battista Alberti. *Opere volgari* (Bari, 1966), 2:107–83.

24. Ibid., pp. 160–61. G. Santinello traces the quotation to Terence in *Leon Battista Alberti. Una visione estetica del mondo e della vita* (Florence, 1962), p. 13. The passage is discussed in Marsh (n. 8), pp. 92–94. I am struck by the similarity between Nicola's assessment and that of Salutati: "Believe me we create nothing new, but like tailors we refashion garments from the oldest and richest fragments which we give out as new" (*Epistolario*, ed. F. Novati, 4 vols. [Rome, 1896], 2:145). The architectural image may be related, once again, to Lucretius (n. 14), II.7–10): "But

nothing is more delightful than to possess well fortified sanctuaries serene, built up by the teachings of the wise, whence you may look down from the height upon others and behold them all astray, wandering abroad and seeking the path of life."

25. Further discussion of the Humanists' dilemma is in N. Struever, *The Language of History in the Renaissance: Rhetorical and Historical Consciousness in Florentine Humanism* (Princeton, N.J., 1970), p. 97. See also S. S. Gravelle, "Humanist Attitudes to Convention and Innovation in the Fifteenth Century," *Journal of Medieval and Renaissance Studies* 11 (1981): 193–209.

26. Although he does accept it elsewhere. See E. Garin's exploration of Alberti's pessimism and cynicism regarding his own culture in "Studi su Leon Battista Alberti," in *Rinascite e rivoluzioni: Movimenti culturali dal XIV al XVII secolo* (Rome, 1975), pp. 133–96. Trinkaus, summing up Alberti's ambivalent attitude toward the role and importance of human action in this world, says that he "takes emphatically each position at various and clearly overlapping times" (C. Trinkaus, "Themes for a Renaissance Anthropology," in *The Scope of Renaissance Humanism* [Ann Arbor, Mich., 1983], p. 377).

27. As Baron has pointed out, the letter shows how Alberti's earlier pessimistic view of modern achievement changed (H. Baron, "The *Querelle* of the Ancients and Moderns as a Problem for Present Renaissance Scholarship," in *In Search of Florentine Civic Humanism*, 2 vols. [Princeton, N.J., 1988], 2:93–94). A similar optimism had already been expressed in *Della Famiglia*, where Alberti says that fortune cannot rob us of our capacity for every great and excellent thing (Grayson [n. 23], prologue, p. 9).

28. *Pros. Lat.*, p. 60.

29. Quint (n. 19), p. 434. The reference in Cicero is *De Oratore*, I.xxxviii.173.

30. Lorenzo Valla, *Gesta Ferdinandi Regis*, in O. Besomi, "Dai 'Gesta Ferdinandi regis Aragonum' del Valla al 'De Ortographia' del Tortelli," *IMU* 9 (1966): 113–14.

31. Lorenzo Valla, *Elegantiae linguae latinae*, in *Laurentius Valla. Opera Omnia* (Basel, 1540), ed. E. Garin (Turin, 1962) 1:4 (trans., M. Baxandall, *Giotto and the Orators: Humanist Observers of Painting in Italy and the Discovery of Pictorial Composition 1350–1450* [Oxford, 1971], p. 197).

32. There is some precedent for the appreciation of originality in sculpture and in rhetoric. Quintilian, for example, says that in the sculptor's art "the very novelty and difficulty of execution is what most deserves our praise" (*Institutio oratoria*, II.xiii.9–11, trans. H. Butler [Cambridge, 1921–1936] p. 295). This passage and its related concepts are discussed in D. Summers, "Contrapposto: Style and Meaning in Renaissance Art," *AB* 59 (1977): 336–61. However, I find no analogous example in reference to architecture.

33. Matteo Palmieri, *Della Vita civile*, I.157–58, ed. G. Belloni (Florence, 1982), p. 44.

34. Nicholas of Cusa, *Idiota de mente*, in *Nicolaus Cusanus: A Fifteenth-Century Vision of Man*, P. M. Watts (Leiden, 1982), pp. 135–36.

35. *PG*, 156: cols. 23–54. My translation of the text is in the Appendix; it is analyzed in chapter 8. From a letter sent by Chrysoloras to Guarino da Verona in January 1412, we learn that the latter had been distributing copies of the work among his friends (*Epistolario*, I.7, ed. R. Sabbadini, 2 vols. [Venice, 1916]).

36. Chrysoloras (n. 35); Appendix.

37. Salmi (n. 3), p. 10.

38. Grayson (n. 23), pp. 160–61, 181–82.

39. Most of the contemporary, and near-contemporary, assessments of Brunelleschi's dome are conveniently gathered in Saalman (n. 3), pp. 11–15. I refer the reader to this work for the original texts; the translations here are mine.

40. Ibid., p. 12. M. Kemp, in his valuable article "From *Mimesis* to *Fantasia*: The Quattrocento

Vocabulary of Creation, Inspiration and Genius in the Visual Arts," *Viator* 8 (1977): 394, notes how exceptional and early is this acclaim of Brunelleschi's "divine genius." In fact, "there appears to be no comparable acclaim of any quattrocento artist by a contemporary humanist."

41. Ibid., p. 12.

42. See C. von Fabriczy, *Filippo Brunelleschi. La vita e le opere* (1892; rpt., Florence, 1979), p. 168, and Biondo Flavio, *Italia illustrata* (Basel, 1531), p. 304.

43. E. Leonard, ed., Ianotii Manetti. *De Dignitate et excellentia hominis* II.36. (Padua, 1975) (trans. and discussion of the passage, Trinkaus [n. 26], pp. 378–79).

44. Sozomeni Pistoriensis presbyteri. *Chronicon universale*, RIS 16, pt. 1, p. 23.

45. Cristoforo Landino, *Carmina omnia*, Book II, no. XV, ed. A. Perosa (Florence, 1939), p. 64

46. Kemp makes this point ([n. 40], p. 388). See also F. D. Prager, "A Manuscript of Taccola, Quoting Brunelleschi on Problems of Inventors and Builders," *Proceedings of the American Philosophical Society* 112 (1968): 131–49. In fact, as Prager has elsewhere shown, not only was there little precedent in fifteenth-century Florentine machinery for Brunelleschi's Great Hoist of 1421, but it was so far ahead of its time that it had little influence on subsequent hoisting mechanisms (F. D. Prager and G. Scaglia, *Brunelleschi: Studies of His Technology and Inventions* [Cambridge, Mass., 1970], pp. 107–9; F. D. Prager, "Brunelleschi's Inventions and the Renewal of Roman Masonry Work," *Osiris* 9 [1950]: 521–25). More recent discussion of the Great Hoist is in Saalman (n. 3), pp. 154ff.

47. Alamanno Rinuccini, *Lettere ed orazioni*, ed. V. Giustiniani (Florence, 1953), pp. 106–7. The text is discussed in E. H. Gombrich, "The Renaissance Conception of Artistic Progress and Its Consequences," in *Norm and Form: Studies in the Art of the Renaissance* (London, 1966), pp. 1–10 (extracts from the text are in an appendix on pp. 139–40).

48. H. Saalman, ed., *The Life of Brunelleschi by Antonio di Tuccio Manetti* (University Park, Pa., 1970).

49. Antonio Filarete, *Antonio Averlino detto il Filarete. Trattato di architettura*, ed. A. M. Finoli and L. Grassi (Milan, 1972); for example, the discussion of San Lorenzo in Book XXV (p. 693) and VIII. fol. 59r; *Giovanni Rucellai ed il suo Zibaldone*, vol. 1: *Il Zibaldone*, ed. A. Perosa, London, 1981, p. 61, where Brunelleschi is described as the "risuscitore delle muraglie antiche alla romanescha." Leonard (n. 43), p. 35.

50. Nemesius of Emesa, *A Treatise on the Nature of Man*, in *Cyril of Jerusalem and Nemesius of Emesa*, ed. W. Telfer (London, n.d.), p. 243; Philo Judaeus, *De Somniis*, II. vii–ix.48–64. Rabanus Maurus seems to follow this text, and perhaps Nemesius, in his discussion of the origin of "oppidum" in *De Universo*, 14.1 (*PL*, 111: col. 375).

51. For detailed discussions of the place of the mechanical arts in Scholastic thought, see E. Whitney, "Paradise Restored: The Mechanical Arts from Antiquity through the Thirteenth Century," *Transactions of the American Philosophical Society* 80 (1990); P. Sternagel, *Die artes mechanicae im Mittelalter* (Kallmunz, 1966); G. Ovitt, "The Status of the Mechanical Arts in Medieval Classifications of Learning " *Viator* 14 (1983): 84–105; D. C. Lindberg, ed., *Science in the Middle Ages* (Chicago, 1978), esp. B. Stock, "Science, Technology and Economic Progress in the Early Middle Ages," pp. 1–51, and S. E. Brown, "The Science of Weights," pp. 179–205; J. A. Weisheipl, "Classification of the Sciences in Medieval Thought," *Medieval Studies* 27 (1965): 54–90; M. Clagett, *The Science of Mechanics in the Middle Ages* (Madison, Wis., 1959). C. Truesdell, *Essay in the History of Mechanics* (New York, 1968) (although primarily concerned with a later period); L. White, *Medieval Religion and Technology: Collected Essays* (Berkeley, 1978); and White, "Cultural Climates and Technological Advance in the Middle Ages," *Viator* 2 (1971): 171–201.

52. C. H. Buttimer, ed., *Hugonis de Sancto Victore. Didascalicon* (Washington, D.C., 1939). (trans., J. Taylor, *The "Didascalicon" of Hugh of St. Victor* [New York, 1961]). See the discussion of Hugh's placement of the *artes mechanicae* in White, *Medieval Religion* (n. 51), pp. 317–38. In fact, as White points out, Hugh's work assigned an unusually high position to the mechanical arts, as one of the four branches of philosophy (the others being theoretical, practical and logical).

53. Alberti (n. 12), prologue, p. 7. The definition is so dense with meaning and, in many ways, innovative that it would merit an essay devoted to it alone. Here I intend to point out one of its most traditional features, the assumption that the first aim of architecture is utility, and that this assumption (rooted in Classical sources, as we have seen) was systematized in the medieval divisions of the sciences.

54. Panella (n. 2).

55. For the aspect of *ingegno* discussed here, of particular importance is Kemp (n. 40).

56. I would be wary, however, of equating this with poetic furor of the kind described by Boccaccio: "Poesis... est fervor quidam exquisite inveniendi atque dicendi, seu scribendi quod inveneris" (*Genealogia deorum gentilium*, quoted in ibid., pp. 356–57).

57. This aspect is discussed in Gombrich (n. 47), who observed a change in attitudes toward the works of art in Florence in the 1420s. The work becomes a *dimostratione*—that is, it shows how to solve a problem—and it makes a "contribution" within a progressive development. Gombrich's conclusion—that the new view of art emphasized its intellectual value, and that "the stronger the admixture of science in art, the more justifiable was the claim to progress" (p. 8)—is particularly relevant to Alberti's letter. Kemp explores the relation between *invenzione* and *inventore* (meaning discoverer) and *excogitare*, implying intellectual and even scientific investigation ([n. 40], p. 349).

58. See R. E. Latham, *Revised Medieval Latin Word-List from British and Irish Sources* (London, 1965); W. H. Maigne D'Arnis, *Lexicon Manuale ad Medie et Infime Latinitatis* (Paris, 1866); and V. Mortet, *Recueil de textes relatifs à l'histoire de l'architecture et à l'histoire des beaux-arts et des belles-lettres pendant le moyen-age*, 2 vols. (Paris, 1911, 1929).

59. White, *Medieval Religion* (n. 51), p. 327. There are, of course, exceptions, perhaps the best known of which is the inscription on Modena Cathedral: "Ingenio clarus Lanfrancus doctus et aptus / est operis princips hujus rectorque magister." The *"ingenierios"* might also build bridges (E. Müntz, "Les Artistes byzantins dans l'Europe latine du Ve au XVe siécle," *Revue de l'art chrétien* 42 [1983]: 187, document of 1281), make a mill for grinding grain (White, *Medieval Religion* [n. 51], p. 129, document of 1322), or consult on the structural engineering of churches (M. Hollingsworth, "The Architect in Fifteenth-Century Florence," *Art History* 7 [1984]: 391, document of 1401).

60. See Ovitt's analysis of the sources and influence of Gundisalvo's work ([n. 51], pp. 97–98).

61. Ibid., p. 98. Apparently, Gundisalvo considered the *scientia de ingeniis* to be a part of mathematics, not physics.

62. White, *Medieval Religion* (n. 51), p. 329.

63. L. R. Shelby, "The Geometrical Knowledge of Medieval Master Masons," *Speculum* 47 (1972): 395–421.

64. J. Ackerman, "Ars sine scientia nihil est." Gothic Theory of Architecture at the Cathedral of Milan," *AB* 31 (1949): 84–111.

65. Mortet, (n. 58), 2:190. Whether this text is evidence for Master Simon's possession of theoretical knowledge is the subject of a forthcoming study by Joseph O'Connor and the author.

66. P. Sanpaolesi, "Ipotesi sulle conoscenze mathematiche, statiche e meccaniche del Brunelleschi," *Belle Arti* 2 (1951): 25–54.

67. Robert Kilwardby, *De Ortu scientiarum*, ed. A. G. Judy (Oxford, 1976), p. 138. Kilwardby taught at Paris from 1237 to 1245, became archbishop of Canterbury, and died in Viterbo in 1279. His work strongly influenced Remigio dei Girolami (Panella [n. 2], pp. 74–75).

68. Ovitt (n. 51), p. 94.

69. For a detailed treatment of the question, see Clagett (n. 51), p. 640ff.

70. Ibid. Clagett notes especially the activity of Giovanni di Casali, Franciscus de Ferrara, Biagio Pelacani di Parma, Messinus, Angelus de Fossambruno, and Jacopo da Forli. The leading minds of the new physics who taught at Padua and Pavia in the early fifteenth century included Paolo Veneto, Gaetano de Thienis, Giovanni da Fontana, and Johannes de Marchanova.

71. R. Fubini and A. Gallorini, "L'Autobiografia di Leon Battista Alberti," *Rinascimento* 12 (1972): 70 (trans., J. B. Ross and M. M. McLaughlin, *The Portable Renaissance Reader* [New York, 1953], p. 482). Of course, Alberti was no longer a university student at Padua by that time. But his introduction to the subjects must have occurred earlier.

72. Giorgio Vasari, *Le Vite*, ed. R. Bettarini and P. Barocchi (Florence, 1971), 3:143. The passage is commented on by E. Garin, *La Cultura filosofica del rinascimento italiano* (Florence, 1961), p. 328.

73. Studied in White, *Medieval Religion* (n. 51), pp. 261–76.

74. L. White, "Medieval Astrologers and Late Medieval Technology," *Viator* 6 (1975): 295–308.

75. Ibid.

76. This system was reiterated as recently as 1415 at the Council of Constance in article XXIX. See J. D. Mansi, *Sacrorum Conciliorum Nova et Amplissima Collectio* (Graz, 1961), 28: cols. 131–32. I thank S. I. Camporeale for this reference.

77. Kemp (n. 40), pp. 392–97, notes that of the twenty-five instances known to him between 1300 and 1500 in which artists are called *ingegni*, more than half apply to architects ([n.40], pp. 392–97). He suggests that one of the reasons for this might be the rediscovery of Vitruvius. My reservations about this suggestion are discussed in the text.

78. That is, following Cicero and, through him, Aristotle. I do not agree with Long that "the influence of Vitruvius was paramount" on the Renaissance ideal of the unification of theory and practice. (P. Long, "The Contribution of Architectural Writers to a 'Scientific' Outlook in the Fifteenth and Sixteenth Centuries," *Journal of Medieval and Renaissance Studies* 15 [1985]: 297).

79. See, on the differences between Vitruvius's and Alberti's architectural treatises, R. Krautheimer, "Alberti and Vitruvius," in *Acts of the XX International Congress of the History of Art*, vol. 2: *The Renaissance and Mannerism*, ed. M. Meiss (Princeton, N.J., 1963), pp. 42–51.

80. P. L. Rose, *The Italian Renaissance of Mathematics* (Geneva, 1975), p. 28; Rose, "Humanist Culture and Renaissance Mathematics: The Italian Libraries of the Quattrocento," *Studies in the Renaissance* 20 (1973): 57; E. B. Fryde, *Humanism and Renaissance Historiography* (London, 1983), p. 217; B. L. Ullman and P. Stadter, *The Public Library of Renaissance Florence: Niccolò Niccoli, Cosimo de'Medici and the Library of San Marco* (Padua, 1972), p. 255; Mancini (n. 1), p. 62. Filelfo's copy was bought by Lorenzo the Magnificent and is now Laurenziana MS. xxviii.45. For the attribution of the work to Aristotle, see Clagett (n. 51), p. 4.

81. See bibliography in n. 80. Manetti's copy is now Vat. Pal. gr. 162.

82. P. Fontana, "Osservazioni intorno ai rapporti di Vitruvio colla teoria dell'architettura del Rinascimento," in *Miscellanea di storia dell'arte in onore di Igino Benvenuto Supino* (Florence, 1933), p. 322, n. 16.

83. On this problem, see Clagett (n. 51), pp. 971–72; Rose, "Humanist Culture" (n. 80), p. 82; and S. Drake and I. E. Drabkin, *Mechanics in Sixteenth-Century Italy: Selections from Tartaglia, Benedetti, Guido Ubaldo and Galileo* (Madison, Wis., 1969), p. 9.

84. On this distinction, see Brown, (n. 51), p. 190.

85. Rose, "Humanist Culture" (n. 80), p. 54; Fryde (n. 80), p. 196.

86. A. P. Treweek, "Pappus of Alexandria. The Manuscript Tradition of the 'Collectio Mathematica,' " *Scriptorium* 11 (1957): 195–233. This late-ninth- or early-tenth-century manuscript is now Vat. gr. 218. Jones maintains that this manuscript was already in the West before 1300 (A. Jones, trans., *Pappus of Alexandria. Book 7 of the 'Collection'* [New York, 1986], pp. 42–55). On this same question, see S. Unguru, "Pappus in the Thirteenth Century in the Latin West," *Archive for the History of the Exact Sciences* 13 (1974): 307–34.

87. C. Stinger, "Ambrogio Traversari and the 'tempio degli Scolari' at S. Maria degli Angeli in Florence," in *Essays Presented to Myron P. Gilmore*, ed. S. Bertelli and G. Ramakus 2 vols. (Florence, 1978), 1:279. This would be an almost unique case of the recovery of ancient engineering devices. See White, "The Medieval Roots of Modern Technology and Science," in *Medieval Religion*, (as in note 51), p. 91. He concludes from his examination of unpublished manuscripts of fifteenth- and sixteenth-century Italian engineers, "I see no trace in them of recovered items of ancient engineering, save the hodometer."

88. F. D. Prager and G. Scaglia, *Mariano Taccola and His Book 'De Ingeneis'* (Cambridge, Mass., 1972), p. 58.

89. Rose, *Italian Renaissance* (n. 80), pp. 31–32. See Clagett (n. 51), p. 9, and Rose, "Humanist Culture" (n. 80), p. 82. Kristeller emphasizes the importance of the Humanists' gathering, translation, and popularization of Archimedes' works for the scientific discoveries of the sixteenth century (P. O. Kristeller, "The Place of Classical Humanism in Renaissance Thought," *JHI* 4 [1943]: 60–61).

90. Ullman and Stadter (n. 80), p. 215. This is now Bib. naz. c.s. I.v.30.

91. Rose, *Italian Renaissance* (n. 80), p. 31.

92. Leon Battista Alberti, *Intercoenales*, in *Pros. Lat.*, p. 642.

93. Prager and Scaglia (n. 88), p. 17. M. Clagett, *Archimedes in the Middle Ages* (Canton, Mass., 1984), 3:319. The manuscript of *De Machinis*, of about 1449, is in the New York Public Library, Spencer Coll., 136.

94. Pappus, *Mathematical Collection*, preface to Book VIII, in *Selections Illustrating the History of Greek Mathematics*, trans. I. Thomas (London, 1951), 2:615.

95. Most of the relevant texts are gathered in Clagett, (n. 93). Before William of Moerbeke, only two complete treatises of Archimedes had been known in the Arabo-Latin tradition. He recovered almost all the works from the Greek tradition, including Archimedes' writings on mechanics, but they had almost no effect until the fifteenth century because they utilized a non-dynamic, statical approach (Rose, *Italian Renaissance* [n. 80], p. 80).

96. Vitruvius, *The Ten Books on Architecture*, ix. Clagett discusses Alberti's knowledge of Archimedes' works from a scientific point of view ([n. 93], pp. 316–18). His lack of interest in the scientific works is in contrast, I believe, with the importance for him of Archimedes as a cultural exemplar, as in the last pages of *Profugiorum*.

97. Clagett (n. 93), 3:1330–35. The first medieval biography of Archimedes is in Vincent of Beauvais's *Speculum historiale*, v.43, where he is presented as a philosopher known for his book on the squaring of the circle. A similar portrait is drawn by John Waleys in *Breviloquium*, II.ii; in the *Dialogus creaturum*, chap. 97, perhaps by Maino de Maineri de'Mediolano; in Walter of Burly's *De Vita et moribus philosophorum*, chap. LXXIV; and in Giovanni Colonna's *De Viribus illustribus*.

98. Cicero, *De Re publica*, I.xiv; Cicero, *De Natura deorum*, II.88; Julius Firmicus Maternus, *Matheseos libri VIII*, VI.30; Ovid, *Fasti*, 6.276–82; Claudian, Poem LI; Lactantius, *Divinae institutiones*, II.18.

99. Manetti (n. 43), pp. 61–62.
100. Livy, *Ab urbe condita*, XXXIV.2–3; Plutarch, *Life of Marcellus*, XIV.7–XVII.7. Plutarch, however, is embarrassed by Archimedes' practical application of theoretical knowledge.
101. Petrarch, *De Viris illustribus*, in the life of Marcellus; Petrarch, *Rerum memorandarum libri*, p. 22 (ed. Billanovich).
102. Pappus (n. 94), p. 619.
103. Another important source for Alberti's view of Archimedes is Diodorus of Sicily, an author frequently drawn on in *De Re aedificatoria*. In Book V.37.4 of his *History*, Diodoros praises Archimedes' invention of the Egyptian screw (a lifting device), and in XXVI.18.1 describes him as a "σοφὸς μηχανητὴς" and "γεωμέτης" who was busy making a "διάγραμμα μηχανικὸν" at the time of his death.
104. Lorenzo Valla, *De Voluptate*, II.xxviii.9 (trans., A. K. Hieatt and M. Lorch, *Lorenzo Valla. On Pleasure* [New York, 1977], p. 200).
105. Translation in Rose, *Italian Renaissance* (n. 80), p. 14.
106. Chrysoloras (n. 35), col. 49C; Appendix, p. 213.
107. Ibid.
108. The work is published in G. L. Huxley, *Anthemius of Tralles: A Study in Later Greek Geometry* (Cambridge, Mass., 1959).
109. See, for example, the study by G. Downey, "Byzantine Architects. Their Training and Methods," *Byzantion* 18 (1948):99–118. This was not the case in the medieval West, where the architect is more usually associated with the master mason. See N. Pevsner, "The Term 'Architect' in the Middle Ages," *Speculum* 17 (1942): 549–62, and S. Kostof, *The Architect: Chapters in the History of the Profession* (New York, 1977), with further bibliography.
110. Lucian, *Hippias, or the Bath*, I, trans. A. Harmon (New York, 1913), p. 35.
111. Brunelleschi's activity as a civil and military engineer is reviewed in L. Olschki, *Geschichte der Neusprachlichen Wissenschaftlichen Literatur*, 3 vols. (Heidelberg, 1919), 1:39–44.
112. Giovanni Villani, *Cronica*, I.xlix, ed. I. Moutier (Florence 1823), p. 70.
113. G. Pochat, "Brunelleschi's 'sacre rappresentazioni': Beginn einer dynamischen Aufführungspraxis," *Daidalos* 14 (1984):14–20.
114. Chrysoloras (n. 35), col. 34C; Appendix, p. 213.

CHAPTER THREE

1. *Petrarch's "Africa,"* trans. T. G. Bergin (New Haven, Conn., 1977), pp. 134–36.
2. Bernard of Clairvaux, *Apologia ad Guillelmum*, XII.28, *PL*, 182: col. 914: "Omitto oratoriorum immensas altitudines, immoderatas longitudines, supervacuas latitudines."
3. J. Onians, "The Last Judgment of Renaissance Architecture," *Royal Society of Arts Journal* 128 (1980): 701–20.
4. For a discussion of the condemnation of luxurious building in Roman literature, see J. Freeman, "The Roof was Fretted Gold," *Comparative Literature* 27 (1975): 254–66.
5. Plato, *Critias*, 116D, trans. R. G. Bury (London, 1929), p. 290.
6. Horace, *Odes*, trans. C. E. Bennet (London, 1914). See also II.xv, II.xviii, and III.i.
7. Sallust, *Bellum Catilinae*, XII.3, trans. J. Rolfe (Cambridge, Mass., 1960), p. 20.
8. Suetonius, *De Vita Caesarum*, trans. J. Rolfe, 2 vols. (Cambridge, Mass., 1915), "The Deified Augustus," 1:259.
9. Ovid, *Fasti*, VI.639–48, trans. J. G. Frazer (London, 1931), pp. 368–69.
10. Plutarch, *Lives*, "Poplicola," 15.5, trans. A. H. Clough (New York, n.d.), p. 126.

NOTES   229

11. Posidonius, in Aetius, *Placita*, I.6 (quoted in W. Tatarkiewicz, *History of Aesthetics*, ed. J. Harrell, 3 vols. [The Hague, 1970], 1:194–95).

12. Aristotle, *Nicomachean Ethics*, IV.iii, trans. H. Rackham (London, 1926), p. 215.

13. Seneca, *Letter* XCI.13, trans. R. Gummere, 3 vols. (London, 1920), 2:441.

14. Prudentius, *Psychomachia*, trans. H. J. Thomson (London, 1949), Book 1, p. 340.

15. Leon Battista Alberti, *De Re aedificatoria*, VII.iii, ed. G. Orlandi (Milan, 1966), p. 545. All references are to this edition; all translations are mine.

16. Saint Antoninus, *Summa major*, pars. III, tit. viii, cap. iv.

17. H. Saalman, "Giovanni di Gherardo da Prato's Designs Concerning the Cupola of S. Maria del Fiore in Florence," *JSAH* 18 (1959): 11–20.

18. A. Lanza, ed., Giovanni Gherardi da Prato. *Il Paradiso degli Alberti* (Rome, 1975), pp. 11–12: "Ogni cosa alfine vola e trapassa, e sol la virtù etterna si giudica al vero. Che giova adunche alzare al cielo le superbissime torri, i magnifici e ampi palazzi colle regali aule di preziosissimi marmi ornate nella istolta openione de'mortali volere edificare per fama etterna seguire?"

19. Coluccio Salutati, *De Seculo et religione*, ed. B. Ullman (Florence, 1957), pp. 60–61.

20. E. R. Curtius, *European Literature and the Latin Middle Ages*, trans. W. Trask (Princeton, N.J. 1953), p. 162.

21. Aristotle, *The Art of Rhetoric*, I.ix.38, trans. J. H. Freese (London, 1939), p. 103.

22. Cicero, *De Partitione oratoria*, VI.19–21, trans. H. Rackham, (London, 1948), p. 326.

23. See T. Buddensieg, "Criticism and Praise of the Pantheon in the Middle Ages and the Renaissance," in *Classical Influences on European Culture, A.D. 500–1500*, ed. R. R. Bolgar (Cambridge, 1971), pp. 259–67.

24. Ammianus Marcellinus, XVI.10.14, trans. J. Rolfe, 3 vols. (London, 1935), 1:249.

25. J. E. Sandys, *A History of Classical Scholarship*, 3 vols. (New York, 1964), 2:29. Niccoli received a copy of the manuscript, which he put in Cosimo's library; the original went to the Vatican.

26. In Buddensieg (n. 23).

27. *The Commentaries of Pius II*, trans. F. Gragg, Smith College Studies in History, 22 (1936–1937), Book II, p. 166.

28. S. De' Conti in *Historia sui temporis*, XVI.2: "operis descriptio omnem antiquitatem pulchritudine et magnitudine superare videtur; in capite enim basilicae testudo futura est latior et altior Templo Pantheon" (quoted in H. Günther, *Das Studium der antiken Architektur in den Zeichnungen der Hoch-renaissance* [Tübingen, 1988], p. 45, n. 234).

29. J. Rasch, "Die Kuppel in der römischen Architektur. Entwicklung, Formgebung, Konstruktion," *Architectura* 15 (1985):117–39.

30. Statius, *Silvae*, IV.ii.30–31, trans. J. H. Mozley (London, 1928), p. 213. For the discovery, see E. Flores, *Le Scoperte di Poggio e il testo di Lucrezio* (Naples, 1980), p. 13. Macdonald suggests that all the vaulting in the Domus Flavia was barrel vaulting and that Statius's description refers to the Aula Regia (W. Macdonald, *The Architecture of the Roman Empire*, vol. 1: *An Introductory Study* [New Haven, 1965], pp. 57–62).

31. Martial, *Epigrams*, VIII.xxxvi, trans. W. Ker (London, 1920), pp. 27–28. Alberti's teacher Barzizza read the epigrams, which were hardly known even though Boccaccio had had a copy, to Francesco Barbaro in Venice in 1407 (R. Sabbadini, *Le Scoperte dei codici latini e greci ne'secoli XIV e XV* [1905; rpt., Florence, 1967], p. 73).

32. *Anthologia Graeca*, ed. H. Beckby (Munich, 1957), Book IX, epigram 58, by Antipater.

33. Ibid., p. 77; R. Pfeiffer, *A History of Classical Scholarship from 1300 to 1850* (Oxford, 1976), p. 48.

34. For Chrysoloras, see Appendix, p. 200.

35. For the tradition of considering as part of the city the walled city and its *contado*, see chapters 6 and 9.

36. Giannozzo Manetti, *De Secularibus et pontificalibus pompis*, ed. E. Battisti, in *Umanesimo e esoterismo: Atti del V convegno internazionale di studi umanistici*, ed. E. Castelli (Padua, 1960), p. 313. See also p. 312: "nam praeter admirabilem eius altitudinem quae cetera orbis terrarum edificia exsuperat."

37. P. O. Kristeller, *Catalogus translationum et commentariorum: Medieval and Renaissance Latin Translations and Commentaries* (Washington, D.C., 1971), 2:61, 3:414.

38. L. Bertalot and A. Campana, "Gli Scritti di Iacopo Zeno e il suo elogio di Ciriaco d'Ancona," *La Bibliofilia* 41 (1939): 368.

39. Pliny, *Natural History*, XXXVI.xvi, trans. D. E. Eichholz, 10 vols. (Cambridge, Mass., 1962), 10:59, 63.

40. Ibid., XXXVI.xix, p. 67.

41. Ibid., XXXVI.xxiv, p. 83.

42. Sextus Julius Frontinus, *De Aquaeductu urbis Romae*, I.64.11–13. The text is partly reprinted in *VZ*, 1:27: "Tot aquarum tam necessariis molibus pyramidas videlicet otiosas compares autcetera, inertia sed fama celebrata opera Graecorum."

43. Philo Byzantinus, *De Septem orbis spectaculis*, ed. L. Allatius (Rome, 1640).

44. For example, Herodotus, *Histories*, II.148.

45. Diodorus Siculus, *The Library of History*, I.64.11, trans. C. Oldfather, 12 vols. (Cambridge, Mass., 1933–1967), 1:223.

46. Appendix, p. 212.

47. Petrarch, *De Remediis utriusque fortinae*, II. 93, "De Tristitia et miseria," discussed in C. Trinkaus, *The Scope of Renaissance Humanism* (Ann Arbor, Mich., 1983), p. 354.

48. Chrysostom, *Homilies on Genesis*, 9.6–8, *PG*, 53: cols. 78–97.

49. The work was thought to be by Gregory of Nyssa until the sixteenth century. It had been translated into Latin in the eleventh and twelfth centuries by Alfano, bishop of Salerno and Burgundione of Pisa (E. Garin, "Traduzioni dal Greco e dall'Arabo," in *Storia della filosofia italiana*, 4 vols. [Turin, 1966], 1:85; C. Trinkaus, *"In Our Image and Likeness": Humanity and Divinity in Italian Humanist Thought* [London, 1970], p. 186).

50. IV.254–55, in Trinkaus (n. 49), p. 186.

51. See the excellent discussion in Trinkaus (n. 47), pp. 378–79.

52. Giannozzo Manetti, *De Dignitate et excellentia hominis*, ed. E. Leonard (Padua, 1975).

53. Manetti (n. 36), p. 313.

54. Trans., E. Gombrich, "Hypnerotomachiana," in *Symbolic Images* (London, 1972), p. 104. G. Pozzi and L. Ciapponi, eds. *Francesco Colonna. Hypnerotomachia Poliphili* (Padua, 1964), p. 17. Similar appreciation for the mind of the architect is in the description of the dome of the Temple of Venus, p. 195.

55. I do not consider the problem of private patronage, especially Medici patronage, which raises different moral issues than public and ecclesiastical building.

56. This attractive hypothesis was forwarded by G. Canfield, "The Florentine Humanists' Concept of Architecture in the 1430's and Filippo Brunelleschi," in *Scritti di storia dell'arte in onore di Federico Zeri*, 2 vols. (Venice, 1984), 1:116.

57. Cicero, *De Inventione*, II.lvi.168, trans. H. M. Hubbell (London, 1949), p. 335. See also Vitruvius, *The Ten Books on Architecture*, Book I, preface, who saw the power of the state enhanced through the "distinguished authority of its public buildings."

58. For the *Laudatio*, see chapter 9, esp. n. 48.

59. Matteo Palmieri, *Della Vita civile*, IV.213–16, ed. G. Belloni (Florence, 1982), p. 151.

60. Manetti (n. 36), p. 313.

61. See chapter 9.

62. Nicholas's deathbed speech was reported by Manetti, *De Vita ac moribus Nicolai Quinti*, RIS, 2: col. 950 (trans., C. Westfall, *In This Most Perfect Paradise: Alberti, Nicholas V, and the Invention of Conscious Urban Planning in Rome, 1447–55* [University Park, Pa., 1974], p. 33). For accounts of opposition to Nicholas's building projects, see M. Miglio, "Una Vocazione in progresso: Michele Canensi, biografo papale," *Studi medievali* 12 (1971): 489.

63. Miglio (n. 62), p. 518: "Quis est enim adeo depravate mentis atque obdurati vir animi quin commoveatur et in bonas artes in probitatem in ipsam quoque religionem exinflectatur quum suum in te animum contorquere ac tuam insuper cum religione sanctimoniam intueri decreverit?"

64. Salutati (n. 19), pp. 129–30.

65. B. Vickers, *In Defence of Rhetoric* (Oxford, 1988), p. 351.

66. H. Baron, "Civic Wealth and the New Values of the Renaissance: The Spirit of the Quattrocento," in *In Search of Florentine Civic Humanism*, 2 vols. (Princeton, N.J., 1988), 1:229.

67. Ibid., p. 233.

68. Poggio Bracciolini, *Historia disceptativa de avaricia*, in *Opera omnia*, ed. R. Fubini (Turin, 1966), 1:13.

69. R. Black, "Ancients and Moderns in the Renaissance: Rhetoric and History in Accolti's 'Dialogue on the Preeminence of Men of his own Time,'" *JHI* 43 (1982): 27. The work, entitled *Dialogus super excellentia curie Romane*, was a defense of the modern church.

70. Ibid., p. 26.

71. Günther (n. 28), p. 45, n. 234. The reference is to *Historia sui temporis*, XVI.2.

72. J. O'Malley, "Fulfillment of the Christian Golden Age under Pope Julius II: Text of a Discourse of Giles of Viterbo," *Traditio* 25 (1969): 322–23. The sermon, "De Ecclesiae incremento," was delivered in 1507 in St. Peter's.

73. Ibid: "unus ipse, Iuli seconde, surrexeris [the temple], qui rem sacram adeo amaveris curaverisque ut sacratissimi templi fastigium ad coelum usque evehere attollereque contenderis, omnium prorsus aedificiorum et miraculorum admirationem in unum istud religionis opus conversus."

74. Francesco Albertini, *Opusculum de mirabilibus novae et veteris urbis Romae*, Book III, in VZ, (n. 42), p. 501: "excedit enim admirationem et magnificentiam Greciae in aedificatione Templi Aephesiae Dianae ducentis viginti annis facti a tota Asia. Templi enim longitudo CCCCXXV pedum, latitudo vero CCXX; columnae CXXVII a singulis regibus factae, LX pedum altitudine ex his XXXVI caelatae, teste Plinio lib. XXXVI. [Then follows a section showing that Florence Cathedral outdoes the Temple at Ephesus.] Quae omnia, Deo dante, sanctitas tua vult superare in supradicta basilica, quod praeclarum et admirabile opus iam ad sidera tendit."

75. Ibid.

76. J. D'Amico, "Papal History and Curial Reform in the Renaissance: Raffaele Maffei's 'Brevis historia' of Julius II and Leo X," *Archivum historiae pontificiae* 18 (1980): 196.

77. M. Fagiolo, "La Basilica Vaticana come tempio-mausoleo 'inter duas metas.' Le Idee e i progetti di Alberti, Filarete, Bramante, Peruzzi, Sangallo e Michelangelo," in *Atti del XXII congresso di storia dell'architettura* (Rome, 1986), p. 208, n. 19.

78. J. O'Malley, *Giles of Viterbo on Church and Reform: A Study in Renaissance Thought* (Leiden, 1968), pp. 105–7. O'Malley was troubled by the anomaly of "the austere and tireless preacher of poverty lending his encouragement to the raising of a monument which from almost

every point of view would seem to embody attitudes and values directly antipathetic to his own" (p. 5).

79. Saint Augustine, *On Music*, VI.44, trans. R. Taliaferro (New York, 1947), p. 367.

CHAPTER FOUR

1. Leon Battista Alberti, *De Re aedificatoria*, VI.iii, ed. G. Orlandi (Milan, 1966), pp. 451–57. All page references are to this edition, all translations into English are mine.

2. R. Krautheimer, "Die Anfänge der Kunstgeschichtsschreibung in Italien," *Repertorium für Kunstwissenschaft* 50(1929): 58 (reprinted as "The Beginnings of Art Historical Writing in Italy," in *Studies in Early Christian, Medieval and Renaissance Art* [New York, 1969], pp. 257–75).

3. P. Frankl, *The Gothic: Literary Sources and Interpretations through Eight Centuries* (Princeton, N.J., 1960), p. 257.

4. C. Westfall, "Society, Beauty and the Humanist Architect in Alberti's *De re aedificatoria*," *Studies in the Renaissance* 16 (1969): 61–81. See chapter 5.

5. Leon Battista Alberti, *Profugiorum ab aerumna libri III*, ed. C. Grayson, in *Leon Battista Alberti. Opere volgari* (Bari, 1960), 2:107.

6. Alberti (n. 1) VI.iii, pp. 455–57, and VI.i, p. 445: "Ex tribus partibus, quae ad universam aedificationem pertinebant, uti essent quidem quae adstrueremus ad usum apta, ad perpetuitatem firmissima, ad gratiam et amoenitatem paratissima, primis duabus partibus absolutis restat tertia omnium dignissima et perquam valde necessaria."

7. Günther has recently observed that Filarete and Francesco di Giorgio also defined the Orders only according to proportions and not according to formal differences (H. Günther, *Das Studium der antiken Architektur in den Zeichnungen der Hoch-Renaissance* [Tübingen, 1988], p. 51).

8. J. Ackerman, " 'Ars sine scientia nihil est.' Gothic Theory of Architecture at the Cathedral of Milan," *AB* 31 (1949): 109.

9. A. M. Finoli and L. Grassi, eds., *Antonio Averlino detto il Filarete. Trattato di architettura* (Milan, 1972). Modern architecture was faulted in regard to "proporzioni e ancora di misure" (XVI, p. 482); the use of the pointed arch (VIII, pp. 226–27); improper proportions for columns, resulting in "colonnette piccole" (VIII, p. 220); and door frames other than right-angled ones (VIII, p. 234).

10. G. Germann, *Gothic Revival in Europe and Britain: Sources, Influences and Ideas*, trans. G. Onn (London, 1972), p. 12; H. Kruft, *Geschichte der Architekturtheorie. Von der Antike bis zur Gegenwart* (Munich, 1986), p. 74.

11. E. Panofsky, "The First Page of Vasari's 'Libro': A Study on the Gothic Style in the Judgment of the Italian Renaissance," in *Meaning in the Visual Arts* (Harmondsworth, (1955), p. 224.

12. Germann (n. 10), p. 13.

13. Ibid.

14. Ibid.

15. E. H. Gombrich, "From the Revival of Letters to the Reform of the Arts. Niccolò Niccoli and Filippo Brunelleschi," in *Essays in the History of Art Presented to Rudolf Wittkower*, ed. D. Fraser, H. Hibbard, and J. Lewine (London, 1967), p. 80. Noting that the windows of the Medici palace are like those of the Bargello, with the "corruption" of the pointed arch removed, he suggested that the real novelty of Quattrocento architecture might lie in its avoidance "of mistakes that would infringe the Classical norm."

H. Klotz, *Die Frühwerke Brunelleschi's und die mittelalterliche Tradition* (Berlin, 1970); M.

Trachtenberg, "Brunelleschi, 'Giotto' and Rome," in *Smyth*, pp. 675–98; G. Morolli, *Firenze e il Classicismo: Un rapporto difficile. Saggi di storiografia dell'architettura del Rinascimento 1977–87* (Florence, 1988), esp. "Architettura del Quattrocento a Firenze. La città immaginata," pp. 145–203.

16. C. Gilbert, "The Earliest Guide to Florentine Architecture, 1423," *Mitteilungen des Kunsthistorisches Institutes in Florenz* 40 (1969): 33–46. Even though Goro Dati had been one of the *operaii* of the Ospedale degli Innocenti in 1422 and 1423, one of the most conspicuous features of his taste is his preference for the Trecento.

17. Leonardo Bruni, *Laudatio florentinae urbis*, trans. in B. Kohl and R. Witt, *The Earthly Republic: Italian Humanists on Government and Society* (Philadelphia, 1978), pp. 138–39.

18. Aeneas Silvius Piccolomini, epistle CLXVI, in Frankl (n. 3), p. 740.

19. English translation is in ibid., pp. 245–46; the Latin is in ibid., appendix 14. Pius also praised York Cathedral, the Duomo of Orvieto, and the churches of Vienna (*The Commentaries of Pius II*, trans. F. Gragg, Smith College Studies in History, 22 [1936–1937], Book I, p. 21 [York], Book IV, p. 337 [Orvieto]); *Historia rerum Friderici Tertii*, in L. Heydenreich, "Pius II als Bauherr von Pienza," *Zeitschrift für Kunstgeschichte* 6 (1937): 116 (Vienna).

20. *Commentaries* (n. 19), Book ix, p. 602.

21. Ibid., ix, p. 602.

22. See Günther (n. 7), p. 26.

23. Giannozzo Manetti, *De Secularibus et pontificalibus pompis*, ed. E. Battisti, in *Umanesimo e esoterismo: Atti del V convegno internazionale di studi umanistici*, ed. E. Castelli (Padua, 1960), p. 312.

24. S. I. Camporeale, "Giovanni Caroli e le 'Vitae fratrum S. M. Novellae.' Umanesimo e crisi religiosa (1460–1480)," *Memorie Domenicane* 12 (1981): 240.

25. Ibid., p. 239.

26. V. Bonito, "The Saint Anne Altar in Sant'Agostino, Rome" (Ph.D. diss., New York University, 1983), p. 251. The poem, by Blosio Palladio, formed part of the *Coryciana*, a collection of poems originally hung on the Saint Anne altar.

27. H. Burns, "Quattrocento and the Antique: Some Problems," in *Classical Influences on European Culture, A.D. 500–1500*, ed. R. R. Bolgar (Cambridge, 1971), p. 276. This view had been advocated earlier by W. K. Ferguson in two essays: "Humanist Views of the Renaissance," *American Historical Review* 45 (1939): 1–28, and *The Renaissance in Historical Thought* (Cambridge, Mass., 1948), chap. 1.

28. Günther (n. 7), p. 22. The first practical guide to the five Orders is in Serlio's treatise, Book IV, of 1537 (Kruft [n. 10], pp. 71, 82).

29. C. von Fabriczy, *Filippo Brunelleschi. La vita e le opere*, ed. A. M. Poma (1892; rpt., Florence, 1979) pp. 66–67. Most recently, this has been supported by Günther (n. 7), p. 10.

30. T. Mommsen, "Petrarch's Conception of the 'Dark Ages,' " *Speculum* 17 (1942): 229.

31. J. Fracassetti, ed., *Francisci Petrarcae epistolae de rebus famialiabus et variae* 3 vols. (Florence, 1859), 1:47.

32. On this, see E. Garin, *Umanesimo italiano. Filosofia e vita civile nel Rinascimento* (1947; rpt., Bari, 1973), pp. 25–46; G. Falco, *La Polemica sul medio evo* (1933; rpt., Naples, 1974); D. Hay, "Flavio Biondo and the Middle Ages," *Proceedings of the British Academy* 45 (1959): 97–125; T. Buddensieg, "Gregory the Great, the Destroyer of Pagan Idols," *JWCI* 28 (1965): 44–65; and Ferguson, "Humanist Views" (n. 27), p. 5.

33. A useful introduction to these systems of periodization is in B. Smalley, *Storici nel Medioevo*, trans. I. Pagani (Naples, 1979); see also B. Guenée, *Histoire et culture historique dans l'Occident médiéval* (Paris, 1980), esp. pp. 148ff.

34. Discussed in Falco (n. 32), p. 52, and Ferguson, "Humanist Views" (n. 27), p. 8, n. 8.

35. Ferguson, "Humanist Views" (n. 27), p. 5.

36. N. Rubinstein, "Il Medio evo nella storiografia italiana del Rinascimento," *Lettere italiane* 24 (1972): 434. On the tripart definition of history, see also M. McLaughlin, "Humanist Concepts of Renaissance and Middle Ages in the Tre- and Quattrocento," *Renaissance Studies* 2 (1988): 131–42.

37. The passage is in *Historiarum florentini populi*, I.i.98 (trans. D. Wilcox, *The Development of Florentine Humanist Historiography in the Fifteenth Century* [Cambridge, Mass.], 1969, p. 72). See the discussion of Bruni's thought in B. Ullman, "Leonardo Bruni and Humanistic Historiography," *Medievalia et Humanistica* 4 (1946): 45–61 (reprinted in Ullman, *Studies in the Italian Renaissance* [Rome, 1955], pp. 321–44). Of interest on this question in the same volume is "Renaissance. The Word and the Underlying Concept," pp. 11–25. Most recently, see H. Baron's updated version of his 1932 article ("Das Erwachsen des Historischen Denkens im Humanismus des Quattrocento," *Historische Zeitschrift* 147 [1932] as "New Historical and Psychological Ways of Thinking: From Petrarch to Bruni and Machiavelli," in *In Search of Florentine Civic Humanism*, 2 vols. (Princeton, 1988), 1:24–42.

38. Biondo Flavio, *Decades. Historiarum ab inclinatione Romanorum imperii libri*. Discussed in Falco (n. 32), pp. 44–45; Hay (n. 32), pp. 102–5; and E. Cochrane, *Historians and Historiography in the Italian Renaissance* (Chicago, 1981), pp. 36–37.

39. Flavio (n. 38), Decade I, Book 3.

40. N. Rubinstein, "Bartolomeo Scala's *Historia Florentinorum*," in *Studi di bibliografia e di storia in onore di Tammaro de'Marinis*, (Verona, 1964), p. 55.

41. Garin (n. 32), p. 25; Garin, "Interpretazioni del Rinascimento," in *Medioevo e Rinascimento* (Bari, 1984), p. 95.

42. Mommsen (n. 30), p. 237.

43. Ferguson, "Humanist Views" (n. 27), p. 17.

44. Ibid.

45. Ullman (n. 37), p. 19.

46. Ibid.; J. Spencer, "Ut Rhetorica Pictura: A Study in Quattrocento Theory of Painting," *JWCI* 20 (1957): 27, where he quotes a letter of Pius II: "After Petrarch letters emerged; after Giotto the hand of the painter arose."

47. See chapter 7.

48. Matteo Palmieri, *Della Vita civile*, ed. G.Belloni (Florence, 1982), pp. 44–46; Biondo Flavio, *Italia illustrata* (Basel, 1531), p. 304.

49. Sicco Polenton, *Scriptorum illustrorum latinae linguae libri XVIII* (1433–1437), discussed in Ullman (n. 37), p. 17. The text was published by Ullman (Rome, 1928), where the relevant passage is on p. 125.

50. Alamanno Rinuccini, preface to Philostratus, *Life of Apollonius* (1474) (trans., E. H. Gombrich, "The Renaissance Conception of Artistic Progress," in *Norm and Form: Studies in the Art of the Renaissance* [London, 1966], p. 2). Vespasiano da Bisticci claimed that when Traversari and Bruni revived the Latin tongue it had been dead and buried for a thousand years or more (Ferguson, "Humanist Views" [n. 27], p. 25).

51. Alfonso of Aragon, letter of 1443, in Hay (n. 32), p. 116; F. Villani, *Liber de civitatis Florentiae famosis civibus*, at the beginning of his life of Dante.

52. Coluccio Salutati, *Epistolario*, ed. F. Novati, 4 vols. (Rome, 1896), 3:83–84. The letter is of 1395.

53. Buddensieg notes that this was first proposed by John of Salisbury in his *Policraticus*,

taken up by Martinus Polonus in the thirteenth century, and then repeated by most chronicles up to the fifteenth century [n. 32], pp. 46–47.

54. Ambrogio Traversari wrote in a letter of 1437 that eloquence had died 600 years before Niccolò Niccoli revived it. (*Ambrosii Traversarii generalis Camaldulensium. Latinae epistulae*, ed. L. Mehus [Florence, 1759], IX.21).

55. Biondo Flavio, *Roma Instaurata*, Book II (Basel, 1559), p. 255: "Solae igitur incusandae et detestandae sunt manus improbae illorum, qui ut privata et quidem sordidissima erigerent aedificia, lapides aut in calcem decoquendos, aut casarum muris adhibendos, ab illa moenium maiestate non sunt veriti asportare."

56. Fazio degli Uberti, *Dittamondo*, II.xvi.11, in Buddensieg (n. 32), p. 50.

57. Palmieri (n. 48), p. 44.

58. Lorenzo Valla, *Elegantiae linguae latinae*, in *Opera omnia*, ed. E. Garin (Turin, 1962), p. 4.

59. Finoli and Grassi (n. 9), VIII, p. 229.

60. Francesco di Giorgio Martini, *Trattati di architettura ingegneria e arte militare*, ed. C. Maltese, 2 vols. (Milan, 1967), 2:297.

61. *The Life of Brunelleschi* by Antonio di Tuccio Manetti, ed. H. Saalman (University Park, Pa. 1970), p. 60.

62. G. Pozzi and L. Ciapponi, eds., *Francesco Colonna. Hypnerotomachia Poliphili* (Padua, 1964; rpt., Padua, 1980), p. 200.

63. Pietro Cataneo, letter of 1554 to Francesco de'Medici, in *Pietro Cataneo, Giacomo Barozzi da Vignola. Trattati*, ed. E. Bassi et al. (Milan, 1985), p. 181.

64. This was first pointed out to me by William Loerke, whom I thank for many stimulating discussions of Alberti's thought.

65. G. Villani, *Cronica*, Book I.lvii, ed. I. Moutier, 5 vols. (Florence, 1823), 1:77, where he says that San Miniato was built in 1013, and I.xlii, p. 60, where he says the baptistery was built by the ancients.

66. Guenée (n. 33), p. 89.

67. The text is in V. Golzio, *Raffaello nei documenti e nelle testimonianze dei contemporanei e nella letteratura del suo secolo* (Vatican City, 1936), pp. 82–92. A particularly interesting analysis is in G. Morolli, *"Le Belle forme degli edifici antichi": Raffaello e il progetto del primo trattato rinascimentale sulle antichità di Roma* (Florence, 1984), p. 65.

68. Morolli comments that Raphael reduced all of Roman architecture to an "eternal present," seeking a norm that was constant and outside of time ([n. 67], p. 65). It is, therefore, not a historical evaluation in the fullest sense.

69. W. Sauerländer, "From Stilus to Style. Reflections on the Fate of a Notion," *Art History*, 6 (1983): 257–59.

70. R. Black, *Benedetto Accolti and the Florentine Renaissance* (Cambridge, 1985), p. 29

71. Frankl (n. 3); Krautheimer (n. 2), p. 266.

72. Black (n. 70), pp. 297–98.

73. Ferguson, *Humanist Views* (n. 27), pp. 8, 15. In the preface to his *The Italian War against the Goths*, of 1441, Bruni says that he has written the work because it is important for people to know about the origins and development of their own country and about what happened to it in earlier times. (G. Griffiths, J. Hankins, and D. Thompson, *The Humanism of Leonardo Bruni* [Binghamton, N.Y., 1987], p. 196).

74. Alberti (n. 5), pp. 160–61. I have not been able to discover the source of Alberti's account. It does not seem to have been Pliny, *Natural History*, LX–LXI, where he gives the history of paved floors.

75. Ibid., pp. 160–61. Alberti's allegory, in which knowledge is conceived as an edifice, may be compared with 1 Corinthians 3:5–8 and Lucretius, *De Rerum natura*, II. 7–10.

76. Alberti (n. 5), p. 182.

77. See especially his role in Leonardo Bruni's dialogue *Ad Petrum Paulum*, in *Pros. Lat.*, pp. 44–99. On the dating of the dialogue, see D. Quint, "Humanism and Modernity: A Reconsideration of Bruni's Dialogues," *Renaissance Quarterly* 38 (1985): 425.

78. Guarino Veronese, *Epistolario*, ed. R. Sabbadini (Venice, 1915), pp. 39–40 (trans. Gombrich [n. 15], p. 78).

79. Rule, for Vasari, consisted in taking measurements from antique works and studying the groundplans of ancient edifices; order was the separation of the Orders one from another so that no interchanging among Doric, Ionic, Corinthian, and Tuscan took place; proportion was the appreciation of universal laws of harmony (G. Vasari, *Le Vite de' più eccellenti pittori scultori e architettori*, ed. P. Barocchi and R. Bettarini [Florence, 1966], 4:3–4, preface to the third part).

80. R. Sabbadini, *Storia del Ciceronianismo e di altre questioni letterarie nell'età della rinascenza* (Turin, 1885). More recent discussion of these trends is in O. Besomi, "Dai 'Gesta Ferdinandi Regis Aragonum' del Valla al 'De Ortographia' del Tortelli," *IMU* 9 (1966): 75–121, esp. 75–76; J. D'Amico, "The Progress of Renaissance Latin Prose: The Case of Apuleianism," *Renaissance Quarterly* 37 (1984): 349–92; and Quint (n. 77), esp. p. 424. According to Gravelle, the Humanists' discussion of convention and innovation focused on three issues: neologisms, the quarrel of the Ancients and Moderns, and the question of whether ancient Latin literature was written in the language of an educated minority or the everyday language of the Roman people (S. S. Gravelle, "Humanist Attitudes to Convention and Innovation in the Fifteenth Century," *Journal of Medieval and Renaissance Studies* 11 [1981]: 194). It is not always sufficiently acknowledged in the older literature that there were two points of view to each of these questions. See, for example, L. Thorndike, "Renaissance or Prenaissance?" *JHI* 4 (1943): 65–74.

81. Gombrich (n. 15), esp. pp. 77ff.

82. Klotz (n. 15).

83. J. Onians, "Brunelleschi: Humanist or Nationalist," *Art History* 5 (1982): 261–62.

84. Manetti (n. 23), p. 50.

85. The rise of Idealism is further discussed in chapters 6 and 10.

86. For Geanakoplos, Argyropoulos "deserves primary credit for the shift in the focus of Florentine humanism from rhetoric, that is eloquence, to metaphysical philosophy, particularly Platonism (D. Geanakoplos, "Italian Renaissance Thought and Learning and the Role of the Byzantine Emigrés Scholars in Florence, Rome and Venice: A Reassessment," *Miscellanea Agostino Pertusi* [*Rivista di Studi Bizantini e Slavi* 3 (1983)] [Bologna, 1984], p. 139). This view has been challenged by A. Field, *The Origins of the Platonic Academy of Florence* (Princeton, N.J., 1988).

87. On this current, see especially C. Trinkaus, "Humanism and Science: Humanist Critiques of Natural Philosophy," in *The Scope of Renaissance Humanism* (Ann Arbor, Mich., 1983), pp. 140–68; Gravelle (n. 80); and N. Struever, *The Language of History in the Renaissance: Rhetorical and Historical Consciousness in Florentine Humanism* (Princeton, N.J., 1970).

88. C. Grayson, "Lorenzo, Machiavelli and the Italian Language," in *Italian Renaissance Studies*, ed. E. F. Jacob (London, 1960), p. 413. Appropriately, the passage is from Bruni's *Vita di Dante* (*Le Vite di Dante e del Petrarca*, ed. A. Lanza [Rome, 1987], p. 49).

89. On the *certame*, see G. Mancini, *Vita di Leon Battista Alberti*, (1882; rpt., Florence, 1911), pp. 228–36, and G. Ponte, *Leon Battista Alberti. Umanista e scrittore* (Genoa, 1981), pp. 182–83.

90. The proem is in Alberti (n. 5), 1:153–56. After the *certame*, in his prefatory letter to

Theogenius, Alberti said that many criticized him for not writing in Latin and that he will answer these charges elsewhere (ibid., p. 55).

91. Ibid., p. 155. As early as 1403, Bruni had praised the perfection of the Florentine vernacular in his *Laudatio Florentinae urbis*; as we saw, he also admired contemporary architecture. See chapter 9.

92. See the discussion in J. Gadol, *Leon Battista Alberti: Universal Man of the Early Renaissance* (Chicago, 1969), p. 218, and Sabbadini (n. 80), p. 127.

93. Summers has shown that Alberti required of painting that it be clear and convincing to learned and unlearned alike (D. Summers, *Michelangelo and the Language of Art* [Princeton, N.J., 1981], p. 88).

94. Alberti (n. 90), p. 155.

95. C. Grayson, ed., *Leon Battista Alberti. La prima grammatica della lingua volgare* (Bologna, 1964).

96. M. Salmi, "La Prima operosità architettonica de Leon Battista Alberti," in *Convegno*, pp. 9–19; A. C. Quintavalle, *Prospettiva e ideologia. Alberti e la cultura del secolo XV* (Parma, 1976), p. 161; C. Chirici, *Il Problema del restauro dal Rinascimento all'età contemporanea* (Milan, 1971), pp. 25ff., 35ff.

97. Salmi (n. 96); R. Wittkower, *Architectural Principles in the Age of Humanism* (London, 1952; rpt., London, 1971), pp. 38–39.

98. Cicero, *De Oratore*, III.ix.39, trans. E. W. Sutton and H. Rackham (London, 1948), p. 32. However, he advises using neologisms only occasionally (III.lii.201, p. 160).

99. Ibid., I.iii.12, p. 11.

100. Ibid., III.xxxvii.149, p. 118; III.lii.201, p. 160.

101. Horace, *Ars poetica*, 58–62, trans. H. R. Fairclough (London, 1926), p. 455.

102. Tacitus, *Dialogus de oratoribus*, 18–24, trans. W. Peterson (London, 1914), pp. 60–78. The work was brought to Rome by Nicholas V's agents in 1455; whether Alberti was familiar with the work at an earlier date is uncertain.

103. Salutati (n. 52), 4:133 (trans., Gravelle [n. 80], p. 196).

104. R. Fubini, ed., *Poggius Bracciolini. Opera omnia* (Turin, 1966), p. 203.

105. Flavio (n. 38), pp. 393–96 (trans., Gravelle [n. 80], pp. 205–6).

106. Valla (n. 58), preface.

107. Ibid.

108. J. S. Nelson, ed. and trans., *Aeneae Silvii. De Liberorum Educatione* (Washington, D.C., 1940), p. 169.

109. A. Poliziano, *Oratio super Fabio Quintiliano et Statii Sylvis* (trans., Gravelle [n. 80], p. 201).

110. C. Trinkaus, *"In Our Image and Likeness": Humanity and Divinity in Italian Humanist Thought* (London, 1970), pp. 279–80.

111. Ibid., pp. 279–80. The quotation is from Cicero.

112. Biondo Flavio, *Roma triumphante* (Basel, 1559), Book IX, p. 187.

113. Lucretius, *De Rerum natura*, trans. W. Rouse (London, 1924).

114. Saint Augustine, *Enarratio in Psalmum 36*, 3.6, and *Enarratio in Psalmum 123*, 8, in H. Marrou, *Saint Augustin et la fin de la culture antique* (Paris, 1938), p. 537.

115. Salmi (n. 96).

116. Alberti (n. 1), IX.x, pp. 855–57. Günther's comment on this passage that the architect might study ancient and some modern buildings "aber nicht gotische Kathedralen" ignores Alberti's enthusiasm for Florence Cathedral ([n. 7], p. 23).

117. Nelson (n. 108), pp. 179–81.

118. Petrarch, *Epistolae de rebus famialibus*, XXIII.19.13–14 (trans., Struever [n. 87], p. 147).

119. Seneca, *Epistulae morales*, LXXXIV, trans. R. Gummere, 3 vols. (Cambridge, Mass., 1917–1953), 2:278.

120. Aristotle, *Rhetoric to Alexander*, 1421A, trans. H. Rackham (Cambridge, Mass., 1937), p. 275.

121. Cicero, *De Inventione*, II.ii, trans. H. Hubbell (Cambridge, Mass., 1949), p. 169.

122. Cicero (n. 98), III.vii.26, p. 23.

123. A similar point is made in regard to the styles of Gentile and Masaccio by R. Panczenko, "La Cultura umanistica di Gentile da Fabriano," *Artibus et Historiae* 8 (1983): 69–70.

124. Benedetto Accolti, *Dialogus de praestanti virorum sui aevi* (trans., Gravelle [n. 80], p. 207).

125. Nelson (n. 108), p. 188.

126. Salutati (n. 52), IV.xviii., pp. 126–45, esp. p. 142.

127. Cicero (n. 98), III.ix.34, p. 29.

128. Krautheimer (n. 2), pp. 263–64.

129. Bruni (n. 88), p. 49.

130. Accolti (n. 124), p. 207.

131. Burns came to the same conclusion, saying that "the imitation of the antique for these writers [Alberti and Francesco di Giorgio] is above all the imitation of antique principles, not antique forms" ([n. 27], p. 276).

CHAPTER FIVE

1. Leon Battista Alberti, *Profugiorum ab aerumna libri III*, ed. C. Grayson, in *Leon Battista Alberti. Opere volgari* (Bari, 1966), 2:107–83. For the Italian, see p. 6.

2. Alberti's literary model for the epithets in *De Re aedificatoria* may be Valerius Maximus (*Facta et dicta memorabilia*), who presents subject headings under which he gives short examples. The specific buildings mentioned in Alberti's treatise most often serve as exampla. The edition of the treatise cited here is *De Re aedificatoria*, ed. G. Orlandi (Milan, 1966).

3. Baxandall also noted the almost total lack of descriptive criticism of works of art in Alberti's writing, as opposed to what he calls "prescriptive criticisms," in which the rules of art are stated (M. Baxandall, "Alberti and Christoforo Landino: The Practical Criticism of Painting," in *Convegno*, pp. 143–54).

4. The *Descriptio* is published in *VZ*, 4:212–22.

5. The letter to Ludovico Gonzaga is in Alberti (n. 1), p. 295.

6. Quoted in C. Grayson, "Il Prosatore Latino e volgare," in *Convegno*, p. 274.

7. This is the definition proposed in A. Nesselrath, "Raffaello e lo studio del'antico nel Rinascimento," in *Raffaello architetto*, ed. C. L. Frommel, S. Ray, and M. Tafuri (Milan, 1984), p. 398. Most studies of the beginnings of art criticism in the Quattrocento have focused on painting and sculpture, rather than architecture. See L. Venturi, "La Critica d'arte in Italia durante i secoli XIV e XV," *L'Arte* 20 (1917): 305–26; R. Krautheimer, "Die Anfänge der Kunstgeschichtsschreibung in Italien," *Repertorium für Kunstwissenschaft* 50 (1929): 49–63; G. Vesco, "L. B. Alberti e la critica d'arte in sul principio del Rinascimento," *L'Arte* 22 (1919): 57–71, 95–104, 136–48; M. Baxandall, *Giotto and the Orators: Humanist Observers of Painting in Italy and the Discovery of Pictorial Composition, 1350–1450* (Oxford, 1971); and Baxandall, *Painting and Experience in Fifteenth Century Italy* (London, 1972).

8. See the useful discussion of art criticism in E. Kleinbauer, *Modern Perspectives in Western*

Art History (New York, 1971), p. 5, and W. Sauerländer, "From Stilus to Style. Reflections on the Fate of a Notion," *Art History* 6 (1983): 253–70.

9. For medieval description, see chapter 9.

10. N. Maraschio, "Leon Battista Alberti, *De Pictura*: Bilinguismo e priorità," *Annali della Scuola Normale superiore di Pisa* 2 (1972): 265–73; D. Summers, "Contrapposto: Style and Meaning in Renaissance Art," *AB* 59 (1977): 336–62; E. Panofsky, " 'Renaissance'—Self-Definition or Self-Deception?" in *Renaissance and Renascences in Western Art* (New York, 1969), pp. 1–41; C. Gilbert, "Antique Frameworks for Renaissance Art Theory: Alberti and Pino," *Marsyas* 3 (1943–1945): 87–106; J. Spencer, "Ut Rhetorica pictura: A Study in Quattrocento Theory of Painting," *JWCI* 20 (1957): 26–44; G. Becatti, "Leon Battista Alberti e l'antico," in *Convegno*, pp. 55–72; M. Baxandall, *Giotto* (n. 7); D.R.E. Wright, "Alberti's 'De pictura': Its Literary Structure and Purpose," *JWCI* 47 (1984): 52–71.

11. H. Mühlmann, "Recht, Retorik und bildende Künste: Uber ihre Beziehungen im literarischen und architektonischen Werk des Leon Battista Alberti" (Ph.D. diss., University of Munich, 1968); Mühlmann, *Aesthetische Theorie der Renaissance. Leon Battista Alberti* (Bonn, 1981); R. Tobin, "Leon Battista Alberti: Ancient Sources and Structure in the Treatises on Art" (Ph.D. diss., Bryn Mawr College, 1979); J. Onians, *Bearers of Meaning: The Classical Orders in Antiquity, the Middle Ages and the Renaissance* (Princeton, N.J., 1988); H. Lorenz, "Studien zum Architektonischen und Architekturtheoretischen Werk L. B. Albertis" (Ph.D. Diss., University of Vienna, 1971).

12. Cicero, *De Optimo genere oratorum*, I.1–4, trans. H. Hubbell (Cambridge, Mass., 1949), pp. 354–56; Cicero, *De Oratore*, I.xi.53, trans. E. W. Sutton and H. Rackham (London, 1948), p. 40.

13. The notion that beauty acts on the emotions has roots in Sophistic thought of the fifth century B.C., in which the beautiful was defined as "that which gives pleasure through hearing and sight." The aesthetic experience was first specifically associated with joy by Democritus: "The great joys are derived from beholding beautiful works" (W. Tatarkiewicz, *History of Aesthetics*, ed. J. Harrell, 3 vols. [The Hague, 1970], 1:97, 90).

14. Ibid., 1:117. See Plato, *Phaedrus*, 268D, and *Gorgias*, 501–2. The irrationality of sight, and the unreliability of sensory evidence was argued by Sextus Empiricus, Lucretius, Vitruvius, and Tertullian (Sextus Empiricus, *Outlines of Pyrrhonism*, I.118–23; Lucretius, *De Rerum natura*, IV.426–32 and IV.380, who notes that "the nature of phenomena cannot be understood by the eyes." Vitruvius, *The Ten Books on Architecture,* VI.ii.2. Tertullian, *Liber de animo*, xvii, in *PL*, 2: col. 674). Almost always, the architectural examples of a square tower that seems round at a distance, or the portico that seems to diminish at its end served as proof.

15. Before 1400, the main rhetorical treatises used in the West were those of Martianus Capella, Cassiodorus, Isidore of Seville, Cicero's *De Inventione*, and the *Rhetorica ad Herennium* wrongly believed to be by Cicero (G. Kennedy, *Classical Rhetoric and Its Christian and Secular Tradition from Ancient to Modern Times* [Chapel Hill, N.C., 1980], p. 24). Perhaps the greatest cache of new rhetorical manuscripts was found in Lodi Cathedral in 1421. Five texts were discovered there: Cicero's *De Inventione* and the *Ad Herrenium* were already known; his *De Oratore* and *Orator* had been previously known in incomplete copies; the *Brutus* had been lost since the fifth century. Quintilian's *Institutio oratoria* had been known in the Middle Ages in a truncated version; its complete text was recovered in the early fifteenth century by Poggio Bracciolini (R. Sabbadini, *Le Scoperte dei codici latini e greci ne'secoli XIV e XV*, 2 vols. [1905; rpt., Florence, 1967], 2:199–200).

16. Dionysius of Halicarnassus, *De Isocrate*, 3, trans. S. Usher (Cambridge, Mass., 1974), p. 113.

17. See chapter 8 and the Appendix.

18. Tacitus, *Dialogus de oratoribus*, trans. W. Peterson (London, 1914), p. 75.

19. For brevity, see E. Curtius, *European Literature and the Latin Middle Ages*, trans. W. Trask (Princeton, N.J., 1953), pp. 487–88.

20. This attitude also existed in pagan writers. Plutarch, for example, said that style gives pleasure, but that the most important part of poetry is content, and its effect on moral character (*How a Youth Should Study Poetry*, esp. I.1–15). Augustine, who taught rhetoric, was ambivalent about ornamented speech (see *De Doctrina christiana*, IV.xiv.31; II.xxxi.48; IV.xix.38; IV.xx.42; *Confessions*, V.vi).

21. Paulinus, *Vita Sancti Ambrosii Mediolanensis Episcopi a Paulino eius notario ad Beatum Augustinum conscripta*, trans. M. Kaneicka (Washington, D.C., 1928), pp. 39–40.

22. These are discussed in H. Caplan, "Classical Rhetoric and the Mediaeval Theory of Preaching," in *Historical Studies of Rhetoric and Rhetoricians*, ed. R. Howes (Ithaca, N.Y., 1961), p. 79.

23. Curtius (n. 19), p. 490.

24. See J. Bialostocki, "The Power of Beauty. A Utopian Idea of Leone Battista Alberti," in *Studien zur toskanischen Kunst. Festschrift für Ludwig Heinrich Heydenreich* (Munich, 1963), pp. 13–19.

25. Alberti (n. 1), 2:127.

26. In fact, Alberti was correct, since it was the fourth-century Sophist Isocrates who first distinguished two types of human product: the useful and the pleasurable (Tatarkiewicz [n. 13], 1:97, 105).

27. See the discussion in C. Trinkaus, *The Scope of Renaissance Humanism* (Ann Arbor, Mich., 1983), p. 19. The five *studia* are grammar, rhetoric, history, poetry, and moral philosophy.

28. For the diffusion of this work in the Quattrocento, see Tobin (n. 11), p. 45.

29. S. I. Camporeale, *Lorenzo Valla. Umanesimo e teologia* (Florence, 1972), p. 111. The quotation is from the *Elegantiae linguae latinae*, 4: preface. The use of rhetorical terms and categories for architectural description is a component of the early Humanists' tendency to decompartmentalization between branches of knowledge explored in chapters 2 and 7.

30. G. Santinello, *Leon Battista Alberti. Una visione estetica del mondo e della vita* (Florence, 1962), pp. 224–28; Maraschio (n. 10), p. 195; L. Vagnetti, " 'Concinnitas': Riflessione sul significato di un termine albertiano," *Studi e documenti di architettura* 2 (1973):139–61.

31. J. Onians, "Style and Decorum in Sixteenth Century Italian Architecture" (Ph.D. diss., University of London, 1968), p. 289.

32. The term *concinnitas* is rare in Classical literature (see Cicero in *Brutus*, XXXVIII.83–84; *Orator*, CXLIX.202; Aulus Gellius I.4, 8 and I.16, 17; and Vitruvius I.2.3.–4).

33. Aulus Gellius, *Noctes Atticae*, XVII.xx.6, trans. J. C. Rolfe (London, 1927), p. 270.

34. Poggio Bracciolini, *Lettere*, epist. xiii, to Andrea Alamanni, 27 June 1455, ed. H. Harth, 3 vols. (Florence, 1984), 3:354.

35. Coluccio Salutati, *Epistolario*, ed. F. Novati, 4 vols. (Rome, 1896), 1:341–2: "Senecam ab eo sententiis equatum, ornatu superatum; Tullium non exundantiorum copia aut gravitate maiorem, veruntamen inventione minorem sine contentione concedet." Discussed in C. Trinkaus, "*Antiquitas* versus *Modernitas*: An Italian Humanist Polemic and Its Resonance," in "Ancients and Moderns: A Symposium," *JHI* 48 (1987): 17.

36. Quoted in N. Struever, *The Language of History in the Renaissance: Rhetorical and Historical Consciousness in Florentine Humanism* (Princeton, N.J., 1970), p. 70.

37. Ibid.

38. These qualities were grandeur, beauty, vigor, ethos, verity, clarity, and gravity (Kennedy [n. 15], p. 103; A. Patterson, *Hermogenes and the Renaissance: Seven Ideas of Style* [Princeton, N.J., 1970]).

39. Demetrius, *On Style*, I.15, trans. W. H. Roberts (London, 1927), p. 309.

40. G. Grube, *The Greek and Roman Critics* (Toronto, 1965), p. 203.

41. J. Nelson, trans., *Aeneae Silvii. De Liberorum educatione* (Washington, D.C., 1940), p. 183.

42. A. Mazzocco, "Leonardo Bruni's Notion of *virtus*," in *Studies in the Italian Renaissance: Essays in Memory of Arnolfo B. Ferruolo*, ed. G. P. Biasin, A. N. Mancini, and N. Perella (Naples, 1988), p. 109. The quotation is from Bruni's *Vita di Dante*.

43. Lorenzo Valla, *On Pleasure. De voluptate*, proem to chapter II.1–3, trans. A. K. Hieatt and M. Lorch (New York, 1977), p. 132.

44. Ibid., proem II.4, p. 134.

45. Ibid., III.19, p. 245.

46. E. De Bruyne, *Etudes d'esthétique médiévale*, 3 vols. (Bruges, 1946), 1:13–14. Hugh of St. Victor, *Didascalicon*, II, 16, in *PL*, 176: col. 757; Hucbald of Saint-Amand, *Musica enchiriadis*, IX. Discussed in U. Eco, *Art and Beauty in the Middle Ages*, trans. H. Bredin (New Haven, Conn., 1986), p. 38.

47. Tatarkiewicz (n. 13) 1:87–88.

48. Onians (n. 11), p. 227. Gaffurio's treatise is entitled *Angelicum ac divinum opus*.

49. V. Zoubov, "Leon Battista Alberti et les auteurs du Moyen Age," *Medieval and Renaissance Studies* 4 (1958): 245–66, saw this passage as directly based on Boethius, *De arithmetica*, II.32: "Quod videlicet non sine causa dictum est, omnis, quae ex contrariis consisterent, armonia quadam coniungi atque componi. Est enim armonia plurimorum adunatio et dissidentium consensio."

50. P. von Naredi-Rainer, "Musikalische Proportionen, Zahlenästhetik und Zahlensymbolik im Architektonischen Werk L. B. Albertis" (Ph.D. diss., University of Graz, 1975), pp. 17–18.

51. Mühlmann, *Aesthetische Theorie* (n. 11), pp. 19–20.

52. For the Pythagorean-Platonic view, see Tatarkiewicz (n. 13), 2:133. See also O. von Simson, *The Gothic Cathedral: Origins of Gothic Architecture and the Medieval Concept of Order* (Princeton, N.J., 1974), p. 191, with special reference to John of Salisbury's *Policraticus*, I, 6, where he says that music, architecture, and the soul are all based on universal laws of harmony. See also L. Spitzer, "Classical and Christian Ideas of World Harmony," *Traditio* 2 (1944): 409–64.

53. Aristotle, *On the Soul*, I.iv, trans. W. S. Hett (London, 1935), p. 43.

54. Macrobius, *Commentary on the Dream of Scipio*, VI.43, ed. W. H. Stahl (New York, 1952), p. 108.

55. Vagnetti noted that in *Pontifex*, Alberti also used the term *concinnitas* to refer to the equilibrium of virtue ([n. 30], p. 145).

56. Ibid., p. 141, referring to *Orator*, 44.149 and 49.164; *Brutus*, 83.287 and 95.325.

57. Paolo Cortesi, *De Hominibus Doctis Dialogus*, trans. M. T. Grazios (Rome, 1973), p. 40. Cortesi is discussing Valla's theory of language.

58. P. von Naredi-Rainer, *Architektur und Harmonie* (Cologne, 1984), p. 24, citing Boethius, *De Institutione musica*, V.2; and the same passages in Cicero's *Orator* and *Brutus* as Vagnetti (n. 56).

59. Giannozzo Manetti, *De Secularibus et pontificalibus pompis*, ed. E. Battisti, in *Umanesimo e esoterismo: Atti del V convegno internazionale di studi umanistici*, ed. E. Castelli (Padua, 1960), pp. 310–20.

60. The motet is published in H. Besseler, ed., *Guillelmi Dufay. Opera Omnia*, vol. 1: *Moteti* (Rome, 1966), pp. 70–75. R. Dammann, "Die Florentiner Domweihmotette Dufays (1436)," in W. Braunfels, *Der Dom von Florenz* (Olten, 1964), pp. 73–85.

61. C. Warren, "Brunelleschi's Dome and Dufay's Motet," *Musical Quarterly* 39 (1973): 92–105. My thanks for this reference to Anthony Cummings, and to him and Graeme Boone for their help with Dufay's motet.

62. Manetti (n. 59), p. 312.

63. Goro Dati: the cathedral "is sixty-six feet wide and two hundred forty feet long" (C. Gilbert, "The Earliest Guide to Florentine Architecture, 1423," *Mitteilungen des Kunsthistorischen Institutes in Florenz* 40 [1969]: 45). F. Albertini, *Memoriale di molte statue et picture di Florentia*, in *Five Early Guides to Rome and Florence*, ed. P. Murray (1972), n.p: "nam templum cathedralis ecclesiae Sancte Mariae floris nuncupatum civitas nostrae Florentinae... excedit praedictae Dianae Templum [at Ephesus]. Latitudo: ipsius est ped. CCCIIII in parte crucis corporis vero CXLVIII. Longitudo totius ecclesiae ped. CCCCC. Altitudo vero CXL. Altitudo autem marmoreae testudinis in medio positae ped. CCCIII."

64. N. Pirrotta, "Dante *Musicus*: Gothicism, Scholasticism and Music," in *Music and Culture in Italy from the Middle Ages to the Baroque* (Cambridge, Mass., 1984), p. 16; on performance in Florence Cathedral, see Pirrotta, "Music and Cultural Tendencies in Fifteenth-Century Italy," in the same volume, p. 383, n. 7.

65. C. Del Bravo, "Il Brunelleschi e la speranza," *Artibus et Historiae* 3 (1981): 70–72.; Pirrotta (n. 64), p. 16.

66. Eriugena, *De Divisione naturae* or *Periphyseon*, Book I, trans. I. Sheldon-Williams, 3 vols. (Dublin, 1968), 1:207: God is the coincidence of opposites: "For He gathers and puts all these things together by a beautiful and ineffable harmony into a single concord: for those things which in the part of the universe seem to be opposed and contrary to one another and to be discordant with one another are in accord and in tune [when] they are viewed in the most general harmony of the universe itself."

Ibid., Book III, 3:69: "The beauty of the whole established universe consists of a marvellous harmony of like and unlike."

67. Ibid., intro., p. 17; P. M. Watts, *Nicolaus Cusanus: A Fifteenth-Century Vision of Man* (Leiden, 1982), p. 49. See also Watts, "Pseudo-Dionysius the Areopagite and Three Renaissance Neoplatonists. Cusanus, Ficino and Pico on Mind and Cosmos," in *Supplementum festivum: Studies in Honor of Paul Oskar Kristeller* (Binghamton, N.Y., 1987), pp. 279–98.

68. Watts, *Cusanus* (n. 67), p. 173; Cassirer believed this new approach to have been stimulated by a new type of mathematical logic that required the possibility of the coincidence of opposites (E. Cassirer, *The Individual and the Cosmos in Renaissance Philosophy*, trans. M. Domandi [1927; rpt., Philadelphia, 1963], p. 14).

69. Watts, *Cusanus* (n. 67), p. 25: "infiniti ad finitum proportionem non esse."

70. Naredi-Rainer recognized the affinity between Alberti's idea of *concinnitas* and Nicholas of Cusa's *coincidentia oppositorum*. ([n. 57], p. 23).

71. Garin, discussing Manetti's account of the Cathedral, noted how important for the Quattrocento was the conviction of "la simmetria strutturale fra il tempio, l'uomo e il cosmo" (E. Garin, *Umanisti, artisti scienziati* [Rome, 1989], p. 156).

72. For the concept of antithesis in nature during the Renaissance, see Summers (n. 10), p. 350.

73. Discussed in L. Batkin, *Die Historische Gesamtheit der italienischen Renaissance* (Dresden, 1979), p. 275.

74. Camporeale (n. 29), p. 46: "il Valla opera il recupero della Patristica con l'esaltazione di ciò che l'antica teologia aveva di proprio e di specifico: il *modus rhetoricus*, anti-'filosofica' e quindi a-logico ed a-metafisico, in diretta contrapposizione all'aristotelismo."

75. Watts, *Cusanus* (n. 67), p. 26.

76. Koenigsberger describes this innovation in Alberti's thought (D. Koenigsberger, *Renais-

sance Man and Creative Thinking: A History of Concepts of Harmony 1400–1700 [Atlantic Highlands, N.J., 1979], p. 9).

CHAPTER SIX

1. Some studies of the ideal city that are especially relevant to my discussion are L. Bek, *Towards Paradise on Earth. Modern Space Conception in Architecture: A Creation of Renaissance Humanism*, Analecta Romana Instituti Danici, no. 9, supplement, (Odense, Denmark, 1980); L. Firpo, *La Città ideale del Rinascimento* (Turin, 1974); V. Franchetti Pardo, *Storia dell'urbanistica dal Trecento al Quattrocento* (Bari, 1982); R. Le Mollé, "Le Myth de la ville ideale à l'époque de la Renaissance italienne," *Annali della Scuola Normale Superiore di Pisa* 2 (1972,): 275–310; P. Marconi, "Una Chiave per l'interpretazione dell'urbanistica rinascimentale. La cittadella come microcosmo," *Quaderni dell'Istituto di Storia dell'Architettura* 15 (1968): 53–94; H. Rosenau, *The Ideal City: Its Architectural Evolution in Europe* (1959; rpt., London, 1983): and P. Zucker, *Town and Square from the Agora to the Village Green* (New York, 1959).

2. Le Mollé (n. 1), pp. 291–92. Ponte ascribed a strong platonizing and idealizing tendency to Albertian city planning (G. Ponte, "Architettura e società nel 'De re aedificatoria' di Leon Battista Alberti," *Giornale italiano di filologia* 21, pt. 2 [1969]: 297). Bruschi also linked Renaissance planning, including that of Alberti, to Platonic thought, emphasizing its scientific and logical qualities that were opposed, he believed, to medieval empiricism (A. Bruschi, "Realtà e Utopia nella città del Manierismo. L'Esempio di Oriolo Romano," *Quaderni dell'Istituto di Storia dell'Architettura* 13 [1966]: 69).

3. Thus Zucker wrote that "from the fifteenth century on, architectural design, aesthetic theory, and the principles of city planning are directed by identical ideas" ([n. 1], p. 99); M. Tafuri, *L'Architettura dell'umanesimo* (Bari, 1969), pp. 309–10.

4. L. Heydenreich, "Pius II als Bauherr von Pienza," *Zeitschrift für Kunstgeschichte* 6 (1937): 106; L. Heydenreich and W. Lotz, *Architecture in Italy, 1400 to 1600* (Harmondsworth, 1974), p. 43; C. L. Frommel, "Francesco del Borgo: Architekt Pius' II und Pauls' II: Der Petersplatz und weitere römische Bauten Pius' II Piccolomini," *Römisches Jahrbuch für Kunstgeschichte* 20 (1983): 110; H. W. Kruft, *Städte in Utopia. Die Idealstadt vom 15. bis zum 18. Jahrhundert* (Munich, 1989), p. 33.

5. P. Murray, *The Architecture of the Italian Renaissance*, rev. 3rd ed. (London, 1986), p. 82. Giovannoni thought the plan displayed a new regularity, without any aim for uniformity (G. Giovannoni, *Saggi sulla architettura del Rinascimento* [Milan, 1935], pp. 270–72).

6. C. Mack, *Pienza: The Creation of a Renaissance City* (Ithaca, N.Y., 1987), pp. 158, 162. An analogous observation, not, however, about Pienza, was made by Rosenau, who commented that while medieval representations of cities often show a circular shape, this idealization had no effect on actual town planning until the sixteenth century ([n. 1], p. 42).

7. L. Gnocchi, "Le Preferenze artistiche di Piero di Cosimo de'Medici," *Artibus et Historiae* 18 (1988): 60–62. Finelli and Rossi were the first to state that Pienza is not an ideal city, although they did not define its aesthetic principles in the context of Early Humanist culture (L. Finelli and S. Rossi, *Pienza tra ideologia e realtà* [Bari, 1979], p. 111).

8. The concept of *varietas* was first discussed in relation to Alberti's thought by M. Gosebruch, " 'Varietà' bei Leon Battista Alberti und der Wissenschaftliche Renaissancebegriff," *Zeitschrift für Kunstgeschichte* 20 (1957), pp. 229–38. His view was contested by H. Klotz, "L. B. Albertis 'De re aedificatoria' in Theorie und Praxis," *Zeitschrift für Kunstgeschichte* 32 (1969): 93–103. Maier's formulation of Alberti's theoretical position as rejecting grand, unified ensembles in favor of

composite assemblies of works that borrowed from many different sources is very close to the aesthetic I propose to apply to Pienza (I. Maier, *Ange Politien. La Formation d'un poète humaniste [1469–80]* [Geneva, 1966], 213–14). Interestingly, the only author who has actually used the term *varietà* to describe the planned differences between the structures at Pienza is Heydenreich, the same scholar who has most insisted on its ideal character ([n. 4], pp. 43, 45).

9. I do not consider here Brunelleschi's project to create a monumental access to Santo Spirito from the Arno, and I reject the suggestion that the Ospedale degli Innocenti formed part of a plan to reorganize Piazza Santissima Annunziata. The only evidence for the latter project is a comment by Arcangelo Giani, of about 1620, that Antonio Manetti Ciaccheri had been commissioned to make a model of the south end of Santissima Annunziata showing a porch and piazza "to be laid out square according to what had already been designed by the illustrious Filippo Brunelleschi." There is no reason to suppose that this source, 200 years later than the actual project, is accurate (E. Battisti, *Filippo Brunelleschi: The Complete Work* [New York, 1981], p. 348, n. 7).

10. J. Ackerman, "The Planning of Renaissance Rome 1450–1580," in *Rome in the Renaissance: The City and the Myth*, ed. P. A. Ramsey, Medieval and Renaissance Texts and Studies, no. 18 (Binghamton, N.Y., 1982), pp. 3–18; K. Bering, *Baupropaganda und Bildprogrammatik der Frührenaissance in Florenz-Rom-Pienza* (Frankfurt am Main, 1984); C. Burroughs, "Below the Angel: An Urbanistic Project in the Rome of Pope Nicholas V," *JWCI* 45 (1982): 94–124; Burroughs, "A Planned Myth and a Myth of Planning: Nicholas V and Rome," in Ramsey, *Rome*, pp. 197–207; C. Frommel, "Papal Policy: The Planning of Rome During the Renaissance," in *Art and History: Images and Their Meaning*, ed. R. Rotberg and T. Rabb (Cambridge, 1988), pp. 39–65; V. Golzio and G. Zander, *L'Arte in Roma nel secolo XV* (Bologna, 1968); C. Mack, "Nicholas the Fifth and the Rebuilding of Rome: Reality and Legacy," in *Light on the Eternal City: Recent Observations and Discoveries in Roman Art and Architecture*, ed. H. Hager and S. Munshower, Papers in Art History from the Pennsylvania State University, no. 2 (University Park, Pa., 1987), pp. 31–56; Mack, "Studies in the Architectural Career of Bernardo di Matteo Gambarelli called Rossellino" (Ph.D diss., University of North Carolina at Chapel Hill, 1972); T. Magnuson, "The Project of Nicholas V for Rebuilding the Borgo Leonino in Rome," *AB* 36 (1954): 89–116; T. Magnuson, *Studies in Roman Quattrocento Architecture* (Rome, 1958); M. E. Müntz, *Les Arts à la cour des Papes*, 3 vols. (Paris, 1882); C. Westfall, *In This Most Perfect Paradise: Alberti, Nicholas V and the Invention of Conscious Urban Planning in Rome, 1447–55* (University Park, Pa., 1974).

11. Mack draws connections between Nicholas V's unsuccessful Roman projects, which failed because of insufficient time and money and the enormity of the enterprises, and Pius II's successful, because much smaller and more manageable, project at Pienza ("Nicholas the Fifth" [n. 10], p. 40).

12. Heydenreich pronounced the square at Pienza to be the first "Ideal-Platz-Anlage der neueren Zeit" and the "Keim zu allen grossen Lösungen der folgenden Epochen, die in den grossen Raum- und Freiraumgestaltungen Michelangelos und Berninis ihre Vollendung finden" ([n. 4], p. 145).

13. The most recent and complete discussion with bibliography is Mack (n. 6). The fundamental study is Heydenreich (n. 4). Other monographic studies are Finelli and Rossi (n. 7); P. Torriti, *Pienza. La Città del rinascimento italiano* (Genoa, 1980); E. Carli, *Pienza. La Città di Pio II* (Rome, 1967); G. B. Mannucci, *Pienza. Arte e storia* (Siena, 1927); and Mannucci, *Pienza. I Suoi monumenti e la sua diocesi* (Montepulciano, 1915). For plans and elevations, see G. Cataldi, L. Di Cristina, F. Formichi, G. Fusco, and L. Marcucci, *Rilievi di Pienza* (Florence, 1977). Some of the more important recent studies on aspects of Pienza are L. Finelli, *L'Umanesimo giovane. Bernardo Rossellino a Roma e a Pienza* (Rome, 1984); N. Adams, "The Acquisition of Pienza 1459–1464," *JSAH* 44 (1985): 99–110; G. Cataldi, "Pienza e la sua Piazza: Nuove ipotesi tipologiche di lettura,"

Studi e documenti di architettura 7 (1978): 75–118; and F. Formichi, "Le Dodici 'case nuove' di Pienza," *Studi e documenti di architettura* 7 (1978): 119–28. Further bibliography is in Mack (n. 6). For Pius II's other urban projects, see Frommel (n. 4); R. Rubinstein, "Pius II as Patron of Art, with Special Reference to the History of the Vatican" (Ph.D. diss., University of London, 1957); and Rubinstein, "Pius II's Piazza S. Pietro and St. Andrew's Head," in *Essays in the History of Architecture Presented to Rudolf Wittkower*, ed. D. Fraser, H. Hibbàrd, and J. Lewine (London, 1967), pp. 22–33, and in *Enea Silvio Piccolomini. Papa Pio II. Atti del Convegno per il Quinto Centenario della Morte e altri scritti raccolti da D.Maffei* (Siena, 1968), pp. 221–45.

14. Opinions range from Forster's complete exclusion of Alberti to Mack's early position that Alberti was directly influential in the choice of Rossellino and in the design of some buildings. At present, most scholars favor the view I am presenting (K. W. Forster, "Discussion: The Palazzo Rucellai and Questions of Typology in the Development of Renaissance Buildings," *AB* 68 [1976]: 111; Mack, *Studies* [n. 10], p. 255).

15. Mack (n. 6), p. 34. Alberti lent Pius a copy of Vitruvius's treatise at the Congress at Mantua in 1459, exactly the moment in which the pope first planned Pienza.

16. F. A. Gragg, trans., *The Commentaries of Pius II*, Smith College Studies in History, vols. 22 (1936–1937); 25 (1939–1940); 30 (1947); 35 (1951); 43 (1957), Book VIII, pp. 546–47.

17. Ibid., Book VIII, p. 547.

18. Rubinstein tried to show that all the major components were built on a system of numerical proportions based on a module derived from the short side of one of the rectangles into which the pavement of the square is divided ("Pius II as Patron" [n. 13], p. 87). However, Mack points out that this grid pattern may not be original. Further, he shows that the Piccolomini Palace does not have the Pythagorean proportions other scholars have claimed to observe ([n. 6], pp. 101, 51).

19. Mack (n. 6), p. 77.

20. Ibid., pp. 77, 83–92.

21. Heydenreich (n. 4), p. 115.

22. Mack (n. 6), p. 82.

23. Ibid., p. 156. Preyer suggested that Pius could have seen the Medici Palace on his visit to Florence in 1459; the palace was but recently finished. In *Giovanni Rucellai ed il Suo Zibaldone*, vol. 2: *A Florentine Patrician and his Palace*, intro. N. Rubinstein (London, 1981), p. 190

24. Mack (n. 6), p. 56; C. Westfall, "Alberti and the Vatican Palace Type," *JSAH* 33 (1974): 101–21; Rubinstein, "Pius II as Patron" (n. 13), p. 92.

25. For the most recent discussion of this problem, with bibliography, see Mack (n. 6), p. 45.

26. See, for example, the houses in the background of Masolino's *The Raising of Tabitha* in the Brancacci Chapel, Santa Maria del Carmine, Florence. The palace is discussed in ibid., p. 104.

27. G. Thiem and C. Thiem, *Toskanische Fassaden-Dekoration in Sgraffito und Fresko* (Munich, 1964), pp. 66–67.

28. Mack (n. 6), pp. 108–12. These windows, an innovation of the time of Nicholas V, are further discussed in Rubinstein, "Pius II as Patron" (n. 13), p. 89. For Roman secular architecture in the Quattrocento, see G. Giovannoni, *I Quartieri romani del Rinascimento* (Rome, 1946).

29. Mack relates the design especially to city halls in Tuscany such as those at Monticiano, Montalcino, and Buonconvento ([n. 6], p. 112). For other comparisions, see the plates in N. Rodolico, *I Palazzi del popolo nei comuni toscani del medio evo* (Milan, 1962).

30. A document of 1462, in which the pope lent Magio 60 ducats to make a house, seems to locate the property on this site. Mack noted that the house, which is a new structure, is of Sienese rather than Roman or Florentine type ([n. 6], p. 143).

31. Pius II (n. 16), Book VIII, p. 598.

32. Mack (n. 6).
33. See chapter 4 for further discussion.
34. Leon Battista Alberti, *De Re aedificatoria*, IX.x, ed. G. Orlandi (Milan, 1966), pp. 855–57. See my discussion of this passage in chapter 4.
35. Heydenreich (n. 4), p. 144.
36. Mack (n. 6), p. 99.
37. Heydenreich (n. 4), p. 144.
38. For the acquisitions and chronology of construction, see Mack (n. 6), pp. 112–13, 41, 142.
39. S. Sinding Larsen, lecture, Institute of Fine Arts, New York University, Spring 1967.
40. Heydenreich (n. 4), p. 142; Zucker (n. 1), p. 111; Heydenreich and Lotz (n. 4), p. 43; Finelli and Rossi (n. 7), p. 17; Mack, (n. 6), pp. 100–101.
41. Mack (n. 6), p. 99.
42. S. Edgerton, *The Renaissance Rediscovery of Linear Perspective* (New York, 1975), p. 122; Edgerton, "Florentine Interest in Ptolemaic Cartography as Background for Renaissance Painting, Architecture and the Discovery of America," *JSAH* 33 (1974): 278.
43. For these, see R. Krautheimer, "The Tragic and Comic Scene of the Renaissance. The Baltimore and Urbino Panels," *Gazette des Beaux-Arts* 33 (1948): 327–46, and P. Sanpaolesi, "Le Prospettive architettoniche di Urbino, di Baltimora e di Berlino," *Bollettino d'arte* 34 (1949): 322–37. These were related to Renaissance planning ideals by F. Kimball, "Luciano Laurana and the 'High Renaissance,' " *AB* 10 (1927–1928): 125–50; N. Gallimberti, "L'Urbanistica della 'rinascenza,' " in *Atti del primo congresso nazionale di storia di architettura* (Florence, 1938), pp. 255–69; and Burroughs, "Below the Angel" (n. 10), p. 124.
44. See Mack (n. 6), p. 160; F. Brunetti, "Le Tipologie architettoniche nel trattato Albertiano," *Studi e documenti di architettura* 1 (1972): 263–92; and C. Westfall, "Society, Beauty, and the Humanist Architect in Alberti's *De Re aedificatoria*," *Studies in the Renaissance* 16 (1969): 61–81. Cf. A. Gunnarsjaa, "Filarete e la gerarchia architettonica," *Acta ad Archeologiam et Artium Historiam Pertinentia* 1 (1981): 229–45.
45. Cicero, *De Oratore*, trans. H. Rackham (London, 1942), p. 167.
46. Quintilian, *Institutio oratoria*, trans. H. Butler (Cambridge, 1921–1936), pp. 154–58. Also, ibid., X.ii.27, p. 88: "We must consider the appropriateness with which those orators handle the circumstances and persons involved in the various cases in which they were engaged, and observe the judgment and powers of arrangement which they reveal."
47. Aristotle, *The Art of Rhetoric*, III.xii, trans. J. H. Freese (London, 1926), p. 419.
48. Horace, *Ars Poetica*, trans. H. R. Fairclough (London, 1947), p. 459.
49. For these trends see R. Goldthwaite, "The Florentine Palace as Domestic Architecture," *American Historical Review* 77 (1972): 977–1012.
50. Spencer relates this to Alberti's *Della Pittura,* where the painter is advised to store up visual memories of nature for future use (J. Spencer, "Ut Rhetorica Pictura. A Study in Quattrocento Theory of Painting," *JWCI* 20 [1957]: 37).
51. E. Garin, "Il Pensiero di L. B. Alberti nella cultura del Rinascimento," in *Convegno*, p. 24.
52. Braunfels recognized that Alberti's recommendation of winding streets revealed a new concern with optical effect and with the relation between spectator and object, but he did not apply this to Pienza (W. Braunfels, *Mittelalterliche Stadtbaukunst in der Toskana* [Berlin, 1953], p. 101).
53. Cusanus, *Idiota de mente. The Layman: About Mind*, chap. IX, trans. C. Miller (New York, 1979), p. 73.
54. As maintained by Finelli and Rossi ([n. 7], p. 17) and Mack ([n. 6], p. 100).

55. This change is examined in H. M. Goldbrunner, "Laudatio Urbis: Zu neueren Untersuchungen über das humanistische Städtelob," *QFIAB* 63 (1983): 321.

56. For Alberti's remark, see chapter 3. The Chain Map is reproduced in G. Boffito and A. Mori, *Piante e vedute di Firenze: studio storico, topografico, cartografico* (Florence, 1926).

57. D. Summers, "Contrapposto: Style and Meaning in Renaissance Art," *AB* 59 (1977): 336–62.

58. Aristotle, *On the Soul*, 424A, trans. W. Hett (London, 1935), p. 135.

59. Quintilian (n. 46), VIII.iii. 52 p. 239.

60. Plutarch, *Moralia*, "Platonic Questions," X.1011, trans. H. Cherniss, 14 vols. (London, (1986), 13:121. Of the extant manuscripts, six out of eleven are fifteenth century. Plutarch, "The Education of Children," 7C, trans. F. C. Babbit (1927), 1:33. The treatise is probably not by Plutarch.

61. Klotz (n. 8), p. 98.

62. Quoted in E. H. Gombrich, "Apollonio di Giovanni: A Florentine Cassone Workshop Seen Through the Eyes of a Humanist Poet," in *Norm and Form: Studies in the Art of the Renaissance* (London, 1966), p. 21.

63. Leon Battista Alberti, *Della Pittura*, trans. J. Spencer (New Haven, Conn., 1966), p. 75.

64. Quoted in M. Baxandall, *Giotto and the Orators: Humanist Observers of Painting in Italy and the Discovery of Pictorial Composition 1350–1450* (Oxford, 1971), pp. 94–95. The quotation is from *De Suavitate dicendi ad Hieronymum Bragadenum*. See also J. Monfasani, *George of Trebizond: A Biography and a Study of His Rhetoric and Logic* (Leiden, 1976), p. 256.

65. Gosebruch (n. 8), esp. p. 233.

66. Quintilian (n. 46), II.x.11–12, p. 277.

67. A. M. Finoli and L. Grassi, eds., *Antonio Averlino detto il Filarete. Trattato di architettura*, I. x, (Milan, 1972), p. 27.

68. P. Watts, ed., *Nicholas de Cusa. De Ludo Globi. The Game of Spheres* (New York, 1986), p. 21. The quotation is from *Idiota de sapientia*, in *Opera omnia*, ed. L. Bauer (Leipzig, 1937), 5:22.

69. A. Field, *The Origins of the Platonic Academy of Florence* (Princeton, N.J., 1988). The demise of early Humanism is thoroughly discussed in G. Holmes, *The Florentine Enlightenment, 1400–1450* (New York, 1969), chap. 8.

70. Discussed in Field (n. 69), pp. 80–83.

71. Ibid., p. 131.

72. Ibid., pp. 3–9, 129, 269–74. For the relative importance of Argyropoulos, see also J. Seigel, "The Teaching of Argyropulos and the Rhetoric of the First Humanists," in Rabb and Seigel, pp. 237–60; E. Garin, "Cultura filosofica toscana e veneta nel Quattrocento," in *Umanesimo Europeo e umanesimo Veneziano*, ed. V. Branca (Venice, 1963), pp. 11–30; and D. Geanakoplos, "Italian Renaissance Thought and Learning and the Role of the Byzantine Emigrés Scholars in Florence, Rome and Venice: A Reassessment," in *Miscellanea Agostino Pertusi* (*Rivista di studi bizantini e slavi* 3 [1983]), pp. 129–57).

73. G. Garfagnini, ed., *Marsilio Ficino e il ritorno di Platone. Studi e documenti* (Florence, 1986); R. Black, "Ancients and Moderns in the Renaissance: Rhetoric and History in Accolti's 'Dialogue on the Preeminence of Men of His Own Time,'" *JHI* 43 (1982): 20; M. L. Mclaughlin, "Histories of Literature in the Quattrocento," in *The Languages of Literature in Renaissance Italy*, ed. P. Hainsworth, V. Lucchesi, C. Roaf, D. Robey, and J. Woodhouse (Oxford, 1988), p. 74.

74. S. I. Camporeale, *Lorenzo Valla tra medioevo e Rinascimento. Encomion S. Thomae—1457* (Pistoia, 1977); H. Gray, "Valla's *Encomion of St. Thomas Aquinas* and the Humanist Conception of Christian Antiquity," in *Essays in History and Literature Presented to Stanley Pargellis*, ed. H. Bluhm (Chicago, 1965), pp. 37–51.

75. Camporeale (n. 74), pp. 8–9.

76. For the following discussion of Platonic aesthetics, see W. Tatarkiewicz, *History of Aesthetics*, ed. J. Harrell, 3 vols. (The Hague, 1970), 1:esp. 116–19.

77. B. Rackusin, "The Architectural Theory of Luca Pacioli. *De Divina proportione*," *Bibliothèque d'Humanisme et Renaissance* 39 (1977): 492.

78. V. Scamozzi, *Dell'Idea della architettura universale*, 2 vols. (Ridgewood, N.J., 1964), 2: chap. X, p. 29.

79. Ibid., chap. X, p. 29.

80. Luca Pacioli, *De Divina proportione* (1497); Vincenzo Scamozzi, *Dell Idea dell'architettura universale* (1615); Giorgio Vasari the Younger, *La Città ideale* (1598) (this is actually a collection of drawings rather than a treatise).

81. Panofsky had noted the difference between Ficino's metaphysical definition of beauty and Alberti's phenomenological one, but he did not extend this observation to characterize early Humanist culture in general (E. Panofsky, *Idea: A Concept in Art Theory*, trans. J. Peake [New York, 1968], p. 54).

82. Plato, *Cratylus*, 439E.

83. The text, of 1535, is in E. Bassi et al., *Pietro Cataneo, Giacomo Barozzi da Vignola, Trattati. Con l'aggiunta degli scritti di architettura di Alvise Cornaro, Francesco Giorgi, Claudio Tolomei, Giangiorgio Trissino, Giorgio Vasari* (Milan, 1985), p. 12.

84. O. von Simson, *The Gothic Cathedral: Origins of Gothic Architecture and the Medieval Concept of Order* (Princeton, N.J., 1974), pp. 201ff., esp. 227–28.

85. P. O. Kristeller, "Humanism and Scholasticism in the Italian Renaissance," *Byzantion* 17 (1944–1945): 368–69.

86. Le Mollé (n. 1), p. 283.

87. Plato, *The Republic*, IX.592, trans. P. Shorey, 2 vols., (Cambridge, Mass., 1956), 2:415–17.

88. In Le Mollé's view, "le *De Re aedificatoria* propose un programme complet pour l'église idéale de la Renaissance" ([n. 1], p. 288). See also W. Eden, "Studies in Urban Theory: The 'De Re aedioficatoria' of L. B. Alberti," *Town Planning Review* 19 (1943): 10–28; C. Westfall, "The Two Ideal Cities of the Renaissance: Republican and Ducal Thought in Quattrocento Architectural Treatises" (Ph.D. diss., Columbia University, 1967), chap. 8; and W. Lotz, "Notizien zum kirchlichen Zentralbau der Renaissance," in *Studien zur Toskanischen Kunst. Festschrift für Ludwig Heydenreich* (Munich, 1964), p. 157.

CHAPTER SEVEN

1. Weiss sums up his importance thus: "L'arrivo del Crisolora a Firenze segna il principio di studi greci, non solo in questa città, ma pure in altre zone dell'Italia rinascimentale. Perche è appunto da questa scuola del Crisolora che è possibile derivare tutta l'attività di studi ellenici, che fiorì cosi rigogliosamente in Italia durante la Rinascenza" (R. Weiss, "Gli Inizi dello studio del Greco a Firenze," in *Medieval and Humanist Greek: Collected Essays* [Padua, 1977], p. 235). Similar assessments are in G. Cammelli, *I Dotti bizantini e le origini dell'Umanesimo*, vol. 1: *Manuele Crisolora* (Florence, 1941); P. O. Kristeller, "Umanesimo italiano e Bisanzio," *Lettere italiane* 16 (1964): 1–14; E. Legrand, *Bibliographie hellènique* (Paris, 1855), vol. i, s.v.; R. Sabbadini, "L'Ultimo ventennio della vita di Manuele Crisolora," *Giornale linguistico* 17 (1890): 321–36; I. Thomson, "Manuel Chrysoloras and the Early Italian Renaissance," *Greek, Roman and Byzantine Studies* 7 (1966): 63–82; A. Vacalopoulos, "The Exodus of Scholars from Byzantium in the Fifteenth Century," *Cahiers d'histoire mondiale* 10 (1967): 463–80; and K. Setton, "The

Byzantine Background to the Italian Renaissance," *Proceedings of the American Philosophical Society* 100 (1956): 1–76.

2. For the loss of Greek in the West during the Middle Ages, see especially B. Bischoff, "The Study of Foreign Languages in the Middle Ages," *Speculum* 36 (1961): 209–24, and J. T. Muckle, "Greek Works Translated Directly into Latin Before 1350," *Medieval Studies* 4 (1942): 33–42. For the restoration of Greek studies, see the works cited in n. 1, as well as standard works such as G. Voigt, *Die Wiederbelebung des Classischen Altertums, oder das erste Jahrhundert des Humanismus* (Berlin, 1859; rpt. Berlin, 1960); J. E. Sandys, *A History of Classical Scholarship* (New York, 1964), vol. 2; R. Sabbadini, *Il Metodo degli umanisti* (Florence, 1920); R. Weiss, *The Renaissance Discovery of Classical Antiquity* (Oxford, 1964); and L. D. Reynolds and N. G. Wilson, *Scribes and Scholars: A Guide to the Transmission of Greek and Latin Literature* (London, 1968). Of particular relevance for this study is P. Ricci, "La Prima cattedra Greco in Firenze," *Rinascimento* 3 (1952): 159–65.

3. Guarino Veronese, *Epistolario di Guarino Veronese*, I.25, ed. R. Sabbadini (Venice, 1916), p. 70.

4. Ibid., II.862, p. 583: "Contigit igitur quod de suis civibus Tullius factum affirmat: 'Post autem auditis oratoribus graecis cognitisque eorum litteris adhibitisque doctoribus incredibili quodam nostri homines dicendi studio flagraverunt.' "

5. Ibid., II.861, pp. 580–81, a letter of 1452: "Quae illius [Chrysoloras] cura et diligentia latas adeo sparsit per Italiae regna radices grandesque et uberes fructus disseminavit, ut Italorum studia immo vero Latinitatis disciplina cuncta quae dudum per inextricabiles vagabantur umbras et errores, Chrysolorae ductu et luminis accensione illustrata et directa perdurent." See also II.863, p. 588, by Guarino's son Battista.

6. Biondo Flavio, *Italia Illustrata* (Basel, 1531), p. 346 (discussed in W. K. Ferguson, "Humanist Views of the Renaissance," *American Historical Review* 45 [1939]: 23).

7. Leonardo Bruni, *Rerum suo tempore gestarum*, in E. Garin, *Storia della filosofia italiana* (Turin, 1966), 1:286. Poggio Bracciolini to Guarino Veronese, 1455, in Poggio Bracciolini, *Lettere*, 18, ed. H. Harth, 3 vols. (Florence, 1984), 3:348.

8. Vespasiano da Bisticci, *Le Vite*, ed. A. Greco, 2 vols. (Florence, 1970), 2:476.

9. A. Corbellini, "Appunti sull'umanesimo in Lombardia," *Bollettino della Società Pavese di Storia Patria* 17 (1915): 20. I thank David Rutherford for this reference.

10. Paolo Cortesi, *De Hominibus Doctis Dialogus*, ed. and trans. M. Grazios (Rome, 1973), p. 14.

11. For example, Vergerio's letter to Chrysoloras of 1400, which closes "ego ubique tuus sum, cui plurimum debere me sentio cum ob doctrinam insignem benevolentiamque tuam in me, tum maxime ratione summe integritatis amplissimeque eruditionis tue" and of 1406, describing him as "the best and most learned man whom your city [Florence] called from the heart of Greece" (Pier Paolo Vergerio, *Epistolario*, lxxxxiiii, ed. L. Smith [Rome, 1934], p. 239, and lxxxxvi, p. 244). Further testimonies are in Guarino's letters (n. 3): I.25, p. 63; 27, p. 73; 54, pp. 112–13; and II.7, p. 20; 862, p. 583; 864, p. 590; and 894, p. 636.

12. The definition was coined by Cato. See also Quintilian, *Institutio oratoria*, I.ix, and Isidore of Seville, *Etymologiarum*, II.iii. A good discussion of paideia as a Renaissance ideal is in C. Trinkaus, "Themes for a Renaissance Anthropology," in *The Scope of Renaissance Humanism* (Ann Arbor, Mich., 1983), pp. 364–403.

13. Guarino (n. 3), I.27, p. 73. Other of Guarino's letters make the same point, for example: I.25, p. 63; 27, p. 73; and II.7, p. 20; 863, p. 587; 864, p. 590; 866, p. 596. See also the letter from Cencio de'Rustici, 1415, in L. Bertalot, "Cincius Romanus und seine Briefe," *QFIAB* 21 (1929–1930): 221.

14. His letter is in *Ambrosii Traversarii generalis Camaldulensium Latinae epistolae*, Book XXIV, epist. 69, ed. L. Mehus, 2 vols. (Florence, 1759; 1: col. 1042.

15. Guarino (n. 3), II.865, p. 592.

16. Ibid., I.54, p. 113 (trans., Thomson [n. 1], p. 73). The connection is repeated in a letter by Battista Guarino (n. 3), II.863, p. 588.

17. Cicero, *De Oratore*, I.viii. 34, trans. E. W. Sutton (London, 1948), pp. 26–27. Similar views are in Cicero, *De Inventione*, I.i.1–2, and I.iv.5; Quintilian, I.10. For the importance of the Ciceronian ideal of civic participation in early Renaissance Florence, see H. Baron, *The Crisis of the Early Italian Renaissance* (Princeton, N.J., 1955), pp. 97–104.

18. Thomson (n. 1), p. 66. His teaching methods are also reflected in Battista Guarino's *De Modo et ordine docendi et discendi* of ca. 1459. For the treatises of Vergerio, Bruni, and Guarino, see H. W. Woodward, *Vittorino da Feltre and Other Humanist Educators* (1897; rpt., New York, 1963). See also E. Garin, *L'Educazione Umanistica in Italia* (1949; rpt., Bari, 1964).

19. Weiss (n. 1), p. 241; Guarino (n. 3), IV. 184–86; H. Baron, *Leonardo Bruni Aretino. Humanistisch-Philosophische Schriften* (Berlin, 1928), pp. 99–100, 160–61.

20. B. Gille, *Les Ingénieurs de la Renaissance* (Paris, 1964), p. 36. See also J. Ackerman, " 'Ars sine scientia nihil est': Gothic Theory of Architecture at the Cathedral of Milan," *AB* 31 (1949): 84–111.

21. The *Comparison* is in *PG*, 156: cols. 24–53.

22. L. Mehus, ed., *Leornardi Arretini epistolarum libri VIII*, epist. VIII.4 (Florence, 1741), p. 111; Baron (n. 17), 1:164–67. See my analysis of his literary sources in chapter 8.

23. Bessarion's *Encomion* was published by S. Lampros, "Bessarionos Enkomion eis Trapezounta," *Neos Hellenomnemon* 13 (1916): 146–204, and is discussed in O. Lampsides, "Datierung des 'Enkomion Trapezountos' von Kardinal Bessarion," *Byzantinische Zeitschrift* 38 (1955): 291–92, and E. J. Storman, "Bessarion Before the Council of Florence: A Survey of His Early Writings," in *Byzantine Papers: Proceedings of the First Australian Byzantine Studies Conference*, ed. E. Jeffreys, M. Jeffreys, and A. Moffatt (Canberra, 1978), pp. 128–56. John Eugenikos's descriptions were published by O. Lampsidis, "Iohannos Eugenikos. Ekphrasis Trapezountos," *Archeion Pontos* 20 (1955): 3–39; see also D. Pallas, "Les 'Ekphrasis' de Marc et de Jean Eugénikos: Le Dualisme culturel vers la fin de Byzance," in *Rayonnement grec: Hommages à Charles Delvoye*, ed. L. Hadermann-Misguich and G. Raepsaet (Brussels, 1982), pp. 505–12. Bryennios is in N. B. Tomadakes, "Rede über die Wiederinstandsetzung der Stadtmauern Konstantinopels," *Epeteris* 36 (1968): 1–16; Apostolis in H. Noiret, ed., *Lettres inédites de Michel Apostolis* (Paris, 1889), pp. 110–12. Many of the texts written about the fall of Constantinople have been gathered in A. Pertusi, *La Caduta di Costantinopoli*, 2 vols. (Verona, 1976).

24. This project, a collaboration between the classicist Joseph O'Connor and me, will offer a corpus of architectural descriptions written in Greek and Latin between 300 and 1500 together with interpretive essays.

25. P. O. Kristeller, *La Tradizione classica nel pensiero del Rinascimento* (1965; rpt., Florence, 1975), pp. 16, 98–99, n. 120. (The work was originally *The Classics and Renaissance Thought*, based on lectures given in 1955.)

26. Kristeller (n. 1), p. 8.

27. D. Geanakoplos, "Italian Renaissance Thought and Learning and the Role of the Byzantine Emigrés Scholars in Florence, Rome and Venice: A Reassessment," in *Miscellanea Agostino Pertusi* (*Rivista di Studi Bizantini e Slavi* 3 [1983]) (Bologna, 1984), p. 130.

28. D. Geanakoplos "The Discourse of Demetrius Chalcondyles on the Inauguration of Greek Studies at the University of Padua in 1463," *Studies in the Renaissance* 21 (1974): 143.

29. Geanakoplos (n. 27), p. 156; Geanakoplos, *Greek Scholars in Venice* (Cambridge, Mass., 1962), p. 1.

30. E. B. Fryde, *Humanism and Renaissance Historiography* (London, 1983), pp. 23–30.

31. Cammelli (n. 1) concentrated on the Italian period of Chrysoloras's life, as did Sabbadini (n. 1). The educational system in Constantinople around 1350 to 1453 is discussed in S. Runciman, *The Last Byzantine Renaissance* (Cambridge, 1970); F. Fuchs, *Die höheren Schulen von Konstantinopel im Mittelalter* (Amsterdam, 1964); and R. Browning, "Byzantine Scholarship," *Past and Present* 28 (1964): 3–20. C. N. Constantinides, *Higher Education in Byzantium in the Thirteenth and Early Fourteenth Century* (Nicosia, 1982), covers earlier Palaiologan education. Some information can be gleaned from studies of Palaiologan culture, such as *Art et societé à Byzance sous les Paléologues: Actes du Colloque organisé par l'Association Internationale des études byzantines* (Venice, 1971); D. Geanakoplos, *Byzantine East and Latin West: Two Worlds of Christendom in Middle Ages and Renaissance* (New York, 1966); A. Heisenberg, "Das Problem der Renaissance in Byzance," *Historische Zeitschrift* 133 (1925–1926): 393–412; H. Hunger, "Von Wissenschaft und Kunst der frühen Palaiologenzeit," *Jahrbuch der Osterreichischen Byzantinischen Gesellschaft* 8 (1959): 123–55; G. Mercati, *Notizie di Procoro e Demetrio Cidone Manuele Caleca e Teodoro Meliteniota ed altri appunti per la storia della teologia e della letteratura bizantina del secolo XIV* (Vatican City, 1931); and A. Pertusi "L'Umanesimo greco della fine del secolo XIV agli inizi del secolo XVI," in *Storia della cultura Veneta*, ed. G. Arnaldi and M. Pastore Stocchi (Venice, 1980), pp. 177–264.

32. I rely here on Runciman (n. 31), p. 54.

33. Ibid, p. 100.

34. Geanakoplos (n. 27), p. 132.

35. E. H. Gombrich, "From the Revival of Letters to the Reform of the Arts. Niccolò Niccoli and Filippo Brunelleschi," in *The Heritage of Apelles: Studies in the Art of the Renaissance* (Ithaca, N.Y., 1976), p. 99.

36. Thomson (n. 1), p. 81.

37. G. Kustas, "The Function and Evolution of Byzantine Rhetoric," *Viator* 1 (1970): 55–73; G. Kennedy, *Classical Rhetoric and Its Christian and Secular Tradition from Ancient to Modern Times* (Chapel Hill, N.C., 1980), pp. 163ff.; R. J. H. Jenkins, "The Hellenistic Origins of Byzantine Literature," *Dumbarton Oaks Papers* 17 (1963): 37–52; Constantinides (n. 31), p. 262.

38. An introduction to the Second Sophistic is in B. P. Reardon, "The Second Sophistic," in *Renaissances Before the Renaissance: Cultural Revivals of Late Antiquity and the Middle Ages*, ed. W. Treadgold (Stanford, 1984), pp. 23–41. For an account of the development of architectural description in Antiquity, see C. Smith, "Christian Rhetoric in Eusebius' Panegyric at Tyre," *Vigiliae Christianae* 43 (1989): 226–47.

39. The fourteen types are fable, tale, chreia, proverb, refutation, confirmation, commonplace, encomion, vituperation, comparison, characterization, description, narrative, and proposal of law (R. Nadeau, "The Progymnasmata of Aphthonius in Translation," *Speech Monographs* 19 [1952]: 264–85).

40. Hermogenes' *Progymnasmata* is in H. Rabe, ed., *Hermogenis. Opera*, Rhetores graeci 6 (Leipzig, 1913); Priscian, *Praeexercitamina*, in *Rhetores Latini minores*, ed. C. Halm (Leipzig, 1867), 7:557, "De Laude"; 10:558, "De Descriptione."

41. These are discussed in Hermogenes' *Peri ideon*, now available in a translation by Wooten, who gives as the English equivalents: clarity, grandeur, beauty, conciseness, ethos, force, and verity C. Wooten, *Hermogenes' "On Types of Style"* [Chapel Hill, N.C., 1989]. Useful discussion of the terms is in A. M. Patterson, *Hermogenes and the Renaissance: The Seven Ideas of Style* (Princeton,

N.J., 1970), and in Kustas (n. 37), p. 65. Comparison between Hermogenes' rhetoric and Latin authorities is in J. Monfasani, "The Byzantine Rhetorical Tradition and the Renaissance," in *Renaissance Eloquence*, ed. J. J. Murphy (Los Angeles, 1983), pp. 174–87.

42. Discussion of the importance of this is in M. Baxandall, *Giotto and the Orators: Humanist Observers of Painting in Italy and the Discovery of Pictorial Composition, 1350–1450* (Oxford, 1971), p. 85.

43. Jenkins had made this observation in regard to writers of the Macedonian renaissance ([n. 37], p. 40); it is also true of the Palaiologan period. See H. Hunger, *Die Hochsprachliche profane Literatur der Byzantiner* (Munich, 1978), vol. 1, esp. pp. 170–88, on ekphrases; K. Krumbacher, *Geschichte der Byzantinischen Litteratur von Justinian bis zum Ende des Oströmischen Reiches (527–1453)* (Munich, 1891); G. Kennedy, *Greek Rhetoric under Christian Emperors* (Princeton, N.J., 1983); Kennedy, *Classical Rhetoric and Its Christian and Secular Tradition from Ancient to Modern Times* (Chapel Hill, N.C., 1980); and T. Burgess, *Epideictic Literature* (Chicago, 1902). Very useful, although concerned with an earlier period, are H. Maguire, "The Art of Comparing in Byzantium," *AB* 70 (1988): 88–103, and Maguire, "Truth and Convention in Byzantine Descriptions of Works of Art," *Dumbarton Oaks Papers* 28 (1974): 113–40.

44. L. Bruni, *Rerum suo tempore gestarum commentarius*, ed. C. di Pierro, *RIS*, 19, pt. 3 (1926), p. 431.

45. Coluccio Salutati, *Epistolario*, IV.xvi, ed. F. Novati, 4 vols. (Rome, 1896), pp. 131–32.

46. Vergerio (n. 11), lxxxxvi, p. 244.

47. Filelfo, epist. V, quoted in Baxandall (n. 42), p. 84.

48. G. Resta, "Filelfo tra Bisanzio e Roma," in *Francesco Filelfo nel quinto centenaria della morte. Atti del XXVII convegno di studi Maceratesi (1981)* (Padua, 1986), pp. 50–51.

49. Woodward (n. 18), p. 49.

50. Bompaire studied the preference for Late Antique and Byzantine authors in Photius's *Library*, concluding that "le premier humanisme byzantin est moins un retour aux sources classiques qu'un authentique prolongement de la Seconde Sophistique" (J. Bompaire, "Photius et la Seconde Sophistique, d'après la *Bibliothèque*," *Travaux et memoires* 8 [1981]: 84).

51. E. Berti, "Alla Scuola di Manuele Crisolora. Lettura e commento di Luciano," *Rinascimento* 27 (1987): 3–74.

52. M. Gigante, *Teodoro Metochites, Saggio critico su Demostene e Aristide* (Milan, 1969).

53. For George of Trebizond, see J. Monfasani, *George of Trebizond: A Biography and a Study of His Rhetoric and Logic* (Leiden, 1976), esp. pp. 17–18. Filelfo's contribution is examined in Geanakoplos (n. 27), p. 143, n. 74; the translation of Aphthonius is noted in Kennedy (n. 43), p. 164.

54. In addition to the works cited in n. 40, see J. Schlosser Magnino, *La Letteratura artistica. Manuale delle fonti della storia dell'arte moderna*, trans. F. Rossi (Florence, 1979), p. 17.

55. For a more detailed discussion of this development, see Smith (n. 38).

56. Hermogenes (n. 40), p. 22.

57. A general comparison between Byzantine and Renaissance culture is in C. Neumann, "Byzantinische Kultur und Renaissance-Kultur," *Historische Zeitschrift* 91 (1903): 215–32. On Byzantine scholarship, see N. G. Wilson, *Scholars of Byzantium* (Baltimore, 1983), and Browning (n. 31).

58. The most recent discussion of this phenomenon, with bibliography, is B. Vickers, *In Defence of Rhetoric* (Oxford, 1988), p. 232. The classic study is R. McKeon, "Rhetoric in the Middle Ages," *Speculum* 17 (1942): 1–32. Also fundamental are J. J. Murphy, *Rhetoric in the Middle Ages: A History of Rhetorical Theory from St. Augustine to the Renaissance* (Berkeley, 1974); Murphy, *Medieval Eloquence: Studies in the Theory and Practice of Medieval Rhetoric* (Berkeley, 1978);

J. A. Schultz, "Classical Rhetoric, Medieval Poetics, and the Medieval Vernacular Prologue," *Speculum* 59 (1984): 1–15; and R. Witt, "Medieval 'Ars Dictaminis' and the Beginnings of Humanism: A New Construction of the Problem," *Renaissance Quarterly* 35 (1982): 1–35. Meyendorff read a most interesting paper comparing medieval and Byzantine education at the 17th International Byzantine Congress, Washington, D.C. (J. Meyendorff, "The Mediterranean World in the Thirteenth Century, Theology: East and West," *The 17th International Byzantine Congress: Major Papers* [New Rochelle, N.Y., 1986], pp. 669–82). An interesting summary of the Western educational curriculum as it was understood in 1415 forms part of the Acts of the Council of Constance, published in J. D. Mansi, *Sacrorum conciliorum nova et amplissima collectio* (Graz, 1961), 28: cols. 131–37.

59. Runciman (n. 31), pp. 28–31; H. Hunger, "Theodoros Metochites als Vorlaufer des Humanismus in Byzance," *Byzantinische Zeitschrift* 45 (1952): 18–19.

60. Guarino (n. 3), I.54, p. 114: "vir doctissimus, prudentissimus optimus, qui tempore generalis concilii Constantiensis diem obiit ea existimatione ut ab omnibus summo sacerdotio dignus haberetur." In his funeral oration for Chrysoloras, Andrea Zulian referred to him as "summam religionis scientiam" (A. Cologerà, *Raccolta d'opuscoli scientifici e filologici XXV* [Venice, 1741], p. 326).

61. See nn. 4, 7, and 11.

62. R. Sinkewicz, "Christian Theology and the Renewal of Philosophical and Scientific Studies in the Early Fourteenth Century: The *Capita 150* of Gregory Palamas," *Medieval Studies* 48 (1986): 334–51; J. Meyendorff, ed., *Grégoire Palamas. Défense des saints hésychastes* (Louvain, 1959), esp. 1:20; Meyendorff, *A Study of Gregory Palamas* (London, 1964).

63. For these scholars, see Runciman (n. 31); Wilson (n. 57); and W. Buchwald, A. Hohlweg, and O. Prinz, *Tusculum Lexikon* (Munich, 1982), s.v.

64. Runciman (n. 31), p. 65; B. Tatakis, *La Philosophie byzantine* (Paris, 1949), p. 246.

65. Tatakis (n. 64), pp. 256, 286.

66. A. Field, *The Origins of the Platonic Academy of Florence* (Princeton, N.J., 1988), pp. 112–13; see also J. Seigel, "The Teaching of Argyropulos and the Rhetoric of the First Humanists," in Rabb and Seigel, pp. 237–60. Argyrpoulus's courses on Aristotle also fostered the tendency toward systematic and complete treatments of material discussed in chapter 6.

67. Field noted that the Neoplatonic character of the scheme probably originated in Late Antique commentaries on Aristotle; George of Trebizond followed the same ascending order of subject matter in his teaching (Monfasani [n. 53], pp. 112–13).

68. Cicero (n. 17), I.xv.64, p. 47. He makes the same point at I.v.21, pp. 16–17.

69. Baxandall (n. 42), p. 26.

70. M. Kemp, "From *Mimesis* to *Fantasia*: The Quattrocento Vocabulary of Creation, Inspiration and Genius in the Visual Arts," *Viator* 8 (1977): 366.

71. Cicero (n. 17), I.viii.32–34, pp. 24–26.

72. See, for example, Leonardo Bruni, *Le Vite di Dante e del Petrarca*, ed. A. Lanza (Rome, 1987), pp. 34–35, and Leon Battista Alberti, *Della famiglia*, in *Opere volgari*, Book I, ed. C. Grayson (Bari, 1960), p. 45.

73. Cammelli (n. 1), pp. 40–41.

74. Libanius, *Antiochikos*, trans. G. Downey, "Libanius' Oration in Praise of Antioch (Oration XI)," *Proceedings of the American Philosophical Society* 103 (1959): p. 675.

75. Weiss (n. 1), p. 237.

76. Vacalopoulos (n. 1), p. 475. He is mistaken in his claim that Chrysoloras taught exclusively in the classroom.

77. Cortesi (n. 10), p. 56.

78. P. Le Merle, *Byzantine Humanism: The First Phase*, trans. H. Lindsay and A. Moffitt, (Canberra, 1986), p. 229.

79. G. Brucker, "Florence and its University, 1348–1434," in Rabb and Seigel, pp. 231–33.

80. Ibid., p. 228.

81. Seven of these orations were published by G. Tanturli in "Cino Rinuccini e la scuola di Santa Maria in Campo," *Studi medievali* 17 (1976): 625–74. Such *controversiae* had served as the final stage of Classical instruction in forensic rhetoric (A. Grafton and L. Jardine, *From Humanism to the Humanities* [London, 1986], p. 7).

82. The debates are discussed in Weiss (n. 1), p. 269, and R. Sabbadini, *Storia del Ciceronianismo e di altre questioni letterarie nell'età della rinascenza* (Turin, 1885), p. 112. See also N. Gilbert, "The Early Italian Humanists and Disputation," in *Renaissance Studies in Honor of Hans Baron*, ed. A. Molho and J. Tedesco (Dekalb, Ill., 1971), pp. 201–26.

83. On the dialogue, the most recent work is P. Burke, "The Renaissance Dialogue," *Renaissance Studies* 3 (1989): 1–12. See also D. Marsh, *The Quattrocento Dialogue: Classical Tradition and Humanist Innnovation* (Cambridge, Mass., 1980); and L. Batkin, *Die Historische Gesamtheit der italienishen Renaissance* (Dresden, 1979).

84. Weiss (n. 1), saw the stimulation of discussion on such topics as one of the essential components of Chrysoloras's teaching activity ([n. 1], pp. 236ff.).

85. Jenkins (n. 37), p. 45.

86. For Niccoli's formation and the importance of late Trecento conversation, see G. Zippel, "Niccolò Niccoli. Contributo alla storia dell'Umanesimo," in *Storia e cultura del Rinascimento italiano* (Padua, 1979), pp. 74–76; Leonardo Bruni, *Ad Petrum Paulum Histrum*, is in *Pros. Lat.*, pp. 44–99.

87. Baron observed that the technique of arguing on both sides of the question (for which he sees Cicero as the source) "has proved so confusing, indeed, that no consensus could ever be established on the author's intentions in writing his work" ([n. 17], pp. 202–4). See my discussion of the text in chapter 2.

88. Bruni (n. 86), p. 48.

89. Ibid., p. 52.

90. Ibid., p. 54.

91. E. Garin, *Umanisti artisti scienziati* (Rome, 1989), p. 53. The letter, of around 1455, was written to oppose the appointment of Argyropoulos to the Studio.

92. The invectives against Niccoli have been most recently studied by M. Davies, "An Emperor Without Clothes? Niccolò Niccoli Under Attack," *IMU* 30 (1987): 95–148. Zippel (n. 86) remains an important source for Niccoli's relations with his fellow Humanists; some penetrating observations are in Gombrich (n. 35).

93. Davies (n. 92), pp. 97–98.

94. Ibid.

95. Cicero (n. 17), I.viii.29–32, pp. 21–24.

96. Libanius (n. 74), p. 668.

97. References are in L. Martines, *Power and Imagination: City-States in Renaissance Italy* (New York, 1980), p. 195.

98. See Baron (n. 17), and J. Seigel, "Civic Humanism or Ciceronian Rhetoric? The Culture of Petrarch and Bruni," *Past and Present* 34 (1966): 3–48. Geanokoplos expressed this sense of the utility of Greek studies, speaking of Demetrius Chalcocondyles: "The Italian Humanists were now seeking to apply the values of Greek civilization especially through what they referred to as 'eloquence,' to the recreation of those values in the life and institutions of their own city-states" ([n. 27], p. 144).

99. See especially Storman (n. 23), and Jenkins (n. 37).

100. Vacalopoulos (n. 1), p. 477

101. C. G. Patrinelis, "An Unknown Discourse of Chrysoloras Addressed to Manuel II Palaiologus," *Greek, Roman and Byzantine Studies* 13 (1972): 497–502.

102. L. Mohler, *Kardinal Bessarion als Theologe, Humanist und Staatsman* 3 vols. (Paderborn, 1967), 3: 479, epist. 30.

103. Ibid., 1:417.

104. Translated by Vacalopoulos (n. 1), p. 466.

105. J. Canabutzes, *Commentary on Dionysius of Halikarnassos' Roman Antiquities*, ed. M. Lehnerdt (Leipzig, 1890), p. 19.

106. Vacalopoulos (n. 1), pp. 470–71; *PG*, 156: cols. 53–56.

107. Guarino (n. 3), I.54, p. 114.

CHAPTER EIGHT

1. The *Comparison* has most recently been discussed in G. Dagron, "Manuel Chrysoloras: Constantinople ou Rome?" *Byzantinische Forschungen* 12 (1987): 281–88. I thank Denis Feissel for this reference. It is also discussed in M. Baxandall, "Guarino, Pisanello and Manuel Chrysoloras," *JWCI* 28 (1965): 183–204, and I. Thomson, "Manuel Chrysoloras and the Early Italian Renaissance," *Greek, Roman and Byzantine Studies* 7 (1966): 69.

2. Guarino da Verona received the work in Florence in October 1411 and had copies made for distribution (R. Sabbadini, ed., *Epistolario di Guarino Veronese*, I.7 [Venice, 1916], pp. 20–21). While the work has always been thought to be addressed to John VIII Palaiologus (who shared the title of emperor with his father), John Barker has convincingly suggested that since Chrysoloras mentions other business correspondence, it was probably sent to Manuel II.

3. R. J. H. Jenkins, "The Hellenistic Origins of Byzantine Literature," *Dumbarton Oaks Papers* 17 (1963): 46.

4. Sabbadini (n. 2), I.7, p. 21.

5. B. Vickers, *In Defence of Rhetoric* (Oxford, 1988), pp. 235, 233.

6. F. Petrarca, *Rerum familiarum libri I–VIII*, VI.2, trans. A. Bernardo (Albany, N.Y., 1975), pp. 290–95; N. Gilbert, "A Letter of Giovanni Dondi Dall'Orologio to Fra Guglielmo Centueri: A Fourteenth-Century Episode in the Quarrel of the Ancients and the Moderns," *Viator* 8 (1977): 299–346. Coluccio Salutati, *Epistolario*, 4 vols. ed. F. Novati (Rome, 1896), 4:480. The letter is from Vergerio after Salutati's death in 1406.

The new epistolography is probably to be interpreted in relation to the Humanists' discovery of Classical models. Until the late fourteenth century, only two letter collections were known: that of Seneca the Younger to Lucilius, and the first 100 letters of Pliny the Younger. Petrarch discovered Cicero's *Epistulae ad Atticum* in 1345, and Salutati instigated the search that led to the recovery of Cicero's *Epistulae ad familiares* in 1392. (P. Grendler, *Schooling in Renaissance Italy: Literacy and Learning, 1300–1600* [Baltimore, 1989], p. 217).

7. See Thomson (n. 1), p. 80.

8. For Bruni's activity in Rome, see G. Griffiths, "Leonardo Bruni and the Restoration of the University of Rome (1406)," *Renaissance Quarterly* 26 (1973): 1–10.

9. Cencio's letter is partly translated in T. Buddensieg, "Gregory the Great, the Destroyer of Pagan Idols," *JWCI* 28 (1965): 53, n. 27.

10. G. Voigt, *Die Wiederbelebung des Classischen Altertums, oder das erste Jahrhundert des Humanismus* (Berlin, 1859; rpt., Berlin, 1960), p. 230.

11. C. Del Bravo, "Il Brunelleschi e la speranza," *Artibus et Historiae* 3 (1981): 63, 66.

12. The text is in *VZ*, 4:110–50

13. Ibid., p. 102.

14. R. Weiss, "Lineamenti per una storia degli studi antiquari in Italia dal dodicesimo secolo al sacco di Roma del 1527," *Rinascimento* 9 (1959): 182.

15. H. Hunger, *Die Hochsprachliche profane Literatur der Byzantiner* (Munich, 1978), 1:172.

16. H. Rabe, ed., *Hermogenis. Opera*, Rhetores graeci 6 (Leipzig, 1913), p. 17.

17. P. Aelius Aristides, *Concerning Concord*, Oration XXIII, in *The Complete Works*, trans. C. Behr, 2 vols. (Leiden, 1981), 2:26–27.

18. See chapter 9 for further discussion.

19. Given the respective dates of the two works, any influence Chrysoloras might have had on Bruni's description came through conversation and not through the example of his *Comparison*. The *Laudatio* is discussed in chapter 9.

20. Dagron (n. 1) p. 284.

21. Ibid., p. 285.

22. That ekphrases really do depend on eyewitness observation is almost never acknowledged in the literature. An exception is A. Cutler, "The 'De Signis' of Nicetas Choniates: A Reappraisal," *American Journal of Archaeology* 72 (1968): 113–18.

23. Dagron (n. 1), pp. 282–83.

24. J. Alsop, *The Rare Art Traditions: The History of Art Collecting and Its Linked Phenomena* (Princeton, N.J., 1982), p. 342.

25. See Appendix, p. 204.

26. G. Dennis, ed., *The Letters of Manuel II Palaeologus*, epist. 16 (Washington, D.C., 1977), p. 44.

27. T. Preger, *Scriptores originum Constantinopolitanarum* (Leipzig, 1901; rpt. 1975); A. Cameron and J. Herrin, *Constantinople in the Eighth Century: The 'Parastaseis syntomoi chronikai'* (London, 1984). This aspect is discussed in G. Dagron, *Constantinople imaginaire. Etudes sur le receuil des "Patria"* (Paris, 1984), pp. 99–125.

28. N. Festa, ed., *Theodori Ducae Lascaris Epistulae CCXVII* (Florence, 1898), p. 107. Baxandall has already noted that Chrysoloras's *Comparison* comes out of this tradition (n. 1, p. 80).

29. N. G. Wilson, *Scholars of Byzantium* (Baltimore, 1983), p. 241, Letters 120 and 55.

30. M. Treu, *Theodori Pediasimi eiusque amicorum quae exstant* (Potsdam, 1899), pp. 14–16.

31. The translation is in Wilson (n. 29), p. 220.

32. See Appendix, p. 200.

33. For Lucian as an art critic, see V. Andò, *Luciano critico d'arte* (Palermo, 1975); for his popularity in the Quattrocento, see, most recently, R. Signorini, *Hoc opus tenue: La Camera dipinta di Andrea Mantegna: lettura storica, iconografica, iconologica* (Parma, 1985). For my connection between this notion and Alberti's architectural criticism, see chapter 1.

34. Lucian, *The Hall*, trans. A. M. Harmon, 8 vols. (New York, 1913), 1:177–79.

35. See Appendix, p. 211.

36. Baxandall supposed that his description focused on the Arch of Constantine ([n. 1], p. 80), whereas Dagron saw evidence of familiarity also with the Column of Trajan and the Arch of Titus (n. 1, pp. 286–87).

37. J. Schlosser, "Die höfische Kunst de Abendlandes in byzantinischen Beleuchtung," *Präludien* (Berlin, 1927), pp. 78–79.

38. For discussion of the concepts of pleasure and necessity in rhetorical theory, see chapter 5.

NOTES

39. R. Foerster, ed., *Libanii Opera*, letter 435 (Leipzig, 1921), 10:425–26. The identification of Jovianus is in O. Seeck, *Die Briefe des Libanius zeitlich geordnet* (Leipzig, 1906), p. 185.

40. J. Chrysostom, Homily XXXII, *PG*, 60: cols. 675–82; Homily XXVI, *PG*, 61: col. 582; Homily III, *PG*, 49: col. 49. For the relation of his style to that of Libanius, see T. E. Ameringer, *The Stylistic Influence of the Second Sophistic on the Panegyrical Sermons of St. John Chrysostom* (Washington, D.C., 1921).

41. Saint Thomas Aquinas, *Summa theologica*, pt. I, q. 73, art. 1, trans. Fathers of the English Dominican Province (New York, 1947), p. 353.

42. For the *Patria*, see n. 27.

43. The use of buildings as evidence for the virtues of magnificence and vice of luxury is examined in chapter 3; use of architecture as historical evidence is considered in chapters 2, 9, and 10.

44. Aelius Aristides, *The Panathenaic Oration*, 8, trans. C. A. Behr (Cambridge, Mass., 1973), p. 15: "For the nature of our country will appear to agree with the nature of its people"; Aristides, *Panegyric in Cyzicus*, 17, Oration XXVII (n. 17), 2:101.

45. Herodotus, *Histories*, II.124–28; III.60; VII.23, trans. A. D. Godley (Cambridge, Mass., 1946): Diodorus Siculus, *The Library of History*, I.47–49; II.7.2–10.6; XIII.82.1–8, trans. C. H. Oldfather (Cambridge, Mass., 1933–1939).

46. See chapters 2 and 3.

47. *VZ*, 4: 142.

48. Aelius Aristides, *On Rome*, 8, Oration XXVI (n. 17), p. 74.

49. The bibliography is gathered in G. Gernentz, *Laudes Romae* (Rostock, 1918).

50. For this device, see E. Curtius, *European Literature and the Latin Middle Ages*, trans. W. Trask (Princeton, N.J., 1953), p. 159.

51. Paralipsis in discussed in Demetrius, *On Style*, with this example from Demosthenes: "I make no mention of Olynthus, Methone, Apollonia and the thirty-two cities in Thrace" (*On Style*, trans. W. H. Roberts [London, 1927], p. 120).

52. *PG*, 60: col. 678.

53. Chrysostom, in *PG*, 61: cols. 581–82.

54. Acts 3:1–10. Is it coincidental that Bruni uses the same biblical passage in his dedicatory letter to John XXIII of April 1411 prefacing his translation of the *Gorgias* (a work cited in the *Comparison*)? (L. Bertalot, "Zur Bibliographie der Uebersetzungen des Leonardus Brunus Aretinus," *QFIAB* 27 (1937): 182)?

55. Much of this section seems to have been taken over by Biondo Flavio in the conclusion of his *Roma Instaurata*, Book III, pp. 271–72.

56. Chrysostom, Homily XXXII, in *PG*, 60: col. 678.

57. The epithet seems to have been used first by Themistius, Discourse 3 (Presbeutikos on Constantinople), probably of 357, then at the Council of Constantinople (381). It appears in a comparison in Paulus Silentiarius's sixth-century poem (lines 164–67), in Corippus's *In Laudem Justini*, and in the *Chronicon Paschale* (mid-seventh century). Constantine Manasses took up the theme in the twelfth century (*Compendium*, verses 2546–48) and Metochites in the fourteenth (*Byzantios*, fol. 301r and v). Fenster is mistaken in identifying the first use of this epithet in Canon 3 of the Second Eucumenical Council of Constantinople in 381 (E. Fenster, *Laudes Constantinopolitanae* [Munich, 1968], p. 55). For Themistius's earlier usage, see G. Downey, ed., *Themistii Orationes*, 3 vols. (Leipzig, 1965–1974), p. 60. W. Hammer, "The Concept of the New or Second Rome in the Middle Ages," *Speculum* 19 (1944): 52, mistakenly believed Corippus to have invented the concept (*MGH.auct.antiq* III, ii [Berlin, 1879], p. 126). The passage from Metochites is in I.

Sevcenko, "The Decline of Byzantium Seen Through the Eyes of Its Intellectuals," in *Society and Intellectual Life in Late Byzantium* (London, 1981), p. 176, n. 37.

58. This population estimate is from J. Meyendorff, *Byzantine Theology: Historical Trends and Doctrinal Themes* (New York, 1983), p. 110, for the year 1439. Alsop supposed that by the fourteenth century there were no Classical works to be seen in Constantinople and no hint of any art collections ([n. 24], p. 343). Chrysoloras's text contradicts that view, not only describing at least three Antique statues, but referring to "many other such things which I have not seen myself, but which I heard are hidden in certain places." See also C. Mango, "Antique Statuary and the Byzantine Beholder," *Dumbarton Oaks Papers* 17 (1963): 70.

59. Nicetas Choniates, *On the Statues*, in *O City of Byzantium: Annals of Niketas Choniates*, trans. H. Magoulias (Detroit, 1984). For the *Patria*, see Cameron and Herrin (n. 27), pp. 97–116.

60. Dagron rightly emphasized the importance and novelty of this decision ([n. 1], p. 283).

61. D.A. Russell and N. G. Wilson, eds., *Menander Rhetor*, Treatise I (Oxford, 1981), pp. 33–75.

62. Ibid., Treatise I, pp. 29–75. That centrality was already a well-worn topos in the fourth century is suggested by Libanius, at the beginning of his *Antiochikos*: "I shall not be persuaded to comply with the usage of most orators, who strain themselves to show that whatever particular place they are praising is the center of the earth" ([n. 39], p. 657).

63. Gregory Nazianzen, *The Last Farewell*, Oration XLII, in *PG*, 36: col. 470, Himerius, Oration XLI, in *Himerii declamationes et orationes*, ed. A. Colonna (Rome, 1951), p. 170.

64. Metochites is quoted in E. Barker, *Social and Political Thought in Byzantium: From Justinian I to the Last Palaeologus* (Oxford, 1957), p. 180. Metochites also claimed the superiority of Constantinople over Rome on the basis of site in his *Byzantios* (Sevcenko [n. 57]. p. 176, n. 5).

65. Aristides (n. 17), 10–11, pp.14–17. However, Aristides uses an almost identical image in *Panegyric in Cyzicus*, 6, pp. 98–99.

66. C. G. Patrinelis, "An Unknown Discourse of Chrysoloras Addressed to Manuel II Palaiologus," *Greek, Roman and Byzantine Studies* 13 (1972): 501–2. Dagron observed that Chrysoloras suggests that *both* Rome and Constantinople arose from the blend of two cultures ([n.1], p. 288).

67. Sources for this argument are probably Themistius (Downey [n. 57]), Discourses 3 (1:58–68) (presbeutikos on Constantinople) and 14 (1:260–65) (presbeutikos to Theodosius), together with the first section of the *Patria*, on the origins of the city.

68. E. Garin, *Science and Civic Life in the Italian Renaissance*, trans. P. Munz (New York, 1965), pp.38–39.

69. F. Dirimtekin, "Adduction de l'eau à Byzance," *Cahiers archéologiques* 10 (1959): 217.

70. Aristides, *Panathenaic Oration* (n. 44), p. 70.

71. Livy, *Ab urbe condita*, trans. F. G. Moore, 13 vols. (Cambridge, Mass., 1935–1940), 1:32–33.

72. Compare this with Alberti's concern with the moral problem of size in chapter 3, and the discussion of beauty versus utility in chapter 5.

73. Constantine of Rhodes is in E. Legrand, "Description des oeuvres d'art et de l'Eglise des Saints Apôtres de Constantinople. Poème en vers iambiques par Constantin le Rhodien," *Revue des études grecques* 9 (1896): 32–103. His list of the Seven Wonders of Constantinople is the statue of Justinian, the column of Constantine, the Senate House and nearby monuments, the column with a cross, the obelisk of Theodosius, the column of Arcadius, and the column of Theodosius.

74. R. Janin, *Constantinople byzantine* (Paris 1950), p. 79. It is not clear whether Chrysoloras refers to Justinian's column in the Augusteon or to the equestrian statue thought to represent him near Hagia Sophia (P. W. Lehmann, "Theodosius or Justinian? A Renaissance Drawing of a Byzantine Rider," *AB* 41 [1959]: 39–57). Might the other monuments near it be the columns of

Leo I and of Eudoxia in the Augusteon (W. Müller-Wiener, *Bildlexikon zur Topographie Istanbuls* [Tübingen, 1977] p. 80)? Two Russian travelers of the late fourteenth and early fifteenth centuries identified three bronze statues of pagan emperors opposite the statue of Justinian (G. Majeska, *Russian Travelers to Constantinople in the Fourteenth and Fifteenth Centuries* [Washington, D.C., 1984], pp. 136, 184).

75. The views are reproduced in G. Gerola, "Le Vedute di Constantinopoli di Cristoforo Buondelmonti," *Studi bizantini e neoellenici* 3 (1931): 249–79.

76. Müller-Wiener (n. 74), pp. 84, 86; C. Mango, *Le Développement urbain de Constantinople (IV–VIIe)* (Paris, 1985).

77. Müller-Wiener (n. 74), p. 247; Mango (n. 76), p. 25.

78. Mango (n. 76), p. 47.

79. Mango says this column is not known from any text (ibid., p. 34). Thus if my identification is correct, Chrysoloras's mention is a unicum. I leave it for Byzantinists to evaluate the meaning of Manuel's residence in the Mangana Palace.

80. Müller-Wiener (n. 74), p. 87; Mango (n. 76), pp. 31, 35. Majeska identifies the figures as an angel and Michael Palaiologus, not Constantine, as Buondelmonti and Zosima thought (Gerola (n. 75), pp. 277–78; Majeska [n. 74], pp. 184, 306).

81. Müller-Wiener (n. 74), p. 81; Mango (n. 76), p. 25. Zosima, who visited in 1423 also thought this was the site of Constantine's palace (Majeska, (n. 74), p. 184).

82. Mango identifies these as representing the sons of Constantine ([n. 76], p. 29). An anonymous Russian traveler of the late fourteenth century describes the Righteous Judges "as large as people and made from red marble. The Franks damaged them; one was split in two and the other had its hands and feet broken and its nose cut off" (Majeska [n. 74], p. 144).

83. Cameron localized the statue as at the Amastrianon, but could not determine whether it corresponded to the reclining Hercules mentioned in chap. 41 of the *Parastaseis* or a reclining Zeus attributed to Phidias (Cameron and Herrin [n. 27], p. 226).

84. Müller-Wiener (n. 74), p. 297. The gate had been rebuilt by John VII Palaiologus in 1389–1390, at which time the first- or second-century A.D. mythological reliefs were added. Thus the ensemble was a new structure, made during Chrysoloras's lifetime.

85. Himerius, Oration XLI (n. 63), p. 171.

86. Longinus, *On the Sublime*, ed. D. A. Russell (Oxford, 1964). Sublimity is discussed throughout the work; variety is recommended at XX.1. The work was only recovered in the West in 1554, according to Vickers (n. 5), p. 307.

87. C. Wooten, trans., *Hermogenes' "On Types of Style"* (Chapel Hill, N.C. 1989).

88. *VZ*, p. 136.

89. Aristides (n. 17), p. 101.

90. Ibid.

91. Ibid., pp. 100–101.

92. Cicero, *Academica*, I.3.9, trans. H. Rackham (London, 1938), p. 418.

93. Sabbadini (n. 2), II.7, p. 21.

CHAPTER NINE

1. Thus Q. Fabius Laurentius Victorinus says in his commentary on Cicero's *Rhetoric* (I.25): "Inter scientiam et artem hoc interest. Ars quidem duplex est, una in scientia, alia in actu. Verum scientia rursus duplex est: est scientia artis, est non artis; neque enim aurum, gemmas probare artis est, aut scire quot milia hominum in illo sint populo, quanto spatio haec urbs ab illa distet;

haec itaque scientia est, sed non artis" (C. Halm, *Rhetores Latini minores* [Leipzig, 1867], p. 219). Emporius Orator in his *Praeceptum demonstrativae materiae* makes a similar point: "Demonstrationes vero urbium locorumque iam non demonstrationes, sed topographiae a plurimus existimantur" (ibid., p. 569).

2. Isidore of Seville, *Etymologiarum sive originum*, XV.2.1, ed. W. M. Lindsay (Oxford, 1911): "Urbs ipsa moenia sunt, civitas non saxa sed habitatores vocantur." Victorinus also identifies the city with its inhabitants: "Dicendum est hoc loco, quid sit civitas. Est autem civitas collecta hominum multitudo ad iure vivendum" (Halm [n. 1], p. 158).

3. Here is one example, from Gregory the Great: "notandum est, quod non dicitur super quem erat aedificium sed quasi aedificium ut videlicet ostenderetur, quod non de corporalis, sed de spiritualis civitatis aedificio cuncta diceruntur. 'Et ipsa est civitas, scilicet sancta Ecclesia' " (H. Rosenau, *The Ideal City: Its Architectural Evolution in Europe* [1959; rpt., London, 1983], p. 27).

4. R. P. McKeon, "Rhetoric in the Middle Ages," *Speculum* 17 (1942): 23. This is also discussed in W. Tatarkiewicz, *History of Aesthetics*, ed. J. Harrell, 3 vols. (The Hague, 1970), 1:265.

5. McKeon (n. 4) p. 27. See also H. Caplan, "Classical Rhetoric and the Medieval Theory of Preaching," in *Historical Studies of Rhetoric and Rhetoricians*, ed. R. F. Howes (Ithaca, N.Y., 1961), pp. 72–74, and B. Vickers, *In Defence of Rhetoric* (Oxford, 1988), p. 222.

6. *VZ*, 1:68.

7. Some of these are in *VZ*; many are collected in G. B. De Rossi, *Inscriptiones christianae urbis Romae*, ed. A. Silvagni (Rome, 1922), 1:xviii–xx.

8. See J. K. Hyde, "Medieval Descriptions of Cities," *Bulletin of the John Rylands Library* 48 (1966):308–40, for further discussion; and G. Tellenbach, "La Città di Roma dal IX al XII secolo vista dai contemporanei d'oltre frontiera," in *Studi storici in onore di Ottorino Bertolini*, 2 vols. (Pisa, 1972), 2:692.

9. C. D'Onofrio, *Visitiamo Roma mille anni fa. La città dei Mirabilia* (Rome, 1988), p. 30. Whether it belongs to the twelfth century, as has usually been thought, and therefore to the climate of revival of the Roman commune and Republican liberty, or to the *renovatio* of Otto III around the year 1000, is not certain. See also the edition in *VZ*, 3:17–65.

10. D'Onofrio suggested that chapters 8, 10, 11, 12, 15, 16, and 17 were copied from earlier collections or based on earlier materials ([n. 9], p. 12).

11. R. Benson, "Political *Renovatio*: Two Models from Roman Antiquity," in *Renaissance and Renewal in the Twelfth Century*, ed. R. Benson and G. Constable (Cambridge, Mass., 1982), pp. 339–86; C. Frugoni, "L'Antichità dai 'Miribilia' alla propaganda politica, " in *Memoria dell'antico nell'arte italiana. I. L'Uso dei classici*, ed. S. Settis (Turin, 1984), p. 5, n. 1. See also P. Jounel, *Le Culte des saints dans les basiliques du Lateran et du Vatican au douzieme siècle* (Rome, 1977), p. 28.

12. Jounel (n.11).

13. R. Maiocchi and F. Quintavalle, eds., *Anonymi Ticinensis. Liber de laudibus civitatis ticinensis* (Città di Castello, 1903–1906); J. Berrigan, "Benzo of Alessandria and the Cities of Northern Italy," *Studies in Medieval and Renaissance History* 4 (1967): 127–92; Bonvicinus de'Ripa, "De Magnalibus urbis mediolani," ed. F. Novati, *Bulletino dell'Istituto Storico Italiano* 20 (1898): 7–176; A. Paredi:, ed., *Bonvesin de la Riva. Grandezze di Milano* (Milan, 1967); I. Moutier, ed., *Cronica di Giovanni Villani*, 5 vols. (Florence, 1823).

14. H. Baron, *The Crisis of the Early Italian Renaissance* (Princeton, N.J., 1955), 2: 515, n. 8.

15. G. Cantino Wataghin, "Archeologia e 'archeologie.' Il Rapporto con l'antico fra mito, arte e ricerca," in Settis (n. 11), p. 191; R. Weiss, "Lineamenti per una storia degli studi antiquari

in Italia dal dodicesimo secolo al sacco di Roma del 1527," *Rinascimento* 9 (1959): 159. The text is in *VZ*, 4:68–73.

16. Although Dondi owned a copy of Vitruvius, his attitudes were formed by a personal library rich in scientific treatises and works by Scholastic philosophers (V. Lazzarini, "I Libri gli argenti le vesti di Giovanni Dondi dall'Orologio," *Bollettino del Museo Civico di Padova* 1 [1925]:11–36).

17. Dondi, in *VZ*, 4:69.

18. Ibid., p. 68.

19. Ibid., pp. 93–100. P. P. Vergerio, *Epistolario*, LXXXVI, ed. L. Smith (Rome, 1934), pp. 211–20. I use the edition in Smith.

20. Vergerio (n.19), p. 211.

21. Ibid., p. 215.

22. Ibid. Weiss quite rightly noted that Vergerio's interest in the Roman monuments reflected influence from Petrarch, of whom he was the biographer and editor ([n. 15], p. 161).

23. Vergerio (n. 19), p. 215.

24. Ibid.: "modo felicitas inveniendi et describendi facilitas adfuisset."

25. The description of Florence is mentioned in Vergerio's letter LXXXXVI of around 1400 to 1404 to an anonymous Florentine (as in note 19, p. 245). For the other descriptions, see D. Robey and J. Law, "The Venetian Myth and the 'De Republica veneta' of Pier Paolo Vergerio," *Rinascimento* 15 (1975): 3–59. The description of Venice is in Robey and Law and in E. A. Cicogna, *Petri Pauli Vergerii senioris Justinopolitani. De republica veneta fragmenta nunc primum edita* (Venice, 1830).

26. Pier Paolo Vergerio, *De Ingenuis Moribus*, trans., J. McManamon, "Innovation in Early Humanist Rhetoric: The Oratory of Pier Paolo Vergerio the Elder," *Rinascimento* 22 (1982): 8

27. Ibid.

28. Ibid., pp. 5, 9, 31.

29. Vergerio tells us that his first attempt to learn Greek, in Padua, bore little fruit and that only as a student of Chrysoloras did he acquire facility in the language ([n. 19], LXXXXVI; XIV).

30. R. Weiss, "Gli Inizi dello studio del Greco a Firenze," in *Medieval and Humanist Greek: Collected Essays* (Padua, 1977), p. 241; J. E. Sandys, *A History of Classical Scholarship*, 2 vols. (New York, 1964), 2:49; R. Sabbadini, "L'Ultimo ventennio della vita di Manuele Crisolora," *Giornale linguistico* 17 (1890): 326; L. Smith, "Note cronologiche vergeriane," *Archivio veneto* 4 (1928): 101. Vergerio was already reading Plutarch, Thucydides, Plato, and Homer in the winter of 1400–1401 (epist. LXXXXV), and expressed hope of perfecting his Greek in a letter to Chrysoloras of 1400 (epist. LXXXXIII) ([n. 19], pp. 241, 237–39).

31. A. Pertusi, "Gli Inizi della storiografia umanistica nel Quattrocento," in *La Storiografia veneziana fino al secolo XVI. Aspetti e problemi* (Florence, 1969), p. 273. Garin suggested that Chrysoloras's translation of the *Republic* is the thread that links all his students, whether in Milan, Pavia, Venice, Ferrara, or Florence (E. Garin, *Umanisti, artisti scienziati* [Rome, 1989], p. 98).

32. Robey and Law (n. 25), p. 28.

33. Ibid., pp. 23–4.

34. See chapter 8, n. 65.

35. Robey and Law (n. 25), p. 43.

36. G. Downey, "Libanius' Oration in Praise of Antioch (Oration XI)," *Proceedings of the American Philosophical Society* 103 (1959): 680.

37. Bruni's statement is in a letter to Francesco Pizolpasso in *Leonardi Aretini epistolarum libri VIII*, epist. VIII, ed. L. Mehus (Florence, 1741), 4:111: "non enim temere neque leviter id opus aggressi sumus, neque vagi, aut incerti per semitas nobis incognitas peregrinantium more nostro

ipsi arbitratu processimus, sed ducem itineris, totiusque laudandi progressus certum indubitatumque habuimus Aristidem celebrem apud Graecos Oratorem eloquentissimum hominem, cuius extat oratio pulcherrima de 'Laudibus Athenarum.' "

Scholars have agreed that Bruni used this oration as his main model, an assumption I will challenge. Sandys (n. 30), 2:47; B. Marx, "Venedig—'altera Roma.' Transformationen eines Mythos," *QFIAB* 60 (1980): 338; H. Baron, *From Petrarch to Leonardo Bruni: Studies in Humanistic and Political Literature* (Chicago, 1968), p. 113; Hyde (n. 8), p. 309. A fourteenth-century manuscript of the oration, now Laurenziana 56, 8, which came from the collection of Niccolò Niccoli and may have once belonged to Chrysoloras, might be the one used by Bruni. For the manuscript, see V. Branca, *Poliziano e l'umanesimo della parola* (Turin, 1983), p. 117.

38. P. Aelius Aristides, *The Complete Works*, trans. C. Behr, 2 vols. (Leiden, 1981), 1:9, 67.

39. B. Kohl and R. Witt, eds., *The Earthly Republic: Italian Humanists on Government and Society* (Philadelphia, 1978), p. 174.

40. Aristides (n. 38), 78, p. 20.; 166–67, p. 38; Bruni (n. 39), pp. 165, 166, 168.

41. Canfield related this image to Plato's *Laws*, Book VI, but I think it is closer to the *Critias* (G. B. Canfield, "The Florentine Humanists' Concept of Architecture in the 1430's and Filippo Brunelleschi," in *Scritti di storia dell'arte in onore di Federico Zeri* 2 vols. [Venice, 1984], 1:114). Garin had also pointed to *Laws* VI as the source for Bruni's notion that architectural planning must correspond to the political and social structure of the city (E. Garin, *Science and Civic Life in the Italian Renaissance*, trans. P. Munz [Garden City, N.Y. 1965], p. 28).

42. Aristides (n. 38), 16, p. 9.

43. Kohl and Witt (n. 39), p. 145.

44. Aristides (n. 38), 49–61, pp. 15–17; Bruni's (n. 39) version is on pp. 159–61; Thucydides, *The History of the Peloponnesian War* II.xxxvi, xl; II..xl.

45. Bruni cites Roman sources in his text (n. 39): Suetonius (p. 152); Lucan (p. 153); Tacitus (p. 154); Cicero (p. 161). Sandys claims that Bruni quotes the beginning of Tacitus's *Histories* in his Panegyric, but I am unable to find this passage ([n. 30], p. 33). It was Chrysoloras who first introduced Italians to this oration, using it in his *Comparison of Old and New Rome*. Evidently, he showed a copy to Bruni much earlier (R. Klein, *Die Romrede des Aelius Aristides* [Darmstadt, 1981], p. 109).

46. Aristides (n. 38), 2:74.

47. Bruni (n. 39), p. 141.

48. Ibid., p. 159; Aristides (n. 38), 2:86.

49. Downey (n. 36), p. 659; Bruni (n. 39), pp. 145–48. The popularity of this author in the period that concerns us is indicated by the fact that of the thirty-eight extant Libanius manuscripts, sixteen are of the fifteenth century and twelve of the fourteenth (R. Foerster, *Libanii Opera* [Leipzig, 1903] 1:412–32).

50. J. Bompaire, "Photius et la Seconde Sophistique, d'après la *Bibliothèque*," *Travaux et memoires* 8 (1981): 81; G. Griffiths, J. Hankins, and D. Thompson, *The Humanism of Leonardo Bruni* (Binghamton, N.Y., 1987), pp. 98–100.

51. Bruni (n. 39), p. 136.

52. Ibid., p. 143.

53. Aristides (n. 38), 1:7; Aristides, *Roman Oration* (n. 38), 2:86.

54. Quintilian, *Institutio oratoria*, III.vii, trans. H Butler, 4 vols. (Cambridge, 1921–1936), 1:477.

55. Bruni (n. 39), p. 149.

56. Leon Battista Alberti, *De Re aedificatoria*, VII.xvii, ed. G. Orlandi (Milan, 1966), p. 663.

57. M. Baxandall, *Giotto and the Orators: Humanist observers of Painting and the Discovery of Pictorial Composition, 1350–1450* (Oxford, 1971), p. 103.

58. Lorenzo Valla, *On Pleasure. De Voluptate*, III.i.1, trans. A. K. Hieatt and M. Lorch (New York, 1977).

59. This aspect of Humanist discourse is examined in H. Gray, "Renaissance Humanism: The Pursuit of Eloquence," *JHI* 24 (1963): 497–514, esp. 506.

60. The text is in *Pros. Lat.*, pp. 7–37.

61. Baron (n. 14), p. 219.

62. For the problem of their dates see G. Tanturli, "Cino Rinuccini e la Scuola di Santa Maria in Campo," *Studi medievali* 17 (1976): 641.

63. *Pros. Lat.*, p. 34 (my translation).

64. Bruni (n. 39), p. 138. A discussion of urban statutes in fifteenth-century Rome is in C. Westfall, *In This Most Perfect Paradise: Alberti, Nicholas V and the Invention of Conscious Urban Planning in Rome, 1447–55* (University Park, Pa., 1974), pp. 78–84; for Tuscany, see W. Braunfels, *Mittelalterliche Stadtbaukunst in der Toskana* (Berlin, 1953), p. 106.

65. Baron (n. 37), p. 158.

66. R. Sabbadini, *La Scuola e gli studi di Guarino Veronese* (Catania, 1986). For his life, see Baxandall (n. 57), p. 87, n. 82; Baxandall, "Guarino, Pisanello and Manuel Chrysoloras," *JWCI* 28 (1965): 190.

67. H. Omont, "Les Manuscrits grecs de Guarine de Vérone et la bibliothèque de Ferrare," *Revue des bibliothèques* 2 (1892): 78. Among the fifty-four Greek manuscripts left at his death in 1460 were works by Isocrates, Lucian, Chysostom, Gregory Nazianzen, and Dionysius of Halicarnassus.

68. I.vii, R. Sabbadini, ed., *Epistolario di Guarino Veronese*, 2 vols. (Venice, 1916), 1:20.

69. Ibid.

70. Ibid., pp. 594–95 (trans., A. Grafton and L. Jardine, "Humanism and the School of Guarino: A Problem of Evaluation," *Past and Present*, 96 [1982]: 69).

71. Baxandall (n. 66), pp. 191–93.

72. Grafton and Jardine (n. 70), p. 69.

73. Sabbadini (n. 68), 2:464.

74. Lucian, *How to Write History*, trans. K. Kilburn (London, 1959), p. 69.

75. R. Avesani, "In Laudem civitatis Veronae," *Studi storici Veronesi* 26–27 (1976–1977), 184.

76. Sabbadini (n. 68), 2:294.

77. Discussed in Avesani, (n. 75), esp. pp. 183–90; H. Goldbrunner, "Laudatio urbis. Zum Untersuchungen über das humanistische Städtelob,," *QFIAB* 63 (1983): 321. Francesco Corna da Soncino mentions Guarino explicitly, "el qual fece più versi in poesia et oratore fu greco e latino" (*Fioretto*, ed. G. P. March [Verona, 1973], p. 89).

78. Sabbadini (n. 68), 1:487. The letter is of 1425: "expecto ut quantum urbem quantum civitatem admireris scribas."

79. In *RIS* 24, pt. 15 (1902).

80. The most recent edition of this portion of the text is in C. D'Onofrio, *Visitiamo Roma nel Quattrocento. La Città degli Umanisti* (Rome, 1989), p. 50.

81. G. Zippel, *Storia e cultura del Rinascimento italiano* (Padua, 1979), p. 82. Most recently, Garin has reaffirmed that Poggio studied with Malpaghini in Florence and not with Chrysoloras ([n. 31], p. 55).

82. I. Thomson, "Manuel Chrysoloras and the Early Italian Renaissance," *Greek, Roman and Byzantine Studies* 7 (1966): 74.

83. Poggio Bracciolini to Guarino Veronese, 1455, in Poggio Bracciolini, *Lettere*, ed. I I. Harth, 3 vols. (Florence, 1984), epist. 18, 3: 348. I follow I. Kajanto's translation in *Poggio Bracciolini and Classicism* (Helsinki, 1987), p. 17.

84. D. Marsh, "Poggio and Alberti. Three Notes," *Rinascimento* 23 (1983): 190.

85. R. Sabbadini, *Il Metodo degli umanisti* (Florence, 1920), p. 23.

86. The dating is that of R. Fubini (Poggius Bracciolini, *Opera omnia*, ed. R. Fubini, 4 vols. [Turin, 1966], 2:499–500). See also Kajanto (n. 83), pp. 36–37, and A. Mazzocco, "Petrarch, Poggio and Biondo: Humanism's Foremost Interpreters of Roman Ruins," in *Francis Petrarch, Six Centuries Later: A Symposium*, ed. A. Scaglione (Chicago, 1975), p. 358. I am citing the edition in Fubini.

87. His discoveries and the sources for the work are discussed in G. Voigt, *Die Wiederbelebung des Classischen Alterthums, oder das erste Jahrhundert des Humanismus*, 2 vols. (1859; rpt., Berlin, 1960) 1:268; Sandys (n. 30), 2:38; R. Sabbadini, *Le Scoperte dei codici latini e greci ne'secoli XIV e XV*, 2 vols. (Florence, 1905; rpt., Florence, 1967), 1:80. Poggio discovered the Einsiedeln Itinerary at either Sankt Gallen or Reichenau; the Ammianus Marcellinus and Statius's *Silvae* were discovered at Sankt Gallen, where, in 1416, he had found a Vitruvius. His knowledge of Greek authors is indicated by his quotation from Dionysius of Halicarnassus, *Roman Antiquities*, IV. 13.5, and Libanius in the text ([n. 86], pp. 508, 525); he translated the first five books of Diodorus Siculus for Nicolas V.

88. Poggio (n. 86), 2:925–37.

89. "Et ab Livanio [*sic*] doctissimo auctore, cum ad amicum scriberet Roman videre cupientem, non urbem, sed quasi quandam caeli partem appellatam." Strangely, Valentini and Zucchetti maintain that the source is Aristides, not Libanius (*VZ*, p. 231).

90. Poggio (n. 86), p. 508.

91. For Petrarch, see A. Mazzocco, "The Antiquarianism of Francesco Petrarca," *Journal of Medieval and Renaissance Studies* 7 (1977): 203–24.

92. Poggio (n. 86), pp. 34–36.

93. The letter is in Bruni (n. 37), VIII. 4, pp. 111ff.; it is discussed in H. Baron, *Leonardo Bruni Aretino. Humanistisch-Philosophische Schriften* (Berlin, 1928), p. 214, no. 4. Valla's attack was in a letter of 1435; Decembrio's dates 1436.

94. O. Besomi and M. Regoliosi, eds., *Laurentii Valle. Epistole* (Padua, 1984), p. 161. I thank S. I. Camporeale for discussing the letter with me.

95. Ibid.

96. Ibid., p. 163.

97. Aristotle, *Politics* (trans., Griffiths, Hankins, and Thompson [n. 50], p. 163).

98. G. Petraglione, "Il 'De laudibus Mediolanensium urbis panegyricus' di P. C. Decembrio," *Archivio Storico Lombardo* 8 (1907): 29.

99. The political motivations of Valla and Decembrio are discussed in Besomi and Regoliosi (n. 94), pp. 152–55.

100. Baron (n. 37), p. 152.

101. A. Lanza, ed., *Leonardo Bruni. Le Vite di Dante e del Petrarca* (Rome, 1987), p. 14.

102. G. Shepherd and T. Tonelli, *Vita di Poggio Bracciolini* (Florence, 1825), appendix 19, pp. xlviii–lii, 25.

103. Bruni (n. 37), p. 111.

104. For the relation between the two panegryics, see pp. 176–77.

105. Bruni (n. 37), p. 112.

106. Ibid. Baron thought that Bruni's source for this distinction was Polybius X. 21.8; ([n. 14], 2: 523).

107. E. Cochrane, *Historians and Historiography in the Italian Renaissance* (Chicago, 1981), p. 504, n. 48. Coluccio Salutati, *Epistolario*, ed. F. Novati, 4 vols. (Rome, 1896), epist. 11.

108. The history of the criticism of rhetoric has most recently been studied by Vickers, (n. 5.)

109. Sabbadini (n. 66), p. 63. The manuscript is British Museum Addit. 12088 (R. Sabbadini, *Carteggio di Giovanni Aurispa* [Rome, 1931], letter LXVI, p. 82). Its reception by the Humanists is discussed in A. M. Brown, "The Humanist Portrait of Cosimo de'Medici, Pater Patriae," *JWCI* 24 (1961): 186–221.

110. Sabbadini (n. 68), 2:462.

111. Ibid., p. 461.

112. Poggio (n. 86), 2:78.

113. Ibid., p. 44.

114. B. Widmer, "Enea Silvio's Lob der Stadt Basel und seine Vorlagen," *Basler Zeitschrift* 58–59 (1959): 111.

115. *Ibid.*, p. 128.

116. Michele Savonarola, *Libellus de magnificis ornamentis regie civitatis Padue*, RIS 24, pt. 15 (1902), p. 4.

117. Philippus de Diversis de Quartigionis Lucensis, *De Situs aedificiorum et laudabilium consuetudinum inclytae civitatis Ragusiae*, ed. V. Brunelli (Zara, 1880); G. Praga, "Ciriaco de Pizzicolli e Marino de'Resti," *Archivio storico per la Dalmazia* 13 (1932–1933): 262–80.

118. Goldbrunner (n. 77), pp. 316–18.

119. See the parallel texts in ibid., p. 319, n. 22. Only Piccolomini picked up the theme of cleanliness, asking in his description of Nuremberg, "quae domorum munditiae? Quis intra nitor platearum?" ("Historia de Europa," in *Aeneae Sylvii Piccolominei senensis. Opera omnia* [Basel, 1555], p. 437).

120. Petraglione (n. 98), p. 36.

121. Ibid., p. 34. For Bruni's text and its sources, see n. 45.

122. Decembrio's letter is in J. Onians, "Style and Decorum in Sixteenth-Century Italian Architecture" (Ph.D. diss., University of London, 1968), p. 232. Pizolpasso's description is in T. Foffano, "La Costruzione di Castiglione Olona in un opusculo di Francesco Pizolpasso," *IMU* 3 (1960): 153–87.

123. Giannozzo Manetti, *Vita Nicolai V summi pontificis*, RIS 3, pt. 2 (1734).

124. This has caused architectural historians considerable difficulty, as Westfall has shown ([n. 64], p. 104). Nonetheless, he and others have tried to reconstruct the real projects of Nicholas V using the text. See G. Urban, "Zum Neubau-Projekt von St. Peter unter Papst Nikolaus V," in *Festschrift für Harald Keller* (Darmstadt, 1963), pp. 131–73; T. Magnuson, "The Project of Nicholas V for Rebuilding the Borgo Leonino in Rome," *AB* 36 (1954): 89–116; Magnuson, *Studies in Roman Quattrocento Architecture* (Rome, 1958); C. Mack, "Nicholas the Fifth and the Rebuilding of Rome: Reality and Legacy," in *Light on the Eternal City: Recent Observations and Discoveries in Roman Art and Architecture*, ed. H. Hager and S. Munshower (University Park, Pa., 1987), pp. 31–56; and C. Burroughs, "A Planned Myth and a Myth of Planning: Nicholas V and Rome," in *Rome in the Renaissance: The City and the Myth*, ed. P. A. Ramsey (Binghamton, N.Y., 1982), pp. 197–207.

125. Manetti (n. 123), col. 931.

126. F. Pagnotti "La Vita di Niccolò V scritta da Giannozzo Manetti," *Archivio della R. Societa Romana di Storia Patria* 14 (1891): 415.

127. Manetti (n. 123), col. 938.

128. L. Onofri, "Sacralità, immaginazione e proposte politiche: La *Vita* di Niccolò V di

Gianozzo Manetti," *Humanistica Lovaniensia*, 28 (1979): 36–37; esp. Cicero, *De Oratore*, II.62; Cicero, *De Legibus*, I.4; Macrobius, *Commentarii in Somnium Scipionis*, I.2.10–11.

129. The influence of Dionysius's modes of rhetoric on historiography is discussed in E. Norden, *La Prosa d'arte antica dal VI secolo a.C. all'età della rinascenza* (1896; rpt., Rome, 1986), 1:90–93. Bessarion owned a copy of his *On Literary Composition*, now in the Marciana (Marcianus 508).

130. The character of the work is discussed in F. Leo, *Die Griechisch-römische Biographie* (Leipzig, 1901), p. 227.

131. Cicero, *Epistulae ad familiares*, V.xii, trans. W. G. Williams (London, 1927), p. 368; G. M. Cagni, "I Manoscritti Vaticani Palatino-latini appartenenti alla biblioteca di Giannozzo Manetti," *La Bibliofilia* 62, pt. 1 (1960): 30.

132. Cochrane (n. 107), pp. 31–32.

133. M. Miglio, "Una Vocazione in progresso: Michele Canensi, biografo papale," *Studi medievali* 12 (1971): 480.

134. G. Ianziti, *Humanistic Historiography under the Sforzas: Politics and Propaganda in Fifteenth-Century Milan* (Oxford, 1988), p. 6.

135. Ibid., p. 6. The *Commentaries* had been known at least since 1427, when Poggio had them copied for Niccoli.

136. Ibid., p. 13.

137. P. Grendler, *Schooling in Renaissance Italy: Literacy and Learning, 1300–1600* (Baltimore, 1989), p. 207.

138. A. Corbinelli, "Note di vita cittadina e universitario pavese nel Quattrocento," *Bollettino della Società Pavese di Storia Patria* 30 (1930): 139 for Guarino; Bruni (n. 37), epist. IX, 3:76–83.

139. Sabbadini (n. 68), p. 460.

140. Poggio, (n. 83), epist. 44, 1:119 (trans., P. Gordon, *Two Renaissance Bookhunters: The Letters of Poggius Bracciolini to Nicholaus de Niccolis* [New York, 1974], p. 176).

141. Corbinelli (n. 138), p. 46. I thank David Rutherford for this reference.

142. See n. 122.

143. Corbinelli (n. 138), p. 256.

144. Hardison showed how the tradition of literary criticism of this text from Late Antiquity through the Middle Ages discussed the poem as an encomion of Aeneas (O. B. Hardison, *The Enduring Moment: A Study of the Idea of Praise in Renaissance Literary Theory and Practice* [Westport, Conn., 1962], p. 33).

145. Ianziti (n. 134), p. 41.

146. For these works, see ibid.

147. F. Tateo, *Alberti, Leonardo e la crisi dell'Umanesimo* (Bari, 1971), p. 94, "Prolusione alla lettura del Petrarca."

148. P. Cortesi, *De Hominibus Doctis Dialogus*, ed. and trans. M. T. Grazios (Rome, 1973), p. 36.

149. R. Black, "Ancients and Moderns in the Renaissance: Rhetoric and History in Accolti's 'Dialogue on the Preeminence of Men of His Own Time,'" *JHI* 43 (1982): 11. The work has usually been dated in the 1440s, but Black showed that it must date 1462–1463 (p. 8).

150. Ibid., p. 12.

151. I develop here a suggestion that Cochrane made in regard to Petrarch (E. Cochrane, "Science and Humanism in the Italian Renaissance," *American Historical Review*, 81 [1976]: 1053).

152. Black (n. 149), p. 31.

153. Alberti (n. 56), prologue, p. 13.

154. A. Finoli and L. Grassi, ed., *Antonio Averlino detto il Filarete. Trattato di architettura*, 2 vols. (Milan, 1972), 1:36.

155. F. Hartt, "Art and Freedom in Quattrocento Florence," in *Essays in Memory of Karl Lehmann*, ed. L. Sandler (New York, 1964), p. 130.

156. G. C. Romby, *Descrizioni e rappresentazioni della città di Firenze nel XV secolo* (Florence, 1976), p. 51.

157. The proem is in *Cristoforo Landino. Scritti critici e teorici*, ed. R. Cardini (Rome, 1974). My view differs from Baron's explanation for the decline in popularity of Bruni's *Laudatio*, which, he thought, was surpassed by "the more sober and objective setting of truly historical presentations" of more fully developed Humanist historiography ([n. 37], p. 153).

158. Polybius, XXIX.12.2–4, trans. W. R. Paton, 6 vols. (London, 1927), 6:67.

159. C. Mack, *Pienza: The Creation of a Renaissance City* (Ithaca, N.Y., 1987), pp. 165–79.

160. Ibid., pp. 170–71.

161. These sources are discussed in relation to Petrarch by E. Fenzi, "Di Alcuni palazzi, cupole e planetari nella letteratura classica e medioevale e nell' 'Africa' del Petrarca," *Giornale storico della letteratura italiana* 153 (1976): 12–59, 186–229, 22–217.

162. R. Hatfield, "Some Unknown Descriptions of the Medici Palace in 1459," *AB* 52 (1970): 232–49.

163. Alberto Avogadro da Vercelli, *De Religione et magnificentia illustris Cosmi Medices Florentini*, in E. H. Gombrich, "Alberto Avogadro's Description of the Badia of Fiesole and the Villa of Careggi," *IMU* 5 (1962): 223.

164. Hatfield (n. 162), p. 234, n. 14.

165. M. Sensi, "Niccolò Tignosi da Foligno: L'Opera e il pensiero," *Annali della Facoltà di Lettere e Filosofia* (Università degli Studi di Perugia) 9 (1973): 359–495, esp. pp. 430, 456 from *Ad Clarissimum virum Ioannem Medicem de laudibus Cosmi Patris eius*; Cristoforo Landino, *Carmina omnia*, Book III, ed. A. Perosa (Florence, 1939), pp. 86–91.

166. Miglio (n. 133), pp. 516–17.

167. "Il *De Republica* di Lauro Quirini," ed. C. Seno and G. Ravegnani, in *Lauro Quirini umanista. Studi e testi a cura di K. Krautter* (Florence, 1977), Book II, pp. 424–25.

168. Ibid. Fornara has discussed this problem in ancient historiography, which was harnessed early on to "higher purposes" and made to serve ethics and patriotism while being conceived of as an instrument of pleasure—all concerns that have the potential to distort. By the fourth century B.C., history was already overtly judgmental, conferring praise and blame (C. W. Fornara, *The Nature of History in Ancient Greece and Rome* [Berkeley, 1988], pp. 105–8). I thank David Rutherford for this reference.

169. Quintilian (n. 54), X.i.29–31, p. 21: "History is very like to poetry and, in a certain sense, a poem in prose."

170. Poggio (n. 86), p. 78; Sabbadini (n. 68), p. 461.

171. Giovanni Pontano, *Dialoghi*, ed. C. Previtera (Florence, 1943), p. 193.

172. Baron (n. 93), pp. 5–19.

173. Corbinelli (n. 138), p. 257.

174. Grendler (n. 137), p. 237.

Bibliography

Primary Sources: Renaissance and Byzantine Authors

Alberti, Leon Battista. *Profugiorum ab aerumna libri III.* In *Leon Battista Alberti. Opere volgari,* edited by C. Grayson. 2 vols. Bari, 1960.
———. *De Re aedificatoria.* Edited by G. Orlandi. 2 vols. Milan, 1966.
———. *On the Art of Building in Ten Books.* Translated by J. Rykwert, N. Leach, and R. Tavernor. Cambridge, Mass., 1988.
———. *On Painting and on Sculpture.* Edited by C. Grayson. London, 1972.
———. *Leon Battista Alberti. On Painting.* Translated by J. Spencer. New Haven, Conn., 1966.
———. *An Autograph Letter from Leon Battista Alberti to Matteo de'Pasti, November 18, 1454.* Edited by C. Grayson. New York, 1957.
   Vagnetti, L. "La 'Descriptio urbis Romae': Uno Scritto poco noto di Leon Battista Alberti." *Quaderno. Università degli Studi di Genova, Facoltà di Architettura* 1 (1967): 25–78.
Anonimo Magliabecchiano. *Tractatus de rebus antiquis et situ urbis Romae.* In VZ, pp. 110–50.
Apostolis, Michael. *Lettres inédites de Michel Apostolis.* Edited by H. Noiret. Paris, 1889.
Bassi, E., et al., eds. *Pietro Cataneo, Giacomo Barozzi da Vignola, Trattati. Con l'aggiunta degli scritti di architettura di Alvise Cornaro, Francesco Giorgi, Claudio Tolomei, Giangiorgio Trissino, Giorgio Vasari.* Milan, 1985.
Bessarion, Cardinal. "Bessarionos Enkomion eis Trapezounta." Edited by S. Lampros. *Neos Hellenomnemon* 13 (1916): 146–204.
Biondo, Flavio. *De Roma triumphante, Romae instauratae, De Origine ac gestis venetorum liber, Italia illustrata, Historiarum ab inclinatio Romano imperio.* Basel, 1559.
———. *Italia illustrata.* Basel, 1531.
———. *De Origine et gestis venetorum.* In *Scritti inediti e rari di Biondo Flavio,* edited by B. Nogara. Rome, 1927.
Bracciolini, Poggio. *Poggius Bracciolini. Opera omnia.* Edited by R. Fubini. 4 vols. Turin, 1966.
———. *Lettere.* Edited by H. Harth. 3 vols. Florence, 1984.
   Gordon, P., trans. *Two Renaissance Book Hunters: The Letters of Poggius Bracciolini to Nicholaus de Niccolis.* New York, 1974.
Bruni, Leonardo. *Le Vite di Dante e del Petrarca.* Edited by A. Lanza. Rome, 1987.

———. *Leonardi Arretini epistolarum libri VIII*. Edited by L. Mehus. Florence, 1741.

———. *Panegyric to the City of Florence*. Translated by B. Kohl and R. Witt. In *The Earthly Republic. Italian Humanists on Government and Society*. Philadelphia, 1978.

———. *Ad Petrum Paulum Histrum*. In *Pros. Lat.*, pp. 44–99.

Bryennios, Joseph.
   Tomadakes, N. B. "Rede über die Wiederinstandsetzung der Stadtmauern Konstantinopels." *Epeteris* 36 (1968): 1–16.

Campano, Giovanni Antonio. *Vita di Pio II*. Edited by G. C. Zimolo. *RIS* 3, pt. 2. 1964.

Canensi, Michele. *Life of Nicholas V*. In M. Miglio, *Storiografia pontificia del Quattrocento*. Bologna, 1975.

Cencio de'Rustici.
   Bertalot, L. "Cincius Romanus und seine Briefe." *QFIAB* 21 (1929–1930): 209–51.

Cennini, Piero.
   Mancini, G. "Il Bel S. Giovanni e le feste patronali di Firenze descritte nel 1475 da Piero Cennini." *Rivista d'arte* 6 (1909): 185–227.

Chrysoloras, Manuel. *Comparison of Old and New Rome*. In *PG*, 156, cols. 23–54.

———. *Comparison of Old and New Rome*. In F. Grabler, *Europa im XV Jahrhundert von Byzantinern Gesehen*. Graz, 1954.

Ciriaco D'Ancona. *Panegyric to the City of Ancona*. In G. Praga, "Ciriaco de'Pizzicolli e Marino de'Resti." *Archivio Storico per la Dalmazia* 13 (1932–1933): 262–80.

Colonna, Franceso. *Francesco Colonna. Hypnerotomachia Poliphili*. Edited by G. Pozzi and L. Ciapponi. 2 vols. Padua, 1964. 1980.

Corna da Soncino, Francesco. *Fioretto de le antiche croniche de Verona e de tutti i soi confini e de le reliquie che se trovano dentro in ditta citade*. Edited by G. P. Marchi. Verona, 1973. Reprint. Verona, 1980.

Cortesi, Paolo. *De Hominibus doctis dialogus*. Edited and translated by M. T. Grazios. Rome, 1973.

Crivelli, Lodrisio. *In Laudem Pientie civitatis*. In L. F. Smith, "Lodrisio Crivelli of Milan and Aeneas Silvius." *Studies in the Renaissance* 9 (1962): 31–63.

Cusanus, Nicholaus. *Idiota de Mente. The Layman: About Mind*. Translated by C. L. Miller. New York, 1979.

Cydones, Demetrius. *Romaiois symbouleutikos*. In *PG*, 153, cols. 961–1008.

Decembrio, Pier Candido. *De Laudibus Mediolanensis urbis panegyricus*. Edited by A. Butti, F. Fossati, and G. Petraglione. *RIS* 20, pt. 1. 1952.

———. "Il '*De Laudibus Mediolanensium urbis panegyricus*' di P. C. Decembrio." Edited by G. Petraglione. *Archivio Storico Lombardo* 8 (1907): 5–45.

———. *Panegyric on Florence*. In G. Shepherd and T. Tonelli, *Vita di Poggio Bracciolini*. Florence, 1825. Appendix 19.

Dei, Benedetto.
   Romby, G. *Descrizioni e rappresentazioni della città di Firenze nel XV secolo con la trascrizione inedita dei manoscritti di Benedetto Dei e un indice ragionata dei manoscritti utili per la storia di Firenze*. Florence, 1976.

Dondi dall'Orologio, Giovanni. *Iter Romanum*. In *VZ*, pp. 68–73.
   Gilbert, N. "A Letter of Giovanni Dondi dall'Orologio to Fra Guglielmo Centueri: A Fourteenth-Century Episode in the Quarrel of the Ancients and Moderns." *Viator* 8 (1977): 299–346.

Eugenikos, John.
　　Lampsides, O. "Iohannos Eugenikos. Ekphrasis Trapezountos." *Archeion Pontos* 20 (1955): 3–39.
Filarete.
　　Antonio Averlino detto il Filarete. Trattato di architettura. Edited by A. M. Finoli and L. Grassi. Milan, 1972.
Filelfo. *Francisci Philelphi. Epistolae breviores et elegantiores.* Venice, 1572.
Francesco di Giorgio. *Francesco di Giorgio Martini. Trattati di architettura ingegneria e arte militare.* Edited by C. Maltese. Milan, 1967.
Frezzi, Federico. *Il Quadriregio.* Edited by E. Filippini. Bari, 1914.
Ghiberti, Lorenzo.
　　Fengler, C. K. "Lorenzo Ghiberti's Second Commentary: The Translation and Interpretation of a Fundamental Renaissance Treatise on Art." Ph. D. diss., University of Wisconsin, 1974.
Giovanni da Prato. *Giovanni Gherardi da Prato. Il Paradiso degli Alberti.* Edited by A. Lanza. Rome, 1975.
———. *Opere complete.* Edited by F. Garilli. Palermo, 1976.
Guarino Veronese. *Epistolario di Guarino Veronese.* Edited by R. Sabbadini. 4 vols. Venice, 1916.
Guidi, Jacopo di Albizzotto.
　　Rossi, V. "Jacopo di Albizzotto Guidi e il suo inedito poema su Venezia." *Nuova archivio Veneto* 5 (1893): 397–451. (Reprinted in V. Rossi. *Scritti di critica letteraria. II. Studi sul Petrarca e sul Rinascimento.* Florence, 1930.)
Landino, Cristoforo. *Carmina omnia.* Edited by A. Perosa. Florence, 1939.
———. *Christoforo Landino. Scritti critici e teorici.* Edited by R. Cardini. Rome, 1974.
———. *Disputationes Camaldulenses.* In *Pros. Lat.*, pp. 716–91.
Leto, Pomponio. *Excerpta.* In *VZ*, pp. 423–36.
Maffei, Timoteo. *In Magnificentiae Cosmi Medicei Florentini detractores.* Edited by G. Lami. In *Deliciae eruditorum.* Vol. 12. Florence, 1742.
Manetti, Antonio. *The Life of Brunelleschi by Antonio di Tuccio Manetti.* Edited by H. Saalman. University Park, Pa., 1970.
Manetti, Giannozzo. *Ianotii Manetti. De Dignitate et excellentia hominis.* Edited by E. Leonard. Padua, 1975. (Also in *Pros. Lat.*, pp. 422–86.)
———. *Vita Nicolai V summi pontificis.* RIS 3, pt. 2. 1734.
———. *De Secularibus et pontificalibus pompis.* Edited by E. Battisti. In *Umanesimo e esoterismo: Atti del V convegno internazionale di studi umanistici*, edited by E. Castelli. Padua, 1960.
Wittschier, H. W. *Giannozzo Manetti. Das Corpus der Orationes.* Cologne, 1968.
Manuel II Palaiologus. *The Letters of Manuel II Palaeologus.* Edited by G. Dennis. Washington, D.C., 1977.
Naldi, Naldo. *Vita Jannotii Manetti.* RIS 20, cols. 527–606.
Palmieri, Matteo. *Città di vita.* Edited by M. Rooke. Smith College Studies in Modern Language, nos. 8 (1926–1927) and 9 (1927–1928).
———. *Della Vita civile.* Edited by G. Belloni. Florence, 1982.
Palmieri, Mattia. *De Temporibus suis.* RIS 26, pt. 1. 1748.
Patrizi, Francesco. *De Istitutione reipublicae.* Paris, 1578.
Piccolomini, Aeneas Sylvius. *De Liberorum educatione.* Edited and translated by J. S. Nelson. Washington, D.C., 1940.

———. *Description of Basel.* In B. Widmer. "Enea Silvio's Lob der Stadt Basel und seine Vorlagen." *Basler Zeitschrift* 58–59 (1959): 111–38.

———. *Aeneae Sylvii Piccolominei senensis. Opera Omnia.* Basel, 1555.

———. *The Commentaries of Pius II.* Translated by F. A. Gragg. Smith College Studies in History, nos. 22 (1936–1937), 25 (1939–1940), 30 (1947), 35 (1951), and 43 (1957).

Wolkan, R. *Der Briefwechsel des Eneas Silvius Piccolomini.* Vienna, 1918.

Pizolpasso, Francesco.

    Foffano, T. "La Costruzione di Castiglione Olona in un opusculo inedito di Francesco Pizolpasso." *IMU* 3 (1960): 153–87.

Platina. *Panegyricus Bessarionis.* In *PG*, 161, cols. iii–xvi.

———. *Vita di Pio II. RIS* 3, pt. 2. 1964.

———. *Liber de vita Christi ac omnium pontificum qui hactenus ducenti fuere et XX. RIS* 3, pt. 1.

Pontano, Giovanni. *I Dialoghi.* Edited by C. Previtera. Florence, 1943.

———. *I Trattati delle virtù sociali.* Translated by F. Tateo. Rome, 1965.

Quartigionis, Filippo de Diversis de. *Situs aedificiorum et laudabilium consuetudinum inclytae civitatis Ragusiae.* Edited by V. Brunelli. Zara, 1880.

Quirini, Lauro. "Il *De Republica* di Lauro Quirini." Edited by C. Seno and G. Ravegni. In *Lauro Quirini umanista. Studi e testi a cura di K. Krautter.* Florence, 1977.

Rinuccini, Alamanno. *Lettere ed orazioni.* Edited by V. Giustiniani. Florence, 1953.

Rucellai, Bernardo. *De Urbe Roma.* In *VZ*, pp. 443–56.

Rucellai, Giovanni. *Giovanni Rucellai ed il suo Zibaldone.* Introduction by N. Rubinstein. Vol. 1: *Lo Zibaldone.* Edited by A. Perosa. London, 1981.

Salutati, Coluccio. *Epistolario.* Edited by F. Novati. 4 vols. Rome, 1896.

———. *De Seculo et religione.* Edited by B. Ullman. Florence, 1957.

Savonarola, Michele. *Libellus de magnificis ornamentis regie civitatis Padue. RIS* 24, pt. 15. 1902.

Signorili, Niccolò. *De Antiquitate et mirabilibus urbis Romae.* In *VZ*, pp. 162–67.

Sozomenus of Pistoia, *Sozomeni Pistoriensis presbyteri. Chronicon universale. RIS* 16, pt. 1.

Tignosi, Niccolò.

    Sensi, M. "Niccolò Tignosi da Foligno: L'Opera e il pensiero." *Annali della Facoltà di Lettere e Filosofia* (Università degli Studi di Perigia) 9 (1973): 359–495.

Tortelli, Giovanni. *Ioannis Tortelii Aretini Ortographia.* Venice, 1501.

Traversari, Ambrogio. *Ambrosii Traversarii generalis Camaldulensium Latinae epistolae.* Edited by L. Mehus. 2 vols. Florence, 1759. Reprint. Florence, 1968.

———. *Hodoeporicon.* Edited and translated by V. Tamburini. Florence, 1985.

———. *Hodoeporicon.* Edited by A. Dini Traversari. Florence, 1912.

Valla, Lorenzo. *De Vero falsoque bono.* Edited by M. Lorch. Bari, 1970.

———. *Lorenzo Valla. On Pleasure. De Voluptate.* Translated by A. K. Hieatt and M. Lorch. New York, 1977.

———. *Laurentii Valle. Epistole.* Edited by O. Besomi and M. Regoliosi. Padua, 1984.

———. *Opera omnia.* Edited by E. Garin. Turin, 1962.

Vergerio, Pier Paolo. *Epistolario.* Edited by L. Smith. Rome, 1934.

    Robey, D., and J. Law. "The Venetian Myth and the *De Republica veneta* of Pier Paolo Vergerio." *Rinascimento* 15 (1975): 3–59.

Vespasiano da Bisticci. *Le Vite.* Edited by A. Greco. 2 vols. Florence, 1970.

Villani, Filippo. *Liber de civitatis Florentiae famosis civibus.* In *Cronica di Matteo Villani*, edited by G. Mazzuchelli. Florence, 1826.

Volpi, G., ed. *Ricordi di Firenze dell'anno 1459 di autore anonimo. RIS* 27, p. 1, cols. 719–5.
Zeno, Iacopo. *Praise of Cyriacus of Ancona.* In L. Bertalot and A. Campana. "Gli Scritti di Iacopo Zeno e il suo elogio di Ciriaco d'Ancona." *La Bibliofilia* 41 (1939): 356–76.

Secondary Sources

Ackerman, J. "The Certosa of Pavia and the Renaissance in Milan." *Marsyas* 5 (1947–1949): 23–37.
———. "'Ars sine scientia nihil est.' Gothic Theory of Architecture at the Cathedral of Milan." *AB* 31 (1949): 84–111.
———. "Architectural Practice in the Italian Renaissance." *JSAH* 13 (1954): 3–11.
"Ancients and Moderns: A Symposium." *JHI* 48 (1987).
Argan, G. C. *The Renaissance City.* New York, 1969.
———. "Il Trattato *De Re aedificatoria.*" In *Convegno,* pp. 43–54.
Assunto, R. *La Critica d'arte nel pensiero medievale.* Milan, 1961.
Atti del V Convegno internazionale di studi sul Rinascimento. Il Mondo antico nel Rinascimento. Florence, 1958.
Avesani, R. "In Laudem civitatis Veronae." *Studi storici Veronesi* 26–27 (1976–1977): 183–97.
Badaloni, N. "La Interpretazione delle arti nel pensiero di L. B. Alberti." *Rinascimento* 24 (1963): 59–113.
Baldwin, E. *Renaissance Literary Theory and Practice: Classicism in the Rhetoric and Poetic of Italy, France and England, 1400–1600.* New York, 1939.
Baron, H. *Leonardo Bruni Aretino. Humanistisch-Philosophische Schriften.* Berlin, 1928.
———. "Franciscan Poverty and Civic Wealth as Factors in the Rise of Humanistic Thought." *Speculum* 13 (1938): 1–37.
———. *The Crisis of the Early Italian Renaissance.* 2 vols. Princeton, N.J., 1955.
———. "The *Querelle* of Ancients and Moderns as a Problem for Renaissance Scholarship." *JHI* 20 (1959): 3–22.
———. *From Petrarch to Leonardo Bruni: Studies in Humanistic and Political Literature.* Chicago, 1968.
———. *In Search of Florentine Civic Humanism.* 2 vols. Princeton, N.J., 1988.
Batkin, L. *Die Historische Gesamtheit der italienischen Renaissance.* Dresden, 1979.
Battisti, E. "Simbolo e classicismo." In *Umanesimo e simbolismo. Atti del IV convegno internazionale di studi umanistici,* edited by E. Castelli, pp. 215–34. Padua, 1958.
———. *Rinascimento e Barocco.* Turin, 1960.
———. *Filippo Brunelleschi: The Complete Work.* New York, 1981.
Bauer, H. *Kunst und Utopie. Studien über das Kunst und Staatsdenken in der Renaissance.* Berlin, 1965.
Baxandall, M. "Bartholomaeus Facius on Painting." *JWCI* 27 (1964): 90–107.
———. "Guarino, Pisanello and Manuel Chrysoloras." *JWCI* 28 (1965): 183–204.
———. *Giotto and the Orators: Humanist Observers of Painting in Italy and the Discovery of Pictorial Composition, 1350–1450.* Oxford, 1971.
———. *Painting and Experience in Fifteenth Century Italy.* London, 1972.
———. "Alberti and Cristoforo Landino: The Practical Criticism of Painting." In *Convegno,* pp. 143–54.
Becatti, G. "Leon Battista Alberti e l'antico." In *Convegno,* pp. 55–72.

Bek, L. *Towards Paradise on Earth: Modern Space Conception in Architecture: A Creation of Renaissance Humanism.* Analecta Romana Instituti Danici, no. 9, supplement. Odense, 1980.

Benevolo, L. *The Architecture of the Renaissance.* 2 vols. London, 1978.

Bering, K. *Baupropaganda und Bildprogrammatik der Frührenaissance in Florenz-Rom-Pienza.* Frankfurt am Main, 1984.

Bertalot, L. "Cincius Romanus und seine Briefe." *QFIAB* 21 (1929–1930): 209–51.

———. "Zur Bibliographie der Uebersetzungen des Leonardus Brunus Aretinus." *QFIAB* 27 (1937): 178–95.

Bertalot, A., and A. Campana. "Gli Scritti di Iacopo Zeno e il suo elogio di Ciriaco d'Ancona." *La Bibliofilia* 41 (1939): 356–76.

Berti, E. "Alla Scuola di Manuele Crisolora. Lettura e commento di Luciano." *Rinascimento* 27 (1987): 3–74.

Besomi, O. "Dai 'Gesta Ferdinandi regis Aragonum' del Valla al 'De Ortographia' del Tortelli." *IMU* 9 (1966): 75–121.

Bettini, M. "Tra Plinio e Sant'Agostino: Francesco Petrarca sulle arti figurativi." In *Memoria*, 1: 221–67.

Bialostocki, J. "The Power of Beauty. A Utopian Idea of Leone Battista Alberti." In *Studien zur toskanischen Kunst. Festschrift für Ludwig Heinrich Heydenreich* (Munich, 1963), pp. 13–19.

———. "The Renaissance Concept of Nature and Antiquity." In *Acts of the XX International Congress of the History of Art.* Vol. 2: *The Renaissance and Mannerism*, edited by M. Meiss. pp. 19–30. Princeton, N.J., 1963.

Black, R. "Ancients and Moderns in the Renaissance: Rhetoric and History in Accolti's 'Dialogue on the Preeminence of Men of his Own Time.'" *JHI* 43 (1982): 3–32.

———. *Benedetto Accolti and the Florentine Renaissance.* Cambridge, 1985.

Bolgar, R. R. *The Classical Heritage and Its Beneficiaries.* Cambridge, 1954.

———, ed. *Classical Influences on European Culture, A.D. 500–1500.* Cambridge, 1971.

Borsi, F. *Leon Battista Alberti.* Milan, 1975.

Branca, V. *Poliziano e l'umanesimo della parola.* Turin, 1983.

———, ed. *Umanesimo Europeo e umanesimo Veneziano.* Venice, 1963.

Brezzi, P., and M. de Panizza Lorch, eds. *Umanesimo a Roma nel Quattrocento. Atti del convegno.* Rome, 1984.

Brown, A. M. "The Humanist Portrait of Cosimo de'Medici, Pater Patriae." *JWCI* 24 (1961): 186–221.

Brucker, G. "Florence and Its University, 1348–1434." In Rabb and Seigel, pp. 220–36.

Brunetti, F. "Le Tipologie architettoniche nel trattato Albertiano." *Studi e documenti di architettura* 1 (1972): 263–92.

Bruschi, A. "Note sulla formazione architettonica dell'Alberti." *Palladio* 25 (1978): 4–44.

Buck, A. "Der Primat der Stadt in der italienischen Geschichte des Mittelalters und der Renaissance." In Buck and Guthmuller, pp. 7–20.

Buck, A., and B. Guthmuller, eds. *La Città italiana del Rinascimento fra utopia e realtà.* Venice, 1984.

Buddensieg, T. "Criticism and Praise of the Pantheon in the Middle Ages and Renaissance." In Bolgar, pp. 259–67.

———. "Gregory the Great, the Destroyer of Pagan Idols." *JWCI* 28 (1965): 44–65.

Bundy, M. *The Theory of Imagination in Classical and Medieval Thought.* Urbana, Ill., 1927.

Burgess, T. *Epideictic Literature.* Chicago, 1902.

Burke, P. *The Italian Renaissance: Culture and Society in Italy.* 1972. Reprint. Oxford, 1986.
———. "The Renaissance Dialogue." *Renaissance Studies* 3 (1989): 1–12.
Burns, H. "Quattrocento Architecture and the Antique: Some Problems." In Bolgar, pp. 269–87.
Cagni, M., "Agnolo Manetti and Vespasiano da Bisticci." *IMU* 14 (1971): 293–312.
Calderini, E. "Intorno alla biblioteca e alla cultura greca del Filelfo." *Studi italiani di filologia classica* 20 (1913): 204–425.
Cammelli, G. *I Dotti bizantini e le origini dell 'Umanesimo.* 2 vols. Florence, 1941.
Camporeale, S. I. *Lorenzo Valla. Umanesimo e teologia.* Florence, 1972.
———. *Lorenzo Valla tra medioevo e Rinascimento. Encomion S. Thomae—1457.* Pistoia, 1977.
———. "Lorenzo Valla e il 'De Falso credita donitione.' Retorica, libertà ed ecclesiologia nel '400." *Memorie Domenicane* 19 (1988): 191–293.
Cancro, C. *Filosofia ed architettura in Leon Battista Alberti.* Naples, 1978.
Canfield, G. B. "The Florentine Humanists' Concept of Architecture in the 1430's and Filippo Brunelleschi." In *Scritti di storia dell'arte in onore di Federico Zeri.* 2 vols., 1:112–21. Venice, 1984.
Cantimore, D. "Rhetoric and Politics in Italian Humanism." *JWCI* 1 (1937): 83–102.
Cantino Wataghin, G. "Roma sotterranea. Appunti sulle origini dell'archeologia cristiana." *Ricerche di storia dell'arte* 10 (1980): 5–14.
———. "Archeologia e 'archeologie.' Il Rapporto con l'antico fra mito, arte e ricerca." In *Memoria*, 1:171–217.
Cantone, G. *La Città di marmo da Alberti a Serlio.* Rome, 1978.
Cassirer, E. *The Individual and the Cosmos in Renaissance Philosophy.* Translated by M. Domandi. 1927. Reprint. Philadelphia, 1983.
Chastel, A. "Un Episode de la symbolique urbaine au XVe siècle: Florence et Rome Cités de Dieu." In *Urbanisme et architecture. Etudes écrites et publiées en l'honneur de Pierre Lavedan*, pp. 75–79. Paris, 1954.
———. *Marsile Ficin et l'art.* Geneva, 1975.
Chastel, A. and J. Guillaume, eds. *Les Traités d'architecture à la Renaissance.* Paris, 1988.
Choay, F. *La Règle et le modèle. Sur la théorie de l'architecture et de l'urbanisme.* Paris, 1980.
Ciapponi, L. "Il 'De Architectura' di Vitruvio nel primo Umanesimo." *IMU* 3 (1960): 59–99.
Clagett, M. *The Science of Mechanics in the Middle Ages.* Madison, Wis., 1959.
———. *Archimedes in the Middle Ages.* 5 vols. Canton, Mass., 1984.
Clarke, M. L. *Rhetoric at Rome.* London, 1953.
Classen, C. J. *Die Stadt im Spiegel der Descriptiones und Laudes urbium in der antiken und mittelalterlichen Literatur bis zum Ende des zwölften Jahrhunderts.* Hildesheim, 1986.
Cochrane, E. "Science and Humanism in the Italian Renaissance." *American Historical Review* 81 (1976): 1039–57.
———. *Historians and Historiography in the Italian Renaissance.* Chicago, 1981.
Corbellini, A., "Appunti sull 'umanesimo in Lombardia." *Bollettino della Società Pavese di Storia Patria*, 15–17 (1915): 327–62, 109–66, 5–51.
———. "Note di vita cittadina e universitario pavese nel Quattrocento." *Bollettino della Società Pavese di Storia Patria* 30 (1930): 1–289.
Costa, G. *La Leggenda dei secoli d'oro nella letteratura italiana.* Bari, 1972.
Coulton, G. C. *Art and the Reformation.* Oxford, 1928.
Curtius, E. R. *European Literature and the Latin Middle Ages.* Translated by W. Trask. Princeton, N.J., 1953.
Dagron, G. "Manuel Chrysoloras: Constantinople ou Rome." *Byzantinische Forschungen* 12 (1987): 281–88.

D'Amico, J. *Renaissance Humanism in Papal Rome.* Baltimore, 1983.
———"The Progress of Renaissance Latin Prose: The Case of Apuleianism." *Renaissance Quarterly* 37 (1984): 349–92.
Da Pisanello alla nascita dei Musie Capitalini. L'Antico a Roma alla vigilia del Rinascimento. Rome, 1988.
Davies, M. "An Emperor Without Clothes? Niccolò Niccoli under Attack." *IMU* 30 (1987): 95–148.
Davis, C. "Topographical and Historical Propaganda in Early Florentine Chronicles and in Villani." *Medioevo e Rinascimento* 2 (1988): 33–52.
De Beer, E. S. "Gothic: Origin and Diffusion of the Term; the Idea of Style in Architecture." *JWCI* 12 (1948): 143–62.
De Bruyne, E. *Etudes d'esthétique médiévale.* 3 vols. Bruges, 1946.
Del Bravo, C. "Il Brunelleschi e la speranza." *Artibus et Historiae* 3 (1981): 57–72.
Diller, A. "The Greek Codices of Palla Strozzi and Guarino Veronese." *JWCI* 24 (1961): 313–21.
Dini Traversari, A. *Ambrogio Traversari e i suoi tempi.* Florence, 1912.
Di Tommaso, A. "Nature and the Aesthetic Social Theory and Practice of L. B. Alberti." *Mediaevalia et Humanistica* 3 (1972): 31–49.
Ditten, H. "βάρβαροι, Ἕλληνες and Ῥωμαῖοι bei den letzten byzantinischen Geschichtsschreibern." In *Actes du XIIe Congrès International d'Etudes Byzantines,* 2:273–99. Belgrade, 1964.
D'Onofrio, C. D. *Visitiamo Roma nel Quattrocento. La Città degli Umanisti.* Rome, 1989.
Doren, A. "Wunschräume und Wunschzeiten." *Vorträge der Bibliothek Warburg* (1924–1925): 158–205.
Drijepondt, H. L. F. *Die Antike Theorie der "varietas."* Hildesheim, 1979.
Eco, U. "Sviluppo dell'estetica medievale." In *Momenti e problemi di storia dell'estetica.* Milan, 1959.
Eden, W. A. "Studies in Urban Theory: The 'De Re aedificatoria' of L. B. Alberti." *Town Planning Review* 19 (1943): 10–28.
Essen, C. van, "I *Commentaria* di Ciriaco d'Ancona." In *Il Mondo antico nel Rinascimento. Atti del V Convegno Internazionale di Studi sul Rinascimento,* pp. 191–194. Florence, 1958.
Falco, G. *La Polemica sul medio evo.* 1933. Reprint. Naples, 1974.
Fasoli, G. "Nascita di un mito." In *Studi storici in onore di Gioacchino Volpe.* 2 vols., 1:445–79. Florence, 1958.
———. "La Coscienza civica nelle'Laudes civitatem'." In *La Coscienza cittadina nei comuni italiani del Duecento,* pp. 9–44. Convegni del Centro di Studi sulla Spiritualità medievale, no. 11. Todi, 1972.
Federici Vescovini, G. *"Arti" e filosofia nel secolo XIV: Studi sulla tradizione aristotelica e i "moderni."* Florence, 1983.
Fenster, E. *Laudes Constantinopolitanae.* Munich, 1968.
Ferguson, W. K. "Humanist Views of the Renaissance." *American Historical Review* 45 (1939): 1–28.
———. *The Renaissance in Historical Thought.* Cambridge, Mass., 1948.
Feuer-Toth, R. "The 'apertionum ornamenta' of Alberti and the Architecture of Brunelleschi." *Acta historiae artium accademiae scientiarum Hungaricae* 24 (1978): 147–52.
Field, A. *The Origins of the Platonic Academy of Florence.* Princeton, N.J., 1988.
Filippo Brunelleschi. La Sua opera e il sue tempo. 2 vols. Atti del convegno internazionale di studi per il sesto centenario della nascita. Florence, 1980.
Finelli, L. *L'Umanesimo giovane. Bernardo Rossellino a Roma e a Pienza.* Rome, 1984.
Finelli, L., and S. Rossi. *Pienza tra ideologia e realtà.* Bari, 1979.
Fiocco, G., "Palla Strozzi e l'umanesimo veneto." In Branca, *Umanesimo,* pp. 349–58.

———. "La Biblioteca di Palla Strozzi." In *Studi di bibliografia e di storia in onore di Tammaro de Marinis*, pp. 289–310. Verona, 1964.
Firpo, L. *La Città ideale del Rinascimento*. Turin, 1974.
Flores, E. *Le Scoperte di Poggio e il testo di Lucrezio*. Naples, 1980.
Fontana, P. "Osservazioni intorno ai rapporti di Vitruvio colla teoria dell'architettura del Rinascimento." In *Miscellanea di storia dell'arte in onore di Igino Benvenuto Supino*, pp. 305–22. Florence, 1933.
Fontana, V. *Artisti e committenti nella Roma del Quattrocento. Leon Battista Alberti e la sua opera mediatrice*. Rome, 1973.
Fornara, C. W. *The Nature of History in Ancient Greece and Rome*. Berkeley, 1988.
Forster, K. W. "Discussion: The Palazzo Rucellai and Questions of Typology in the Development of Renaissance Buildings." *AB* 68 (1976): 109–13.
Franchetti Pardo, V. *Storia dell'urbanistica dal Trecento al Quattrocento*. Bari, 1982.
Fraser Jenkins, A. "Cosimo de'Medici's Patronage of Architecture and the Theory of Magnificence." *JWCI* 33 (1970): 162–70.
Freeman, J. "The Roof Was Fretted Gold." *Comparative Literature* 27 (1975): 254–66.
Frommel, C. L. "Papal Policy: The Planning of Rome during the Renaissance." In *Art and History: Images and Their Meaning*, edited by R. Rotberg and T. Rabb, pp. 39–65. Cambridge, 1988.
Frugoni, C. "L'Antichità dai 'Miribilia' alla propaganda politica." In *Memoria*, 1:3–72.
Fubini, R., and A. Gallorini. "L'Autobiografia di Leon Battista Alberti." *Rinascimento* 12 (1972): 21–78.
Fryde, E. B. *Humanism and Renaissance Historiography*. London, 1983.
Gabel, L. "The First Revival of Rome 1420–1484." In *The Renaissance Reconsidered: A Symposium*, edited by N. Hoyt, pp. 13–25. Northampton, Mass., 1964.
Gadol, J. *Leon Battista Alberti: Universal Man of the Early Renaissance*. Chicago, 1969.
Gaeta, F. "Sull'idea di Roma nell'Umanesimo e nel Rinascimento." *Studi Romani* 25 (1977): 169–86.
Gallimberti, N. "L'Urbanistica della 'rinascenza.'" In *Atti del primo congresso nazionale di storia di architettura*, pp. 255–69. Florence, 1938.
Garfagnini, G. C., ed. *Marsilio Ficino e il ritorno di Platone. Studi e documenti*. 2 vols. Florence, 1986.
Garin, E. "La 'Dignitas hominis' e la letteratura patristica." *Rinascita* 1 (1938): 102–46.
———. *Umanesimo Italiano. Filosofia e vita civile nel Rinascimento*. 1947. Reprint. Bari, 1973.
———. *L'Educazione Umanistica in Italia*. 1949. Reprint. Bari, 1964.
———. *Medioevo e Rinascimento*. 1954. Reprint. Bari, 1980.
———. "Cultura filosofica toscana e veneta nel Quattrocento." In Branca, *Umanesimo*, pp. 11–30.
———. *La Cultura filosofica del rinascimento italiano*. Florence, 1961.
———. *Science and the Civic Life in the Italian Renaissance*. Translated by P. Munz. New York, 1965.
———. *Prosatori Latini del Quattrocento*. Milan, and Naples, 1965.
———. *Storia della filosofia italiana*. 4 vols. Turin, 1966.
———. "Il Pensiero di L. B. Alberti nella cultura del Rinascimento." In *Convegno*, pp. 21–44.
———. "Studi su Leon Battista Alberti." In *Rinascite e rivoluzioni: Movimenti culturali dal XIV al XVII secolo*, pp. 133–96. Rome, 1975.
———. *Umanisti, artisti scienziati*. Rome, 1989.
Geanakoplos, D. *Greek Scholars in Venice*. Cambridge, Mass., 1962.

———. *Byzantine East and Latin West: Two Worlds of Christendom in the Middle Ages and Renaissance.* New York, 1966.
———. "La Colonia greca di Venezia e il suo significato per il Rinascimento." In Pertusi, *Venezia*, pp. 183–203.
———. "The Discourse of Demetrius Chalcondyles on the Inauguration of Greek Studies at the University of Padua in 1463." *Studies in the Renaissance* 21 (1974): 118–44.
———. "Italian Renaissance Thought and Learning and the Role of the Byzantine Emigrés Scholars in Florence, Rome and Venice: A Reassessment." In *Miscellanea Agostino Pertusi (Rivista di studi Bizantini e Slavi* 3 [1983]), pp. 129–57. Bologna, 1984.
Gengaro, M. L. "L'Architettura romana nella interpretazione teorica di Leon Battista Alberti." *Bollettino del Reale Istituto di Archeologia e Storia dell'Arte* 9 (1940): 37–42.
Gerland, E. *Das Studium der byzantinischen Geschichte vom Humanismus bis zur Jetztzeit.* Athens, 1934.
Germann, G. *Einführung in die Geschichte der Architekturtheorie.* Darmstadt, 1980.
Gernentz, G. *Laudes Romae.* Rostock, 1918.
Gilbert, C. "Antique Frameworks for Renaissance Art Theory: Alberti and Pino." *Marsyas* 3 (1943–1945): 87–106.
———. "The Archbishop on the Painters of Florence, 1450." *AB* 41 (1959): 75–87.
———. "The Earliest Guide to Florentine Architecture, 1423." *Mitteilungen des Kunsthistorischen Institutes in Florenz* 40 (1969): 33–46.
———. "On Castagno's Nine Famous Men and Women: Sword and Book as the Basis for Public Service." In *Life and Death in Fifteenth-Century Florence*, edited by M. Tetel, R. Witt, and R. Goffen, pp. 176–92. Durham, N.C., 1989.
Gill, J. *Personalities of the Council of Florence.* Oxford, 1964.
Gille, B. *Les Ingénieurs de la Renaissance.* Paris, 1964.
Giovannoni, G. *Saggi sulla architettura del Rinascimento.* Milan, 1935.
Gnocchi, L. "Le Preferenze artistiche di Piero di Cosimo de' Medici." *Artibus et Historiae* 18 (1988): 41–78.
Goebel, G. *Poeta Faber. Erdichtete Architektur in der italienischen, spanischen und französischen Literatur der Renaissance und des Barock.* Heidelberg, 1971.
Goldschmidt, E. P. "The First Edition of Lucian of Samosota." *JWCI* 14 (1951): 7–20.
Goldbrunner, H. M. "Laudatio Urbis: Zu neueren Untersuchungen über das humanistische Städtelob." *QFIAB* 63 (1983): 313–28.
Gombrich, E. H. "Botticelli's Mythologies. A Study in the Neoplatonic Symbolism of His Circle." *JWCI* 8 (1945): 7–60.
———. "Icones symbolicae: The Visual Image in Neoplatonic Thought." *JWCI* 12 (1948): 163–92.
———. "Apollonio di Giovanni. A Florentine Cassone Workshop Seen Through the Eyes of a Humanist Poet." *JWCI* 18 (1955): 6–34. (Reprinted in *Norm and Form: Studies in the Art of the Renaissance.* London, 1966.)
———. "A Classical Topos in the Introduction to Alberti's *Della Pittura.*" *JWCI* 20 (1957): 173.
———. "Alberto Avogadro's Description of the Badia of Fiesole and the Villa of Careggi." *IMU* 5 (1962): 217–29.
———. "From the Revival of Letters to the Reform of the Arts. Niccolò Niccoli and Filippo Brunelleschi." In *Essays in the History of Art Presented to Rudolf Wittkower*, edited by D. Frazer, H. Hibbard, and J. Lewine, pp. 71–82. London, 1967. (Reprinted in *The Heritage of Apelles: Studies in the Art of the Renaissance.* Ithaca, N.Y., 1976.)

Gosebruch, M. " 'Varietà' bei Leon Battista Alberti und der Wissenschaftliche Renaissancebegriff." *Zeitschrift für Kunstgeschichte* 20 (1957): 229–38.
Grafton, A. "Renaissance Readers and Ancient Texts: Comments on Some Commentaries." *Renaissance Quarterly* 38(1985): 615–49.
Grafton, A., and L. Jardine. "Humanism and the School of Guarino: A Problem of Evaluation." *Past and Present* 96 (1982): 51–80.
———. *From Humanism to the Humanities.* London, 1986.
Grassi, L. "Una Utopia sociologica e pedagogica in A. Averlino detto il Filarete." In *Arte in Europa. Scritti di storia dell'arte in onore di W. Arslan*, pp. 377–403. Milan, 1965.
Gravelle, S. S. "Humanist Attitudes to Convention and Innovation in the Fifteenth Century." *Journal of Medieval and Renaissance Studies* 11 (1981): 193–209.
Gray, H. "Renaissance Humanism: The Pursuit of Eloquence." *JHI* 24 (1963): 497–514.
———. "Valla's *Encomion of St. Thomas Aquinas* and the Humanist Conception of Christian Antiquity." In *Essays in History and Literature Presented to Stanley Pargellis*, edited by H. Bluhm, pp. 37–51. Chicago, 1965.
Grayson, C. "The Composition of L. B. Alberti's 'Decem libri de re aedificatoria.' " *Münchner Jarhbuch der Bildenden Kunst* 11 (1960): 152–61.
———. "Lorenzo, Machiavelli and the Italian Language." In *Italian Renaissance Studies*, edited by E. F. Jacob, pp. 410–32. London, 1960.
———. *Leon Battista Alberti: La Prima grammatica della lingua volgare.* Bologna, 1964.
———. "The Text of Alberti's 'De Pictura.' " *Italian Studies* 23 (1968): 71–92.
———. "Il Prosatore Latino e volgare." In *Convegno*, pp. 273–86.
Green, L. *Chronicle into History: An Essay on the Interpretation of History in Florentine Fourteenth-Century Chronicles.* Cambridge, 1972.
Greene, M. *The Arts and the Art of Criticism.* Princeton, N.J., 1940.
Grendler, P. *Schooling in Renaissance Italy: Literacy and Learning, 1300–1600.* Baltimore, 1989.
Griffiths, G., J. Hankins, and D. Thompson. *The Humanism of Leonardo Bruni.* Binghamton, N.Y., 1987.
Grubb, J. "When Myths Lose Power: Four Decades of Venetian Historiography." *Journal of Modern History* 58 (1986): 43–94.
Guenée, B. *Histoire et culture historique dans l'Occident médiéval.* Paris, 1980.
Guidoni, E. "Trasformazioni urbanistiche e teoria della città nell'età brunelleschiana." In *Fil/Brunelleschi*, 1:65–77.
———. *La Città dal Medioevo al Rinascimento.* Rome, 1981.
Günther, H. *Das Studium der antiken Architektur in den Zeichnungen der Hoch-Renaissance.* Tübingen, 1988.
Gunnarsjaa, A. "Filarete e la gerarchia architettonica." *Acta ad Archaeologiam et Artium Historiam Pertinentia* 1 (1981): 229–45.
Hager, H., and S. Munshower, eds. *Light on the Eternal City: Recent Observations and Discoveries in Roman Art and Architecture.* Papers in Art History from the Pennsylvania State University, no. 2. University Park, Pa., 1987.
Hammerstein, N. "Die Utopie als Stadt. Zu Italienischen Architektur. Traktaten der Renaissance." In Buck and Guthmuller, pp. 37–53.
Hardison, O. B. *The Enduring Monument: A Study of the Idea of Praise in Renaissance Literary Theory and Practice.* Westport, Conn., 1962.
Hatfield, R. "Some Unknown Descriptions of the Medici Palace in 1459." *AB* 52 (1970): 232–49.
Hay, D. "Flavio Biondo and the Middle Ages." *Proceedings of the British Academy* 45 (1959): 97–125.

Heckscher, W. S. *Die Romruinen. Die Geistigen Voraussetzungen ihrer Wertung im Mittelalter und in der Renaissance.* Wurzburg, 1936.

Heninger, S. K. *Touches of Sweet Harmony: Pythagorean Cosmology and Renaissance Poetics.* San Marino, Calif. 1974.

Heydenreich, L. "Pius II als Bauherr von Pienza." *Zeitschrift für Kunstgeschichte* 6 (1937): 105–46.

Hollingsworth, M. "The Architect in Fifteenth-Century Florence." *Art History* 7 (1984): 385–410.

Holmes, G. *The Florentine Enlightenment, 1400–1450.* New York, 1969.

———. *Florence, Rome and the Origins of the Renaissance.* Oxford, 1986.

Horster, M. "Brunelleschi und Alberti in ihrer Stellung zur Römischen Antike." *Mitteilungen des Kunsthistorischen Institutes in Florenz* 17 (1973): 29–64.

Huizinga, J. *The Waning of the Middle Ages.* 1924. Reprint. Garden City, N.Y., 1954.

Hunger, H. *Die Hochsprachliche profane Literatur der Byzantiner.* Munich, 1978.

Hyde, J. K. "Medieval Descriptions of Cities." *Bulletin of the John Rylands Library* 48 (1966): 308–40.

Ianziti, G. "From Flavio Biondo to Lodrisio Crivelli: The Beginnings of Humanistic Historiography in Sforza Milan." *Rinascimento* 20 (1980): 3–39.

———. *Humanistic Historiography Under the Sforzas: Politics and Propaganda in Fifteenth-Century Milan.* Oxford, 1988.

Il Cardinale Bessarione nel V centenario della morte (1472–1972). Conferenze di studio, 1972. Rome, 1974. (*Miscellanea Francescana* 73 [1973]: 249–86.)

Jenkins, R. J. H. "The Hellenistic Origins of Byzantine Literature." *Dumbarton Oaks Papers* 17 (1963): 37–52.

Joost-Gaugier, C. L. "The Early Beginnings of the Notion of *Uomini famosi* and the 'De Viris illustribus' in Greco-Roman Literary Tradition." *Artibus et Historiae* 3 (1982): 97–115.

———. "Poggio and Visual Tradition: *Uomini famosi* in Classical Literary Description." *Artibus et Historiae* 12 (1985): 57–74.

Kajanto, I. *Poggio Bracciolini and Classicism.* Helsinki, 1987.

Kauffmann, H. "Über 'rinascere,' 'Rinascita,' und einige Stilmerkmale der Quattrocentobaukunst." In *Concordia Decennalis. Festschrift der Universität Köln zum 10 jährigen Bestehen des Deutsch-Italienischen Kulturinstituts Petrarcahaus,* pp. 123–46. Cologne, 1941.

Keller, A. G. "A Byzantine Admirer of 'Western' Progress: Cardinal Bessarion." *Cambridge Historical Journal* 11 (1955): 343–48.

Kelley, D. *Foundations of Modern Scholarship: Language, Law and History in the French Renaissance.* New York, 1970.

Kemp, M. "From *Mimesis* to *Fantasia*: The Quattrocento Vocabulary of Creation, Inspiration and Genius in the Visual Arts." *Viator* 8 (1977): 347–98.

Kennedy, G. *Classical Rhetoric and Its Christian and Secular Tradition from Ancient to Modern Times.* Chapel Hill, N.C., 1980.

———. *Greek Rhetoric Under Christian Emperors.* Princeton, N.J., 1983.

Kennedy, R. "The Contribution of Martin V to the Rebuilding of Rome 1420–31." In *The Renaissance Reconsidered. A Symposium,* pp.27–52. Northampton, Mass., 1964.

Klette, Th. *Beiträge zur Geschichte und Literatur der Italienischen Gelehrten-renaissance.* Greifswald, 1888.

Klotz, H. "L. B. Albertis 'De re aedificatoria' in Theorie und Praxis." *Zeitschrift für Kunstgeschichte* 32 (1969): 93–103.

———. *Die Frühwerke Brunelleschi's und die mittelalterliche Tradition.* Berlin, 1970.

Koenigsberger, D. *Renaissance Man and Creative Thinking: A History of Concepts of Harmony, 1400–1700.* Atlantic Highlands, N.J., 1979.

Kohl, B., and R. Witt. *The Earthly Republic: Italian Humanists on Government and Society.* Philadelphia, 1978.

Kostof, S. *The Architect: Chapters in the History of the Profession.* New York, 1977.

Krautheimer, R. "Die Anfänge der Kunstgeschichtsschreibung in Italien." *Repertorium für Kunstwissenschaft* 50 (1929): 49–63.

———. "The Tragic and Comic Scene of the Renaissance: The Baltimore and Urbino Panels." *Gazette des Beaux-Arts* 33 (1948): 327–46.

Kristeller, P. O. "The Place of Classical Humanism in Renaissance Thought." *JHI* 4 (1943): 59–63.

———. "Augustine and the Early Renaissance." *Review of Religion* 8 (1943–1944): 339–58.

———. "Humanism and Scholasticism in the Italian Renaissance." *Byzantion* 17 (1944–1945): 346–74.

———. *Studies in Renaissance Thought and Letters.* Rome, 1956.

———. "Umanesimo italiano e Bisanzio." *Lettere italiane* 16 (1964): 1–14.

———. *Renaissance Thought II: Papers on Humanism and the Arts.* New York, 1965.

———. *La Tradizione classica nel pensiero del Rinascimento.* 1965. Reprint. Florence, 1975.

———. *Medieval Aspects of Renaissance Learning.* Durham, N.C., 1974.

———. *Renaissance Thought and Its Sources.* New York, 1979.

Kruft, H. W. *Geschichte der Architekturtheorie. Von der Antike bis zur Gegenwart.* Munich, 1986.

———. *Städte in Utopia. Die Idealstadt vom 15. bis zum 18. Jahrhundert.* Munich, 1989.

Kustas, G. "The Function and Evolution of Byzantine Rhetoric." *Viator* 1 (1970): 55–73.

Labowsky, L. *Bessarion's Library and the Biblioteca Marciana.* Leiden, 1979.

La Capra, D. *History and Criticism.* Ithaca, N.Y., 1985.

Lanza, A. *Polemiche e berte letterarie nella Firenze del primo Quattrocento.* Rome, 1971.

Lavedan, P. *Histoire de l'urbanisme.* 1941. Reprint. Paris, 1959.

———. *Urbanisme et architecture. Etudes écrites et publiées en l'honneur de Pierre Lavedan.* Paris, 1954.

Lazzarini, V. "I Libri gli argenti le vesti di Giovanni Dondi dall'Orologio." *Bollettino del Museo Civico di Padova* 1 (1925): 11–36.

Legrand, E. *Bibliographie hellènique.* Paris, 1855.

Lehmann-Brockhaus, O. "Alberti's 'Descriptio urbis Romae.'" *Kunstchronik* 13 (1960): 345–47.

Le Mollé, R. "Le Mythe de la ville ideale à l'epoque de la Renaissance italienne." *Annali della Scuola Normale Superiore di Pisa* 2 (1972): 275–310.

Liborio, M. "Contributi alla storia dell' *Ubi sunt.*" *Cultura neolatina* 20 (1961): 141–209.

Logan, G. "Substance and Form in Renaissance Humanism." *Journal of Medieval and Renaissance Studies* 7 (1977): 1–34.

Long, P. "The Contribution of Architectural Writers to a 'Scientific' Outlook in the Fifteenth and Sixteenth Centuries." *Journal of Medieval and Renaissance Studies* 15 (1985): 265–98.

Lorenz, H. "Studien zum Architektonischen und Architekturtheoretischen Werk L. B. Albertis." Ph. D. Diss., University of Vienna, 1971.

Lotz, W. *Studies in Italian Renaissance Architecture.* Cambridge, Mass., 1977.

Mack, C. *Pienza: The Creation of a Renaissance City.* Ithaca, N.Y., 1987.

Magnuson, T. "The Project of Nicholas V for Rebuilding the Borgo Leonino in Rome." *AB* 36 (1954): 89–116.

———. *Studies in Roman Quattrocento Architecture.* Rome, 1958.

Maltese, C. "Colore, luce e movimento nello spazio albertiano." *Commentari* 27 (1976): 238–48.

Mancini, G. *Vita di Leon Battista Alberti*. 1882. Reprint. Florence, 1911.
———. *Giovanni Tortelli, cooperatore di Niccolò V.* Florence, 1921.
Maraschio, N. "Aspetti del bilinguismo albertiano nel 'De pictura.' " *Rinascimento* 12 (1972): 183–228.
———. "Leon Battista Alberti, *De pictura*: Bilinguismo e priorità." *Annali della Scuola Normale Superiore di Pisa* 2 (1972): 265–73.
Marconi, P., ed. *La Città come forma simbolica. Studi sulla teoria dell'architettura nel rinascimento.* Rome, 1973.
Mariani, G. B. "Brunelleschi e Roma: Un incontro fra mito e realtà." *Studi romani* 25 (1977): 187–210.
Mariani, M. "La Favola di Roma nell'ambiente fiorentino dei secoli XIII–XV." *Archivio della Società Romana di Storia Patria* 81 (1958): 1–54.
Marsh, D. "Grammar, Method and Polemic in Lorenzo Valla's Elegantiae." *Rinascimento* 19 (1979): 91–116.
———. *The Quattrocento Dialogue: Classical Tradition and Humanist Innovation.* Cambridge, Mass., 1980.
———. "Poggio and Alberti. Three Notes." *Rinascimento* 23 (1983): 189–215.
Martindale, A. *The Rise of the Artist in the Middle Ages and Early Renaissance.* New York, 1972.
Martines, L. *Power and Imagination: City-States in Renaissance Italy.* New York, 1980.
Marx, B. "Venedig—'altera Roma.' Transformationen eines Mythos." *QFIAB* 60 (1980): 325–73.
Mazzocco, A. "Petrarch, Poggio and Biondo: Humanism's Foremost Interpreters of Roman Ruins." In *Francis Petrarch, Six Centuries Later: A Symposium*, edited by A. Scaglione, pp. 354–63. Chicago, 1975.
———. "The Antiquarianism of Francesco Petrarca." *Journal of Medieval and Renaissance Studies* 7 (1977): 203–24.
———. "Leonardo Bruni's Notion of *Virtus*." In *Studies in the Italian Renaissance: Essays in Memory of Arnolfo B. Ferruolo*, edited by G. P. Biasin, A. N. Mancini, and N. Perella, pp. 105–15. Naples, 1988.
McKeon, R. "Rhetoric in the Middle Ages." *Speculum* 17 (1942): 1–32.
McLaughlin, M. L. "Histories of Literature in the Quattrocento." In *The Languages of Literature in Renaissance Italy*, edited by P. Hainsworth, V. Lucchesi, C. Roaf, D. Robey, and J. Woodhouse, pp. 63–80. Oxford, 1988.
———. "Humanist Concepts of Renaissance and Middle Ages in the Tre- and Quattrocento." *Renaissance Studies* 2 (1988): 131–42.
McManamon, J. "The Ideal Renaissance Pope: Funeral Oratory from the Papal Court." *Archivum historiae pontificae* 14 (1976): 5–70.
———. "Innovation in Early Humanist Rhetoric: The Oratory of Pier Paolo Vergerio the Elder." *Rinascimento* 22 (1982): 3–32.
Mehus, L. *Historia litteraria florentina*. 1769. Reprint. Munich, 1968.
Meltzoff, S. *Botticelli, Signorelli and Savonarola: 'Theologia poetica' and Painting from Boccaccio to Poliziano.* Florence, 1987.
Mercati, G. *Scritti d'Isidoro il Cardinale Ruteno e codici a lui appartenuti.* Rome, 1926.
Michel, A. "Architecture et rhetorique chez Alberti: La Tradition humaniste a Pienza." In *Présence de l'architecture et de l'urbanisme romains: Actes du colloque des 12, 13, decembre 1981. Hommage à Paul Dufournet*, edited by R. Chevallier, pp. 171–77. Paris, 1983.
Miglio, M. "Una Vocazione in progresso: Michele Canensi, biografo papale." *Studi medievali* 12 (1971): 463–524.
———. *Storiografia pontificia del Quattrocento.* Bologna, 1975.

———. "Et rerum facta est pulcherrima Roma." In *Aspetti culturali della società italiana nel periodo del papato Avignonese. Atti del XIX convegno di studi*, pp. 313–69. Todi, 1981.

———. "Tradizione storiografica e cultura umanistica nel 'liber de vita Christi ac omnium pontificum.'" In *Bartolomeo Sacchi, Il Platina (Piadena 1421–Roma 1481)*, edited by S.A. Campana and P. Medioli Masotti, pp. 63–89. Padua, 1986.

———. "Roma dopo Avignone. La Rinascita Politica dell'antico." In *Memoria*, 1:75–111.

Momigliano, A. "Ancient History and the Antiquarian." In *Studies in Historiography*, pp. 1–39. London, 1966.

Mommsen, T. "Petrarch's Conception of the 'Dark Ages.'" *Speculum* 17 (1942): 226–42.

———. "St. Augustine and the Christian Idea of Progress." *JHI* 12 (1951): 346–74.

Monfasani, J. *George of Trebizond: A Biography and a Study of His Rhetoric and Logic*. Leiden, 1976.

———. "The Byzantine Rhetorical Tradition and the Renaissance." In *Renaissance Eloquence*, edited by J. J. Murphy, pp. 174–87. Los Angeles, 1983.

Morisani, O. "Art Historians and Art Critics. III. Cristoforo Landino." *Burlington Magazine* 95 (1953): 267–70.

Morolli, G. *Firenze e il Classicismo: Un rapporto difficile. Saggi di storiografia dell'architettura del Rinascimento: 1977–87*. Florence, 1988.

A. Morrogh, F. Superbi Gioffredi, P. Morselli, and E. Borsook, eds. *Renaissance Studies in Honor of Craig Hugh Smyth*. Florence, 1985.

Mühlmann, H. "Recht, Rhetorik und bildende Künste: Über ihre Beziehungen im litterarischen und architektonischen Werk des Leon Battista Alberti." Ph.D. diss., University of Munich, 1968.

———. "Über den humanistischen Sinn einiger Kerngedanken der Kunsttheorie seit Alberti." *Zeitschrift für Kunstgeschichte* 33 (1970): 127–42.

———. *Aesthetische Theorie der Renaissance. Leon Battista Alberti*. Bonn, 1981.

Munoz, A. "Alcune fonti letterarie per la storia dell'arte bizantina." *N. Bulletino di archeologia cristiana* (1904): 220–32.

———. "Le ἐκφράσεις nella letteratura bizantina e i loro rapporti con l'arte figurativa." In *Recueil d'études dédiées à la memoire de N. P. Kondakov*, pp. 139–42. Prague, 1926.

Müntz, E. *Les Arts à la cour des papes*, 3 vols. Paris, 1882.

———. *Les Collections des Médicis au XVe siècle*. Paris, 1888.

Muratore, G. *La Città rinascimentale. Tipi e modelli attraverso i trattati*. Milan, 1975.

Murphy, J. J. *Rhetoric in the Middle Ages: A History of Rhetorical Theory from St. Augustine to the Renaissance*. Berkeley, 1974.

Murru, F. "Poggio Bracciolini e la riscoperta dell' Institutio oratoria' di Quintiliano." *Critica storica* 20 (1983): 621–25.

Murray, P. "Studies in Tuscan Sources for the History of Art." Ph.D. diss., University of London, 1956.

———. "Art Historians and Art Critics. IV. 'XIV uomini singhulari in Firenze.'" *Burlington Magazine* 99 (1957): 330–36.

———. *The Architecture of the Italian Renaissance*. Rev. 3rd ed. London, 1986.

———, ed. *Five Early Guides to Rome and Florence*. 1972.

Naredi-Rainer, P. von. "Musikalische Proportionen, Zahlenästhetik und Zahlensymbolik im architektonischen Werk L. B. Albertis." Ph.D. diss., University of of Graz, 1975.

———. *Architektur und Harmonie*. Cologne, 1984.

Neumann, C. "Byzantinische Kultur und Renaissance-Kultur." *Historische Zeitschrift* 91 (1903): 215–32.

———. "Ende des Mittelalters? Die Legende der Ablösung des Mittelalters durch die Renaissance." *Deutsche Vierteljahrsschrift für Literaturwissenschaft und Geistesgeschichte* 12 (1934): 124–71.

Nicol, D. *Byzantium: Its Ecclesiastical History and Relations with the Western World.* London, 1972.

Norden, E. *La Prosa d'arte antica dal VI secolo a.C. all' età della rinascenza.* 1896. Reprint. Rome, 1986.

O'Connell, R. J. *Art and the Christian Intelligence in St. Augustine.* Oxford, 1978.

Olschki, L. *Geschichte der Neusprachlichen wissenschaftlichen Literatur.* 3 vols. Heidelberg, 1919.

Omaggio ad Alberti. Studi e documenti di architettura. Florence, 1972.

O'Malley, J. "Giles of Viterbo: A Reformer's Thought on Renaissance Rome." *Renaissance Quarterly* 20 (1967): 1–11.

———. "Fulfillment of the Christian Golden Age under Pope Julius II: Text of a Discourse of Giles of Viterbo." *Traditio* 25 (1969): 265–338.

———. *Giles of Viterbo on Church and Reform: A Study in Renaissance Thought.* Leiden, 1968.

———. "Man's Dignity, God's Love and the Destiny of Rome. A Text of Giles of Viterbo." *Viator* 3 (1972): 389–416.

———. "Some Renaissance Panegyrics of Aquinas." *Renaissance Quarterly* 27 (1974): 174–92.

———. "The Vatican Library and the Schools of Athens: A Text of Battista Casali, 1508." *Journal of Medieval and Renaissance Studies* 7 (1977): 271–87.

———. *Praise and Blame in Renaissance Rome: Rhetoric, Doctrine and Reform in the Sacred Orators at the Papal Court, c. 1450–1521.* Durham, N.C., 1979.

Omont, H. "Les Manuscrits Grecs de Guarine de Vérone et la bibliothèque de Ferrare." *Revue des bibliothèques* 2 (1892): 78–81.

Onians, J. "Style and Decorum in 16th Century Italian Architecture." Ph.D. diss., University of London, 1968.

———. "Alberti and Φιλαρήτε. A Study in Their Sources." *JWCI* 34 (1971): 96–115.

———. "Filarete and the 'qualità': Architectural and Social." *Arte Lombarda* 38–39 (1973): 116–28.

———. "The Last Judgment of Renaissance Architecture." *Royal Society of Arts Journal* 128 (1980): 701–20.

———. "Brunelleschi: Humanist or Nationalist?" *Art History* 5 (1982): 259–72.

———. "On How to Listen to High Renaissance Art." *Art History* 7 (1984): 411–37.

———. *Bearers of Meaning: The Classical Orders in Antiquity, the Middle Ages and the Renaissance.* Princeton, N.J., 1988.

Onofri, L. "Sacralità, immaginazione e proposte politiche: La *Vita* di Niccolò V di Gianozzo Manetti." *Humanistica Lovaniensia* 28 (1979): 27–77.

Ovitt, G. "The Status of the Mechanical Arts in Medieval Classifications of Learning." *Viator* 14 (1983): 84–105.

Pagliara, P. N. "Vitruvio da testo a canone." In *Memoria*, 3:5–85.

Pagnotti, F. "La Vita di Niccolò V scritta da Giannozzo Manetti." *Archivio della R. Società Romana di Storia Patria* 14 (1891): 411–36

Pallas, D. "Les 'Ekphrasis' de Marc et de Jean Eugénikos: Le Dualisme culturel vers la fin de Byzance." In *Rayonnement grec: Hommages à Charles Delvoye*, edited by L. Hadermann-Misguich and G. Raepsaet, pp. 505–12. Brussels, 1982.

Panczenko, R. "La Cultura umanistica di Gentile da Fabriano." *Artibus et Historiae* 8 (1983): 27–75

Panofsky, E. "The First Page of Vasari's 'Libro': A Study on the Gothic Style in the Judgment of the Italian Renaissance." In *Meaning in the Visual Arts*, pp. 206–76. Harmondsworth, 1955.

———. " 'Renaissance'—Self-Definition or Self-Deception?" In *Rennaissance and Renascences in Western Art*, pp. 1–41. New York, 1969.

Patera, B. *La Letteratura sull'arte nell'antichità*. Palermo, 1975.

Patterson, A. M. *Hermogenes and the Renaissance: Seven Ideas of Style*. Princeton, N.J., 1970.

Pellegrin, E. "Bibliothèques d'humanistes lombards de la cour des Visconti et Sforza." *Bibliothèque d'Humanisme et de la Renaissance* 17 (1955): 218–45.

———. *La Bibliothèque des Visconti et des Sforza ducs de Milan au XV siècle*. Paris, 1955.

Pellizzari, A. *I Trattati attorno le arti figurative in Italia e nella penisola Iberica, dall'antichità classica al secolo XIII*. Naples, 1915.

Pertusi, A. *Storiografia umanistica e mondo bizantino*. Palermo, 1967.

———. *La Storiografia veneziana fino al secolo XVI. Aspetti e problemi*. Florence, 1969.

———. *La Caduta di Constantinopoli*. Verona, 1976.

———. "L'Umanesimo greco della fine del secolo XIV agli inizi del secolo XVI." In *Storia della cultura Veneta*, edited by G. Arnaldi and M. Pastore Stocchi, pp. 177–264. Venice, 1980.

———, ed. *Venezia e l'Oriente fra tardo medioevo e Rinascimento*. Venice, 1966.

Pfeiffer, R. *History of Classical Scholarship from 1300 to 1850*. Oxford, 1976.

Piccolomini, E. "Inventario della libreria medicea privata compilato nel 1495." *Archivio storico Italiano* 20 (1874): 51–94.

Pigman, G. W. "Imitation and the Renaissance Sense of the Past: The Reception of Erasmus' Ciceronianus." *Journal of Medieval and Renaissance Studies* 9 (1979): 155–77.

Pintor, F. "Per la storia della libreria medicea nel Rinascimento. Appunti d'archivio." *IMU* 3 (1960): 189–210.

Pollitt, J. J. *The Ancient View of Greek Art: Criticism, History and Terminology*. New Haven, Conn., 1974.

Pomaro, G. "Censimento dei manoscritti della biblioteca di S. Maria Novella." *Memorie Domenicane* 11 (1980): 325–470, and 13 (1982): 203–353.

Ponte, G. "Architettura e società nel 'De Re aedificatoria' di Leon Battista Alberti." *Giornale Italiano di filologia* 21, pt. 2 (1969): 297–312.

———. *Leon Battista Alberti. Umanista e scrittore*. Genoa, 1981.

Prager, F. D. "Brunelleschi's Inventions and the Renewal of Roman Masonry Work." *Osiris* 9 (1950): 427–554.

———. "A Manuscript of Taccola, Quoting Brunelleschi on Problems of Inventors and Builders." *Proceedings of the American Philosophical Society* 112 (1968): 131–49.

Prager, F. D., and G. Scaglia. *Brunelleschi: Studies of His Technology and Inventions*. Cambridge, Mass., 1970.

———. *Mariano Taccola and His Book "De Ingeneis."* Cambridge, Mass. 1972.

Preston, J. "Was There an Historical Revolution?" *JHI* 38 (1977): 353–64.

Quint, D. *Origin and Originality in Renaissance Literature*. New Haven, Conn., 1983.

———. "Humanism and Modernity: A Reconsideration of Bruni's Dialogues." *Renaissance Quarterly* 38 (1985): 423–45.

Quintavalle, A. C. *Prospettiva e ideologia. Alberti e la cultura del secolo XV*. Parma, 1976.

Rabb, T. "The Historian and the Art Historian Revisited." *Journal of Interdisciplinary History* 14 (1984): 647–55.

Rabb, T., and J. Seigel, eds. *Action and Conviction in early Modern Europe: Essays in Memory of E. H. Hardison*. Princeton, N.J., 1969.

Ralph, P. L. *The Renaissance in Perspective*. New York, 1973.
Ramsey, P. A., ed. *Rome in the Renaissance: The City and the Myth*. Medieval and Renaissance Texts and Studies, no. 18. Binghamton, N.Y., 1982.
Resta, G. "Filelfo tra Bisanzio e Roma." In *Francesco Filelfo nel quinto centenaria della morte. Atti del XVII convegno di studi Maceratesi*, pp. 1–87. Padua, 1986.
Reynolds, L. D. *Texts and Transmission: A Survey of the Latin Classics*. Oxford, 1983.
Reynolds, L. D., and N. G. Wilson. *Scribes and Scholars: A Guide to the Transmission of Greek and Latin Literature*. London, 1968.
Robathan, D. "Flavio Biondo's 'Roma instaurata.'" *Medievalia et Humanistica* 1 (1970): 203–16.
Robey, D., and J. Law. "The Venetian Myth and the 'De Republica veneta' of Pier Paolo Vergerio." *Rinascimento* 15 (1975): 3–59.
Romby, G. C. *Descrizioni e rappresentazioni della città di Firenze nel XV secolo*. Florence, 1976.
Rose, P. L., and S. Drake. "The Pseudo-Aristotelian *Questions of Mechanics* in Renaissance Culture." *Studies in the Renaissance* 18 (1971): 65–104.
———. "Humanist Culture and Renaissance Mathematics: The Italian Libraries of the Quattrocento." *Studies in the Renaissance* 20 (1973): 46–105.
———. *The Italian Renaissance of Mathematics*. Geneva, 1975.
Rosenau, H. *The Ideal City: Its Architectural Evolution in Europe*. 1959. Reprint. London, 1983.
Rossi, S. *Dalle Botteghe alle accademie: Realtà artistiche a Firenze del XIV al XVI secolo*. Milan, 1980.
Rovetta, A. "Cultura e codici vitruviani nel primo umanesimo milanese." *Arte Lombarda* 60 (1981): 9–14.
———. "Filarete e l'umanesimo greco a Milano: Viaggi, amicizie e maestri." *Arte Lombarda* 66 (1983): 89–102.
Rubinstein, N. "Bartolomeo Scala's *Historia Florentinorum*." In *Studi di bibliografia e di storia in onore di Tammaro de Marinis*, pp. 49–59. Verona, 1964.
———. "Il Medio evo nella storiografia italiana del Rinascimento." *Lettere italiane* 24 (1972): 431–47.
Rubinstein, R. "Pius II as a Patron of Art, with Special Reference to the History of the Vatican." Ph.D. diss., University of London, 1957.
———. "Pius II and Roman Ruins." *Renaissance Studies* 2 (1988): 197–203.
Runciman, S. *The Last Byzantine Renaissance*. Cambridge, 1970.
Russell, D. A., and M., Winterbottom. *Ancient Literary Criticism*. Oxford, 1972.
Saalman, H. "Early Renaissance Architectural Theory and Practice in Antonio Filarete's *Trattato di architettura*." *AB* 41 (1959): 89–106.
Sabbadini, R. *Storia del Ciceronianismo e di altre questioni letterarie nell'età della rinascenza*. Turin, 1885.
———. "L'Ultimo ventennio della vita di Manuele Crisolora." *Giornale linguistico* 17 (1890): 321–36.
———. *La Scuola e gli studi di Guarino Veronese*. Catania, 1896.
———. *Le Scoperte dei codici latini e greci ne'secoli XIV e XV*. 2 vols. 1905. Reprint. Florence, 1967.
———. *Il Metodo degli umanisti*. Florence, 1920.
Salmi, M. "La Prima operosità architettonica di Leon Battista Alberti." In *Convegno*, pp. 9–19.
Salomone, S. "Fonte greche nel 'De Equo animante' di Leon Battista Alberti." *Rinascimento* 26 (1986).

Sandys, J. E. *A History of Classical Scholarship*. New York, 1964.
Sanpaolesi, P. "Le Prospettive architettoniche di Urbino, di Baltimora e di Berlino." *Bollettino d'arte* 34 (1949): 322–37.
———. "Ipotesi sulle conoscenze mathematiche, statiche e meccaniche del Brunelleschi." *Belle arti* 3 (1951): 25–54.
Santinello, G. *Leon Battista Alberti. Una visione estetica del mondo e della vita.* Florence, 1962.
Sauerländer, W. "From Stilus to Style. Reflections on the Fate of a Notion." *Art History* 6 (1983): 253–70.
Schalk, F. "Aspetti della vita contemplativa nel rinascimento Italiano." In Bolgar, (1954) pp. 225–38.
Schlosser, J. von. *Präludien.* Berlin, 1927.
Schmidt, D. *Untersuchungen zu den Architektur-Ekphrasen in der 'Hypnerotomachia Poliphili.' Die Beschreibung des Venus Tempels.* Frankfurt, 1978.
Schmidt, P. G. "Mittelalterliches und humanistisches Städtelob." In *Die Rezeption der Antike: Zum Problem der Kontinuität zwischen Mittelalter und Renaissance*, edited by A. Buck. Hamburg, 1981.
Schmitt, C. *Aristotle and the Renaissance.* Cambridge, Mass. 1983.
Sciolla, G. C., ed. *La Città ideale nel Rinascimento. Scritti di Alberti, Filarete, Francesco di Giorgio Martini, Cataneo, Palladio, Vasari il Giovane, Scamozzi.* Turin, 1975.
Scott, I. *Controversies over the Imitation of Cicero.* New York, 1910.
Seigel, J. "Civic Humanism or Ciceronian Rhetoric? The Culture of Petrarch and Bruni." *Past and Present* 34 (1966): 3–48.
——— "The Teaching of Argyropoulos and the Rhetoric of the First Humanists." In Rabb and Seigel, pp. 237–60.
Semprini, G. *L. B. Alberti.* Milan, 1928.
Settis, S., ed. *Memoria dell'antico nell'arte italiana.* Vol. 1: *L'Uso dei classici* (Turin, 1984); vol. 2: *I Generi e i temi ritrovati* (Turin, 1985); vol. 3: *Dalla Tradizione all'archeologia* (Turin, 1986).
———. "Von *auctoritas* zu *vetustas*: Die Antike Kunst im mittelalterlicher Sicht." *Zeitschrift für Kunstgeschichte* 51 (1988): 157–79.
Setton, K. "The Byzantine Background to the Italian Renaissance." *Proceedings of the American Philosophical Society* 100 (1956): 1–76.
———. *Europe and the Levant in the Middle Ages and the Renaissance.* London, 1974.
Sevcenko, I. "Intellectual Repercussions of the Council of Florence." *Church History* 24 (1955): 291–323.
———. "The Decline of Byzantium Seen Through the Eyes of Its Intellectuals." In *Society and Intellectual Life in Late Byzantium.* London, 1981.
Shugar, D. "The Christian Grand Style in Renaissance Rhetoric." *Viator* 16 (1985): 337–66.
Simone, F. "La Coscienza della Rinascita negli Humanisti." *La Rinascita* 3 (1940): 163–86.
Simoncini, G. "Architettura e musica." *Palladio* 14 (1966): 67–79.
Simson, O. von *The Gothic Cathedral: Origins of Gothic Architecture and the Medieval Concept of Order.* Princeton, N.J., 1974.
Smith, L. "Note cronologiche vergeriane." *Archivio veneto* 4 (1928): 92–141.
Spencer, J. "Ut Rhetorica Pictura: A Study in Quattrocento Theory of Painting." *JWCI* 20 (1957): 26–44.
Spitzer, L. "Classical and Christian Ideas of World Harmony." *Traditio* 2 (1944): 409–64.
———. "The Problem of Latin Renaissance Poetry." *Studies in the Renaissance* 2 (1955): 118–38.

Stein, O. *Die Architekturtheoretiker der italienischen Renaissance*. Karlsruhe, 1944.
Stinger, C. *Humanism and the Church Fathers: Ambrogio Traversari and Christian Antiquity in the Italian Renaissance*. Albany, N.Y., 1977.
———. "Greek Patristics and Christian Antiquity in Renaissance Rome." In Ramsey, pp. 153–69.
Struever, N. *The Language of History in the Renaissance: Rhetorical and Historical Consciousness in Florentine Humanism*. Princeton, N.J., 1970.
Summers, D. "Contrapposto: Style and Meaning in Renaissance Art." *AB* 59 (1977): 336–62.
———. *The Judgment of Sense: Renaissance Naturalism and the Rise of Aesthetics*. New York, 1987.
Summerson, J. "Antitheses of the Quattrocento." In *Heavenly Mansions*. New York, 1948.
Tafuri, M. *L'Archittetura dell'Umanesimo*. Bari, 1969.
Tanturli, G. "Cino Rinuccini e la scuola di Santa Maria in Campo." *Studi medievali* 17 (1976): 625–74.
———. "Rapporti del Brunelleschi con gli ambienti letterari Fiorentini." In *Filippo Brunelleschi*, 1:125–44.
———. "La Cultura fiorentina volgare del Quattrocento davanti ai nuovi testi greci." *Medioevo e rinascimento* 2 (1988): 217–44.
Tatarkiewicz, W. *History of Aesthetics*. Edited by J. Harrell. 3 vols. The Hague, 1970.
Tateo, F. "Le Virtù sociali e l'immanità nella trattatistica pontaniana." *Rinascimento* 5 (1965): 119–54.
———. *Alberti, Leonardo e la crisi dell'Umanesimo*. Bari, 1971.
Thoenes, C. and H. Günther. "Gli Ordini architettonici: Rinascita o invenzione?" In *Roma e l'antico nell'arte e nella cultura del Cinquecento*, edited by M. Fagioli, pp. 261–310. Rome, 1985.
Thomson, I. "Manuel Chrysoloras and the Early Italian Renaissance." *Greek, Roman and Byzantine Studies* 7 (1966): 63–82.
Thorndike, L. *Science and Thought in the Fifteenth Century*. New York, 1929.
———. "Renaissance or Prenaissance?" *JHI* 4 (1943): 65–74.
Tietze, H. "Romanische Kunst und Renaissance." *Vorträge der Bibliothek Warburg* (1926–1927): 43–57.
Tobin, R. "Leon Battista Alberti: Ancient Sources and Structure in the Treatises on Art." Ph. D. diss., Bryn Mawr College, 1979.
Toker, F. "Alberti's Ideal Architect: Renaissance—or Gothic?" In *Smyth*, pp. 667–74.
Trachtenberg, M. "Brunelleschi, 'Giotto' and Rome." In *Smyth*, pp. 675–98.
"Tradition and Innovation in Fifteenth Century Italy: A Symposium." *JHI* 4 (1943).
Trinkaus, C., *Adversity's Noblemen: The Italian Humanists on Happiness*. New York, 1965.
———. *"In Our Image and Likeness": Humanity and Divinity in Italian Humanist Thought*. 2 vols. London, 1970.
———. *The Scope of Renaissance Humanism*. Ann Arbor, Mich., 1983.
———. "*Antiquitas* versus *Modernitas*: An Italian Humanist Polemic and Its Resonance." In *JHI* 48 (1987): 11–21.
Trinkaus, C., and H. Oberman, eds. *The Pursuit of Holiness in Late Medieval and Renaissance Religion*. Leiden, 1974.
Ullman, B. L. "Leonardo Bruni and Humanistic Historiography." *Mediavalia et Humanistica* 4 (1946): 45–61.
———. *Studies in the Italian Renaissance*. Rome, 1955.
———. *The Humanism of Coluccio Salutati*. Padua, 1963.
Ullman, B. L., and P. Stadter. *The Public Library of Renaissance Florence: Niccolò Niccoli, Cosimo de' Medici and the Library of San Marco*. Padua, 1972.

Urban, G., "Zum Neubau-Projekt von St. Peter unter Papst Nikolaus V." In *Festschrift für Harald Keller*, pp. 131–73. Darmstadt, 1963.

Vacalopoulos, A. "The Exodus of Scholars from Byzantium in the Fifteenth Century." *Cahiers d'histoire mondiale* 10 (1967): 463–80.

Vagnetti, L. "Concinnitas: Riflessione sul significato di un termine albertiano." *Studi e documenti di architettura* 2 (1973): 139–61.

———. "Lo Studio di Roma negli scritti albertiani." In *Convegno*, pp. 73–140.

Valentini, R., and G. Zucchetti. *Codice topografico della città di Roma*. 4 vols. Rome, 1953.

Vasoli, C. "L'Estetica dell'Umanesimo e Rinascimento." *Momenti e problemi di storia dell'estetica'* (1959): 325–434.

———. *Studi sulla cultura del Rinascimento*. Manduria, 1968.

Vecchio, A. "The Library and Moral Personality of Leon Battista Alberti." Ph.D. diss., New York University, 1966.

Venturi, L. "La critica d'arte in Italia durante i secoli XIV e XV." *L'Arte* 20 (1917): 305–26.

———. "Ragione e Dio nell'estetica di L. B. Alberti." *Esame* 1 (1922): 325–29.

———. "La Critique d'art en Italie a l'époque de la Renaissance. I: Léon Battista Alberti." *Gazette des beaux-arts* 5 (1922): 321–31.

———. *History of Art Critism*. Translated by C. Marriott. 1936. Reprint. New York, 1964.

Verpeaux, J., "Byzance et l'Humanisme." *Bulletin de l'Association Guillaume Budé* 3 (1952): 25–38.

Vesco, G., "L. B. Alberti e la critica d'arte in sul principio del Rinascimento." *L'Arte* 22 (1919): 57–71, 95–104, 136–48.

Vickers, B. *In Defence of Rhetoric*. Oxford, 1988.

Vickers, B., ed. *Arbeit, Musse, Meditation. Betrachtungen zur 'vita activa' und 'vita contemplativa.'* Zurich, 1985.

Voigt, G. *Die Wiederbelebung des Classischen Alterthums, oder das erste Jahrhundert des Humanismus*. 1859. Reprint. Berlin, 1960.

Waswo, R. "The 'Ordinary Language Philosophy' of Lorenzo Valla." *Bibliothèque d'Humanisme et Renaissance* 41 (1979): 255–71.

Watts, P. *Nicolaus Cusanus: A Fifteenth-Century Vision of Man*. Leiden, 1982.

———. "Pseudo-Dionysius the Areopagite and Three Renaissance Neoplatonists. Cusanus, Ficino and Pico on Mind and Cosmos." In *Supplementum festivum: Studies in Honor of Paul Oskar Kristeller*, pp. 279–98. Binghamton, N.Y., 1987.

Webb, D. "The Decline and Fall of Eastern Christianity: A Fifteenth-Century View." *Bulletin of the Institute of Historical Research. University of London* 49 (1975): 198–216.

Weisheipl, J. A. "Classification of the Sciences in Medieval Thought." *Medieval Studies* 27 (1965): 54–90.

Weisinger, H. "The Renaissance Theory of the Reaction Against the Middle Ages as a Cause of the Renaissance." *Speculum* 20 (1945): 461–67.

———. "Ideas of History During the Renaissance." *JHI* 6 (1945): 415–35.

———. "Renaissance Accounts of the Revival of Learning." *Studies in Philology* 45 (1948): 105–18.

Weiss, R. "Per gli studi greci di Coluccio Salutati." In *Atti del V Convegno internazionale di studi sul Rinascimento. Il Mondo antico nel Rinascimento*, pp. 49–53. Florence, 1958.

———. "Il Primo Rinascimento e gli studi archeologici." *Lettere italiane* 11 (1959): 89–94.

———. "Lineamenti per una storia degli studi antiquari in Italia dal dodicesimo secolo al sacco di Roma del 1527." *Rinascimento* 9 (1959): 141–201.

———. "Biondo Flavio archeologo." *Studi romagnoli* 14 (1963): 335–41.

———. "Un Umanista antiquario: Cristoforo Buondelmonti." *Lettere italiane* 16 (1964): 105–16.
———. *The Renaissance Discovery of Classical Antiquity*. Oxford, 1969.
———. *Medieval and Humanist Greek: Collected Essays*. Padua, 1977.
Westfall, C. "The Two Ideal Cities of the Renaissance: Republican and Ducal Thought in Quattrocento Architectural Treatises." Ph.D. diss., Columbia University, 1967.
———. "Society, Beauty, and the Humanist Architect in Alberti's *De Re aedificatoria*." *Studies in the Renaissance* 16 (1969): 61–81.
———. "Biblical Typology in the *Vita Nicolai V* by Giannozzo Manetti." In *Acta Conventus Neolatini Lovaniensis (Proceedings of the First International Congress for Neo-Latin Studies)*, edited by J. Ijsewijn and E. Kessler, pp. 701–10. Munich, 1973.
———. "Alberti and the Vatican Palace Type." *JSAH* 33 (1974): 101–21.
———. *In This Most Perfect Paradise: Alberti, Nicholas V and the Invention of Conscious Urban Planning in Rome, 1447–55*. University Park, Pa., 1974.
White, L. "Cultural Climates and Technological Advance in the Middle Ages." *Viator* 2 (1971): 171–201.
———. "Medical Astrologers and Late Medieval Technology." *Viator* 6 (1975): 295–308.
———. *Medieval Religion and Technology: Collected Essays*. Berkeley, 1978.
Wilcox, D. *The Development of Florentine Humanist Historiography in the Fifteenth Century*. Cambridge, Mass., 1969.
———. "The Sense of Time in Western Historical Narratives from Eusebius to Machiavelli." In *Classical Rhetoric and Medieval Historiography*, edited by E. Breisach, pp. 167–237. Kalamzoo, Mich., 1985.
Wilkins, E. H. "On Petrarch's Appreciation of Art." *Speculum* 36 (1961): 299–301.
Wilson, N. G. *Scholars of Byzantium*. Baltimore, 1983.
Witt, R. "Cino Rinuccini's 'Risponsiva alla invettiva di Messer Antonio Lusco.'" *Renaissance Quarterly* 23 (1970): 133–49.
———. "Salutati and Contemporary Physics." *JHI* 38 (1977): 667–72.
———. "Medieval 'Ars Dictaminis' and the Beginnings of Humanism: A New Construction of the Problem." *Renaissance Quarterly* 35 (1982): 1–35.
Wittkower, R. *Architectural Principles in the Age of Humanism*. London, 1952. Reprint. London, 1971.
Woodward, W. H. *Vittorino da Feltre and Other Humanist Educators*. 1897. Reprint. New York, 1963.
Wright, D. R. E. "Alberti's 'De Pictura': Its Literary Structure and Purpose." *JWCI* 47 (1984): 52–71.
Zevi, B. "L'Operazione linguistica di Leon Battista Alberti." *L'Architettura. Cronache e storia* 18 (1972): 142–43.
Zippel, G. *Storia e cultura del Rinascimento italiano*. Padua, 1979.
Zoubov, V. "Leon Battista Alberti et les auteurs du Moyen Age." *Medieval and Renaissance Studies* 4 (1958): 245–66.
Zurko, E. R. de. "Alberti's Theory of Form and Function." *AB* 39 (1957): 142–45.

Index

Acciaioli, Donato, 65, 78
Accolti, Benedetto, 52, 78, 127, 193
Active versus contemplative life, 14–17, 36–37
Albert of Saxony, 32
Alberti, Leon Battista, 3–97 passim, 118, 137, 193
  architectural works by or connected with: campanile at Ferrara, 73; Rucellai Chapel, 73, 76; Santa Maria Novella (facade), 73, 76; Sant'Andrea, 81; SS. Annunziata (tribune), 76; Tempio Malatestiano, 76. *See also* Pienza
  concinnitas, concept of, 87–89, 91, 94, 96–97
  literary works cited: *Autobiography*, 32; *Della Famiglia*, 72, 222 n.8, 223 n.27, 253 n.72; *Della Pittura*, 125, 246 n.50; *De Ludi matematici*, 36, 72; *De Motibus ponderis*, 13; *De Re aedificatoria*, 9, 11, 17–18, 29, 42, 46–47, 50, 57–59, 60, 68, 70, 73, 75–76, 78, 81, 82–84, 86, 88–89, 91–93, 110, 117, 119, 127, 129, 179, 193; *Descriptio urbis Romae*, 81; *Grammatica della lingua volgare*, 73; *Intercoenales*, 35; letter to Brunelleschi prefacing *Della Pittura*, 12, 19–53, 179, 188, 191, 193; *Profugiorum ab aerumna*, 3–18, 60, 67, 69–70, 73, 76, 78, 80–97, 218 n.1; *Theogenius*, 4, 236 n.90
Albertini, Francesco, 52, 94
Albizzi family, 4
Ammianus Marcellinus, 43, 185
Ancients and Moderns
  architectural practice of, 59, 61, 64
  Byzantine view of, 140
  Ciceronian versus eclectic position in early Humanism, 68–74, 78–79
  imitation of, versus invention, 22–25, 28, 38–39
  influence on Humanist historiography, 185–88, 191–93, 197
Angeli da Scarperia, Jacopo, 140
Anonimo Magliabecchiano, 152, 160, 167, 173
Anthemios of Tralles, 35, 37–38
Antithesis
  literary device of: in *Comparison of Old and New Rome*, 156, 167; in *Profugiorum ab aerumna*, 6, 12, 60, 81, 84–89, 111. *See also* Active versus contemplative life; Beauty; Epistemology; Rhetorical theory

in musical composition, 91–99
as principle of perception, 124–25
Antoninus, Saint, bishop of Florence, 42, 64
Antonio da Barga, 5
Aphthonius, 139, 141, 182
Apostolis, Michael, 137, 143, 148
Aquinas, Saint Thomas, 127, 139, 158
Aragazzi, Bartolomeo, 151
Archaeology and antiquarianism, 153–56, 158
Archimedes, 13–14, 16–18, 34–38, 227 n.95
Argyropoulus, Johannes, 71, 126–27, 143, 145
Aristides, Aelius
  as Second Sophistic orator, 171, 182
  source for: Leonardo Bruni's *Laudatio florentinae urbis* 140–41, 176–79, 187; Manuel Chrysoloras's *Comparison of Old and New Rome* 136, 152–53, 159, 160, 164, 168; Pier Candido Decembrio's *De Laudibus Mediolanensium urbis panegyricus*, 189
Aristotle
  on beauty, 41, 83, 90–93, 128–29, 158
  on cognition, 10, 125
  courses on, by Argyroupolus, 143–44, 158
  on emotion, 11, 91
  hierarchy of knowledge, 15–16, 29, 37–38
  as part of canon of Greek classics, 136, 178
  principles of oratorical style, 84, 87, 117, 196
  on virtue, 26, 38, 79, 87, 196, 219 n.11
  works attributed to: *Economics*, 51; *Mechanical problems*, 34–35
  works by, cited: *Art of Rhetoric*, 9, 26, 43, 77, 84, 87, 117, 196; *Eudemian Ethics*, 83; *Metaphysics*, 15; *Nicomachean Ethics*, 11, 15, 26, 38, 41, 87, 219 n.8, 219 nn.9, 11; *On Memory and Recollection*, 10; *On the Soul*, 10, 91, 125; *Physics* 35, 169; *Politics*, 186
Augustine, Saint, 53, 76, 91–92, 220 n.24
  works by: *City of God*, 7, 10–11, 16–17, 96; *De Doctrina christiana*, 60, 240 n.20; *On Music*, 53; *Soliliquies*, 83
Aurispa, Giovanni, 35, 44, 134, 188
Avogadro da Vercelli, Alberto, 195, 197

Barduzzi, Bernardino, 184
Basil, Saint, 77, 136
Beauty
  definition of, 90, 92, 95, 127–28, 158, 164, 167
  emotional and sensual experience of, 83, 92, 95
  importance of, 46, 86–87
  intellectual perception of, 92, 95, 127–28
  verbal appreciation of, 9–10, 167
  versus necessity, 5, 46, 85–91, 94, 156, 167–68, 186–88, 193. *See also* Antithesis; Rhetorical theory
Beauvais, Vincent of. *See* Vincent of Beauvais
Benzo of Alessandria, 173
Bernard of Clairvaux, Saint, 10, 40
Bessarion, Cardinal, 34, 137, 145, 148, 266 n.129
Biondo, Flavio. *See* Flavio, Biondo
Bisticci, Vespasiano da, 134
Boccaccio, Giovanni, 8, 65
Bracciolini, Poggio, 4, 50–51, 78, 137, 145–47, 266 n.35
  De Varietate fortunae, 185–86, 196
  his discoveries of texts, 23, 43–46
  historiographer, 191, 196
  ideas on language, 74, 78, 90, 134

relation to Chrysoloras, 145, 151, 184–86
Bradwardine, Thomas, 32
Bramante, Donato, 44, 53, 63, 66
Brunelleschi, Filippo, 19–39, 47–48, 63, 66
Bruni, Leonardo, 4, 15, 51, 125
  Ancients and Moderns, views on, 11, 23–25, 146–47, 193
  as historiographer, 49–50, 64, 65, 124, 185–88, 191–92, 267 n.157
  ideas on language, 72, 78, 90, 186–87, 193, 196, 237 n.91
  as panegyrist, 62, 153, 176–80, 185–87, 189, 191–92, 197
  relation to Chrysolaras, 136–37, 140, 144, 146–47, 151, 171, 185
  as translator, 144, 185–86, 257 n.54
  works by, cited: *Ad Petrum Paulum Histrum*, 11, 23–24, 146, 179–81, 188, 193; *De Interpretatione recta*, 144; *De Studiis et litteris*, 136, 196; *Laudatio Florentinae urbis*, 49–50, 62, 124, 153, 176–82, 186–190, 188, 192, 194, 197, 237 n.91, 267 n.157; letters, 186; translation of Aristotle's *Politics*, 186; *Vita di Dante e del Petrarca*, 72, 187
Bryennios, Joseph, 137, 142
Buondelmonti, Cristoforo, 165

Canabutzes, Johannes, 149
Canensi, Michele, 51, 191, 195
Caroli, Giovanni, 63
Cassiodorus, 66, 239 n.15
Certame coronario, 4, 72
Cesariano, Cesare, 62
Chalcondyles, Demetrius, 138, 254 n.98
Chortasmenus, John, 143
Chrysoloras, John, 133–34, 149
Chrysoloras, Manuel, 133–49
  circle of, 171–72, 174–86

Comparison of Old and New Rome, 150–70, 199–215
  criteria for architectural description, 26, 34, 37–38, 43, 45, 47, 49, 85, 176, 179, 182–83
  ideas on language, 90,
  portrait of, fig. 23
  teacher of the early Humanists, 129, 176, 179, 182–83, 223 n.35
Chrysostom, John, 48, 140, 157, 160–61, 263 n.67
Cicero, 129
  on dignity of man, 36, 219 n.15
  on ethics, 8, 15–16, 196
  historiography, 8, 188, 190–91
  ideal of early Humanists, 129, 133, 135, 144–46, 169–70, 175–76, 262 n.45
  ideas on language, 43–44, 73–75, 77–79, 82–84, 86, 88–91, 93–94, 118, 126, 128, 135, 144–46, 190–91, 196, 254 n.87,
  politics, 15–16, 49, 147
  Rhetorica ad Herrenium, mistakenly attributed to, 239 n.15, 259 n.1
  views on architecture, 29, 33, 49
  works by, cited: *Academica*, 170; *Brutus*, 78, 84, 239 n.15, 240 n.32; *De Inventione*, 49, 77, 239 n.15; *De Natura deorum*, 29, 33, 219 n.15, 227 n.98; *De Officiis*, 88; *De Optimo genere oratorum*, 82, 89, 90; *De Oratore*, 24, 82–85, 89–90, 117, 133–35, 144, 147, 188, 191, 196, 222 n.8, 239 n.15, 266 n.128; *De Partitione oratoria*, 43, 89; *De Re publica*, 227 n.98; *Epistulae familiares*, 190–91, 255 n.6; *In Somnium Scipionis*, 93; *Orator*, 239 n.15, 240 n.32; *Tusculanarum disputationum*, 4, 15–16, 220 n.31
Clairvaux, Saint Bernard of. *See* Bernard of Clairvaux, Saint

Classical orders. *See* Orders, the classical
Claudian, 66, 227 n.98
Colonna, Francesco, 49, 66
Constantine of Rhodes, 153, 165
Constantinople, 34, 64, 137, 150–70, 199–215
  institutions: university of (Catholicon Mouseion), 139, 142; Patriarchal Academy, 139, 142
  sites: aqueducts, 164, 209; churches 139, 165; columns, 165–66, 210–11; forums, 165–66, 210–11; Hagia Sophia, 26, 37–38, 43, 47, 49, 85, 144, 156, 165–67, 211–13, 195; walls, 164–66, 207
  plan of, fig. 24
Conti, Stefano de', 52
Corna da Soncino, Francesco, 184
Cortesi, Paolo, 94, 134–35, 145, 193
Crivelli, Lodrisio, 192, 195
Cusanus, Nicholas. *See* Nicholas of Cusa
Cydones, Demetrios, 139, 142, 149, 154
Cyriacus of Ancona, 45, 189

Dante, 65–66, 68, 91, 178, 194
Dati, Goro, 94, 233 n.16
Decembrio, Pier Candido, 135, 186–87, 189, 192
Decorum, concept of, 78, 88, 117–19, 141
Dei, Benedetto, 194
Demosthenes, 136, 140, 141, 149, 178, 257 n.51
Dignity of man, concept of, 4–5, 8–9, 28, 36, 38, 43, 48, 97, 167–68. *See also* Man as second creator
Diodorus Siculus, 47, 159, 185, 228 n.103
Dionysius of Halicarnassus, 84, 185, 196, 263 n.67
Dionysius the Areopagite, 91, 95, 149
Domenico da Corelli, Fra, 28
Dondi, Giovanni, 32, 151, 173, 174
Dufay, Guillaume, 94

Eclecticism, concept of, 70–71, 74, 76, 77, 79, 111. *See also* Variety, concept of
Epistemology, 10–11, 43, 82, 92, 124–25, 158, 169
Eriugena, Johannes Scotus, 95
Euclid, 7
Eugenikos, John, 137, 156

Facio, Bartolomeo, 4, 179, 191
Federico da Montefeltro, 28
Ficino, Marsilio, 127, 248 n.81
Ferrara/Florence Council, 38, 71, 137
Filarete, 58, 61, 66, 126–27, 193, 195
Filelfo, Francesco, 34, 37, 140–41, 192
Filelfo, Giovan Maria, 183
Filostratus, 155
Fioravanti, Bartolomeo, 34
Flavio, Biondo, 28, 43, 65–66, 74–75, 127, 134
Florence
  cathedral, 3–80
  churches, 63, 67, 73, 76, 145, 180, 195, 235 n.65, 244 n.9
  palaces, 109, 114, 180, 195, 232 n.15
  Studium, 71, 126–27, 143, 146–47, 197
Fontana, Giovanni, 32
Fortune, 6–7, 13–14, 17, 35, 169, 171, 185, 213–14
Francesco da Fiano, 151
Francesco di Giorgio, 49, 66, 71, 127, 232 n.9
Frontinus, 10, 46, 127

Gellius, Aulus, 90, 240 n.32
Gennadius II Scholarius, 143
George of Trebizond, 126, 141
Ghiberti, Lorenzo, 78, 125
Giles of Viterbo, 52–53
Giorgi, Francesco, 128
Giovanni da Prato, 42
Giovanni da San Miniato, 136
Girolami, Remigio de', 19, 26, 30

Gothic architecture, 27, 30–31, 39, 57–79, 101, 109, 125
Gregoras, Nicephoras, 139, 143
Gregory Nazianzus, 45, 163, 263 n.67
Guarino Veronese
    Chrysoloras, relation with, 133–35, 138, 150, 170, 182–83, 253 n.60
    historiography of, 65, 146, 188, 191, 196
    ideas on language, 133–35, 138, 146–47, 150, 182–84, 191
    Niccolò Niccoli, relation with, 70,
    writings on architecture, 70, 182–84
Guido da Vigevano, 32

Hermogenes, 136, 139–41, 152, 166–67, 182
Herodotus, 47, 138, 140, 159, 185
Heron, 35
Himerius, 163, 166
Historiography, 68–69, 150–97
Homer, 22, 91, 140, 146, 177–78, 189, 192
Horace, 7–8, 40–41, 79, 86, 117
Hugh of Saint Victor, 29, 31

Isidore of Seville, 60, 86, 239 n.15, 249 n.12, 260 n.2

John Chrysostom. *See* Chrysostom, John
John XXIII (pope), 150–51
Jordanus de Nemore, 34
Julius II (pope), 52
Juvenal, 66

Kilwardby, Robert, 31
Kyeser, Conrad, 32

Lactantius, 36, 66, 227 n.98
Landino, Cristoforo, 28, 81, 192, 194, 197
Lapo da Castiglionchio, 52
Lascaris, Theodore, 154–55
Leo X (pope), 52–53

Libanius
    as source for Chrysoloras, 136, 145, 148, 157, 160, 199–200, 258 n.62
    as source for early Humanists, 171, 176, 178, 182, 185, 262 n.49, 264 n.89
Lippomano, Marco, 34
Livy, 165, 228 n.100
Loschi, Antonio, 134, 180–81, 185
Lucian, 9, 37–38, 141, 155, 183–85, 263 n.67
Lucretius, 8, 22–23, 25, 39, 75, 219 n.15, 222 n.24, 236 n.75, 239 n.14

Macrobius, 93, 190, 266 n.128
Man as second creator, 8–9, 14, 17, 23, 48. *See also* Dignity of man
Manetti, Antonio, 28, 66, 71
Manetti, Giannozzo
    on dignity of man, 5, 9, 28, 36, 48–50
    as historiographer, 190–91
    interest in mechanics, 34, 48–50
    writings on architecture, 28, 34, 45, 48–50, 94, 137, 184, 190–91, 242 n.71
Manuel II Palaiologus, 139, 165
    correspondence with Chrysoloras, 148, 150, 161, 164, 255 n.2
    education of, 142, 154
    writings on architecture, 154, 156
Marcellinus, Ammianus. *See* Ammianus Marcellinus
Marsigli, Luigi, 145
Marsuppini, Carlo, 28, 126–27
Martial, 41, 44–45
Mathematics
    in architectural writing, 7, 13
    in architecture, 30–32, 36–37, 128, 245 n.18
    as highest object of knowledge, 15–16
    liberal art of, 21, 30–32
    music, based on, 92–94, 245 n.18
    painting, based on, 13

See also Mechanical arts; Mechanics, science of
Mechanical arts, 19–39, 136. *See also* Mechanics, science of
Mechanics, science of, 13, 16–18, 19–39, 167. *See also* Weights, science of
Medici family
   Cosimo, 35, 195, 229 n.25
   Filippo, 65
   Lorenzo, 35
   Nicola di Vieri, 4–5, 7, 10, 12, 24, 69–70, 218 n.3
Menander Rhetor, 141, 162, 164
Metochites, Theodore, 141, 163, 178, 257 n.57
Mirabilia urbis Romae, 151–52, 173, 175
Morandi, Benedetto, 75

Nemesius of Emesa, 29, 48, 230 n.49
Neoplatonism, 49, 71, 92–93, 128–29. *See also* Plato
Niccoli, Niccolò, 4, 184
   attitude toward contemporary culture, 24–25, 70–72, 79, 146–47
   books owned or read by, 35, 229 n.25, 262 n.37, 266 n.135
   criticism of, 70, 147, 191
Nicholas V (pope), 35, 51, 100, 183, 190–91, 195, 231 n.62
Nicholas of Cusa, 8, 26, 95–96, 121, 126

Orders, the classical, 13, 27, 57–58, 60–63, 67–71, 128
Originality, concept of, 19–39, 76, 167–68, 223 n.32
Ovid, 41, 195, 227 n.98

Pacioli, Luca, 127–28
Palladio, 58, 71
Palmieri, Matteo, 50, 65–66, 148
Pandolfini, Agnolo, 3–17, 27, 70, 218 n.2
Pandoni, Porcellio, 194–95
Panormita, Antonio, 192
Paolo Veneto, 32

Pappus, 34–37
Patria, the, 154, 158, 162, 165, 258 n.67
Patrizi, Francesco, 127, 148
Paulus Silentiarius, 153, 257 n.57
Peruzzi family, 4
Petrarch, 36, 40, 77, 178, 255 n.6
   attitude toward Middle Ages, 64–65
   on dignity of man, 48
   on *fortuna*, 185
   in revival of letters, 90, 146, 151
   writing on architecture, 64, 175–76, 195, 261 n.22
Philo (architect), 50, 190
Philo Judaeus, 29
Photius, 141, 145, 178, 252 n.50
Piccolomini, Aeneas Silvius. *See* Pius II (pope)
Pico della Mirandola, 5
Pienza, 62–63, 98–129, 194
Piero da Noceto, 50
Pierozzi, Antonio, 64
Pius II (pope), 37, 115
   ideas on language, 75, 77–78, 91, 134, 188–89
   patron of Pienza, 100, 125, 194. *See also* Pienza
   writing on architecture, 44, 62–63, 188–89, 194
Pizolpasso, Francesco, 186–89, 192, 261 n.37
Planudes, Manuel, 154
Plato, 96
   and city planning, 98–99, 101, 116, 129, 164
   definition of beauty, 83, 92–93, 98–99, 128
   Humanists' knowledge of, 126–27, 136, 140, 146
   on idealism, 71, 81, 98–99, 101, 116, 128
   ideas on language, 82–83, 89
   literary style of, 90
   works by, cited: *Cratylus*, 128; *Critias*, 40, 177; *Gorgias*, 82–83, 161, 257

n.54; *Laws*, 93, 181, 196, 262 n.41; *Phaedrus*, 6, 83; *Protagoras*, 196; *Republic*, 16, 129, 164, 176, 196, 261 n.31
   writing on architecture, 40, 129, 177, 181, 262 n.41
   See also Neoplatonism
Platonic philosophy, 7, 71, 93, 143
Plethon, Gemistos, 143
Pliny, 46, 52, 60, 77, 235 n.74
Pliny the Younger, 22–23, 188, 255 n.6
Plutarch, 41, 44, 125, 135, 140, 190, 228 n.100, 240 n.20, 247 n.60
Poliziano, Agnolo, 75
Polybius, 138, 194, 264 n.106
Pontano, Giovanni, 196
Ponteo, Giovanni Antonio, 184
Porcari, Stefano, 51
Priscian, 140
Proclus, 7
Progress, concept of, 19–39, 58, 144, 162
Prudentius, 42, 195
Ptolemy, 157
Pythagorean philosophy, 91–93, 99, 101, 116, 128

Quartigiani, Filippo de'Diversi de', 189
Quintilian, 75, 77, 86, 90–91, 117–18, 125, 175–76, 179, 181–82, 196, 223 n.32, 249 n.12
Quirini, Lauro, 189, 196

Raphael, 67, 71
Rhetorical theory, 26, 43, 45, 82–96, 117–18, 125, 133–97. *See also* Cicero; Quintilian; Second Sophistic rhetoric
Rinuccini, Alamanno, 28, 65, 127
Rinuccini, Cino, 145–47
Rinnuccio da Castiglione, 35
Ripa, Bonvesin della, 173
Romanesque architecture, 59, 61, 63, 67
Rome, 47, 66, 109, 150–70, 199–215
   sites: aqueducts, 46; Capitoline, 100, 185; churches, 52, 63, 101, 161, 174, 190; Colosseum, 67; columns, 156, 160; palaces, 114; Pantheon, 43–44, 52, 60, 65, 67, 168, 174
Rossellino, Bernardo, 49, 100–101
Rossi, Roberto de', 147
Rucellai, Giovanni, 28
Rustici, Cencio de', 151, 249 n.13

Sallust, 41
Salutati, Coluccio, 51–52, 140, 147, 151, 255 n.6
   on active and contemplative lives, 15
   admiration for, 24, 145
   historiography of, 51–52, 187–88, 196
   ideas on language, 10–11, 65, 74, 78, 90, 171
   writings on architecture, 42, 176, 180–82
Sangallo, Giuliano da, 63
Savonarola, Michele, 184, 189
Scala, Bartolomeo, 65
Scamozzi, Vincenzo, 128
Scarperia, Jacopo Angeli da. *See* Angeli da Scarperia, Jacopo
Scholasticism, 127–28, 181, 261 n.16
   dialectic, 12, 96, 127–28, 175, 186
   division of the arts, 29, 31, 33, 35, 39
   See also Aquinas, Saint Thomas; Bradwardine, Thomas; Kilwardby Robert; Vincent of Beauvais
Second Sophistic rhetoric, 136, 138–42, 144, 148, 155, 157, 171, 178, 195. *See also* Aelius Aristides; Aphthonius; Hermogenes; Libanius; Lucian; and Rhetorical theory
Seneca, 4, 9, 14, 29, 41–42, 77, 90
Serlio, Sebastiano, 43, 71
Seven Wonders, 37, 44–48, 52, 157
   Colossus of Rhodes, 157, 200
   hanging gardens of Babylon, 47
   mausoleum of Halicarnassus, 47, 154, 157
   Pharos lighthouse, 157
   pyramids, 42, 47, 157, 200

temple of Artemis at Ephesus, 47, 69–70, 73, 76, 78, 157, 195, 200
walls and gates of Thebes, 157, 200
walls of Babylon, 47, 157, 165
Signorili, Nicola, 152
Silvestri, Domenico, 65
Sozomeno of Pistoia, 28
Statius, 41, 44
Stoic philosophy, 7, 9, 41
Strabo, 183
Strozzi family, 3–4, 145
Suetonius, 41

Taccola, Il, 35
Tacitus, 74, 86, 190
Themistius, 153, 257 n.57, 258 n.67
Tignosi, Niccolò, 195
Thucydides, 138, 140
Tortelli, Giovanni, 127
Toscanelli, Paolo dal Pozzo, 32, 35
Traversari, Ambrogio, 35, 65, 78, 134

Uberti, Fazio degli, 66

Valla, Lorenzo
  on active and contemplative lives, 15–17, 37
  attitude toward contemporary culture, 25, 65–66, 74
  on dignity of man, 8
  historiography of, 65–66, 186
  ideas on language, 74, 87, 91, 127, 148, 186
Variety, concept of, 58, 69–70, 78–79, 98–129. *See also* Eclecticism

Varro, 77, 169–70
Vasari, Giorgio, 62, 70–71
Vegio, Maffeo, 192, 196
Vergerio, Pier Paolo, 135, 140, 191, 255 n.6
  relation to Chrysoloras, 142, 149, 174–76, 249 n.11
  writings on architecture, 137, 151, 171, 174–76, 185
Vernacular language, 3–4, 57, 71–73, 78–79, 177. *See also* Eclecticism, concept of; Rhetorical theory
Vespasiano da Bisticci. *See* Bisticci, Vespasiano da
Vignola, Giacomo, 71
Villani, Filippo, 65, 68
Villani, Giovanni, 173
Vincent of Beauvais, 127, 187
Virgil, 91, 146, 192, 197
Visconti family
  Filippo Maria, 187, 192
  Giangaleazzo, 177, 180
Vitruvius, 36, 44, 67, 77, 239 n.14
  architectural criteria, 85
  Classical Orders, 62–63
  copies of treatise, 185, 245 n.15, 261 n.16
  on invention, 31
  on status of architecture, 8, 29, 33, 219 n.15, 220 n.24, 230 n.57
Vittorino da Feltre, 140

Weights, science of, 13, 18, 30, 34–35
William of Moerbeke, 35–36

Zeno, Iacopo, 45